AUTOMATA
Theory, Trends, and Applications

AUTOMATA

Theory, Trends, and Applications

Alexander Meduna
Tomáš Kožár
Brno University of Technology, Czech Republic

World Scientific
NEW JERSEY • LONDON • SINGAPORE • BEIJING • SHANGHAI • HONG KONG • TAIPEI • CHENNAI • TOKYO

Published by

World Scientific Publishing Co. Pte. Ltd.
5 Toh Tuck Link, Singapore 596224
USA office: 27 Warren Street, Suite 401-402, Hackensack, NJ 07601
UK office: 57 Shelton Street, Covent Garden, London WC2H 9HE

Library of Congress Cataloging-in-Publication Data
Names: Meduna, Alexander, 1957– author | Kožár, Tomáš, 1998– author.
Title: Automata : theory, trends, and applications / Alexander Meduna, Tomáš Kožár, Brno University of Technology, Czech Republic.
Description: New Jersey : World Scientific, [2024] | Includes bibliographical references and index.
Identifiers: LCCN 2023021908 | ISBN 9789811278129 (hardcover) | ISBN 9789811278136 (ebook for institutions) | ISBN 9789811278143 (ebook for individuals)
Subjects: LCSH: Machine theory.
Classification: LCC QA267 .M428 2024 | DDC 511.3/5--dc23/eng/20230816
LC record available at https://lccn.loc.gov/2023021908

British Library Cataloguing-in-Publication Data
A catalogue record for this book is available from the British Library.

For any available supplementary material, please visit
https://www.worldscientific.com/worldscibooks/10.1142/13464#t=suppl

Desk Editors: Balasubramanian Shanmugam/Amanda Yun

Typeset by Stallion Press
Email: enquiries@stallionpress.com

To Alex, my co-author.

— **Tomáš Kožár**

To Tomáš, my co-author.

— **Alexander Meduna**

Preface

Subject

In essence, *automata* are mathematical models that formalize procedures in the intuitive sense, and as such, they play a crucially important role in computer science as a whole. Indeed, they allow us to study procedures by using the invaluable apparatus of mathematics as a systematized body of unshakable knowledge obtained by precise and infallible reasoning. The fundamentals obtained by this study are summarized in *automata theory*. This theory, in turn, underlies the *theory of computation*, whose principal goal is to establish what is computable and what is not. Viewed from this standpoint, automata represent the principal subject of the current book.

First and foremost, this book gives an account of the classical automata theory while paying special attention to the consequences following from this theory in terms of computability. Second, it provides an overview of currently active trends in automata theory. Finally, apart from these theoretically oriented topics, the book also describes real practical uses of automata in a variety of scientific areas, ranging from language processing to musicology.

Primarily, this book represents a theoretically oriented treatment of automata and computability. All the formalisms concerning automata are introduced with enough rigor to make all results quite clear and valid. Every complicated mathematical passage is preceded by its intuitive explanation so that even the most complex parts of the book are easy to grasp. Secondarily, by presenting many useful algorithms, the current book also maintains an emphasis on their applications.

Use of the Book

This book is useful to everybody who wants to become familiar with the theory of automata and computation. It can also be used as a textbook for an advanced course in theoretical computer science at the senior level; the text allows the flexibility needed to select some of the discussed topics.

Structure of the Book and its Synopsis

The entire text contains 14 chapters, which are organized into Parts 1–5.

Part 1, which consists of Chapters 1–2, gives an introduction to the subject. Chapter 1 recalls all the necessary mathematical notions used later in the book. Notably, it gives the basic terminology of formal language theory, which is often used in automata theory. Chapter 2 explains the reason why this book is important to computer science by giving its motivation, aim, and major objectives as clearly and intuitively as possible.

Part 2, consisting of five chapters, represents a key part of the entire book in many respects because it covers the classical automata theory in detail. Indeed, this theory is based on three fundamental types of automata: *finite automata*, *pushdown automata*, and *Turing machines*, discussed in Chapters 3, 4, and 5, respectively. Finite automata are strictly finitary models of computation, whose components are of fixed and finite sizes, whereas none of them can be extended during the course of computation. Pushdown automata are, in essence, finite automata extended by a potentially infinite workspace organized as stacks, which the automata theory calls pushdowns, hence their name. Turing machines represent ultimately general models of computation. Chapter 6 covers selected *grammatical models* while concentrating their principal attention on grammars with the same power as automata discussed earlier in Part 2, so the selected grammars actually represent counterparts to the equally powerful automata. Chapter 7 revisits the discussion of Chapter 5 by demonstrating how Turing machines, as general computational models, underlie the *theory of computation* in its entirety. Based on these machines, it builds up a *metaphysics of computation* that states what is computable and what is not. Analogously, it shows that some problems are algorithmically decidable while others are not. Concerning decidable problems, the theory of computation takes a finer look at some of them by studying their time and space computational complexity. Based on the results concerning this complexity, it distinguishes problems whose computation takes a reasonable amount of time from intractable problems whose solutions require an unmanageable amount of time. Considering all these results, it comes as no surprise that Chapter 7 is by far the most important chapter of Part 2 as a whole.

Part 3, which consists of Chapters 8–10, covers selected modern trends in automata theory. Chapter 8 deals with regulated automata, whose behavior is controlled by additional simple languages. Chapter 9 changes classical finite automata to jumping automata as adequate models of discontinuous computation. Chapter 10 covers generalized versions of pushdown automata, referred to as deep pushdown automata, which can modify their pushdowns deeper than on their tops.

Part 4, which consists of Chapters 11 and 12, demonstrates applications of the formal models studied earlier in this book. Chapter 11 describes the applications of finite and pushdown automata in programming language processing in detail. Chapter 12 sketches grammatical applications in natural language processing and musicology.

Part 5 consists of two brief chapters. Chapter 13 sums up the book, after which Chapter 14 places its subject into a historical context; in addition, it recommends more books for further reading.

Support

The text makes use of a great number of mathematical notions. Therefore, the end matter of this book contains the subject index, which should help the reader with a quick orientation to the text. For additional support on the Internet, please visit

http://www.fit.vutbr.cz/~meduna/books/atta.

Here, one will find further backup materials related to this book, such as suggestions for further reading as well as errata as we learn of them. To support the use of this book as a textbook, this website also contains several teaching tips as well as lectures to download in the PDF. Some parts of this website are protected by a password. Read About the Supplementary Website to learn all you need, including the password, to access this website in its entirety. The e-mail address for correspondence to this book is meduna@fit.vut.cz.

About the Authors

Alexander Meduna (born 1957 in Olomouc, Czech Republic) is a theoretical computer scientist and expert on the theory of computation. He is a full professor of computer science at the Brno University of Technology in the Czech Republic. Formerly, he taught theoretical computer science at various American, Asian, and European universities, including the University of Missouri, where he spent a decade teaching advanced topics of the theory of computation, with a principal focus on automata, and Kyoto Sangyo University, where he spent several months teaching automata theory as well. Concerning the subject of this book, he is the author of over 90 papers and several books, listed at http://www.fit.vutbr.cz/~meduna/work.

Tomáš Kožár (born 1998 in Prešov, Slovakia) is a distinguished computer science PhD student supervised by Alexander Meduna at the Brno University of Technology in the Czech Republic. His main research area is the theory of computation, including automata and formal languages. He is also interested in applications of automata in such areas as language processing, including compiler writing.

Acknowledgments

We wish to acknowledge our indebtedness to several people for their assistance in various aspects of this book. Perhaps most significantly, we have benefited from conversations with Jozef Makiš and Zbyněk Křivka. Many more students, friends, and colleagues — too many to name them all — have contributed a good deal to the production of this book, both by setting us thinking about questions concerning automata as well as by helping us to see the answers. We are also grateful to Amanda Yun and Andrea Wolf at the World Scientific Publishing Co. Pte. Ltd. for their encouragement and patience when we failed to meet the original deadline. To all of the above, we render our thanks.

Alexander Meduna's special thanks go to Professor Masami Ito for his strong support, wholehearted friendship, and, most importantly, true humaneness; *arigatō*, Masami.

This work was supported by two Brno University of Technology grants, namely S-20-6293 and FIT-S-23-8209.

Contents

Part 1

Introduction

This part consists of Chapters 1 and 2. The former introduces all the terminology, including the basic notions of formal language theory, needed to follow the rest of the book. The latter sketches the subject of this book in an intuitive way. That is, it explains the fundamental reason why the automata theory and its applications are crucially important to computer science as a whole.

Chapter 1

Terminology

Mathematical Background

This chapter reviews rudimentary concepts from logic (Section 1.1), set theory (Section 1.2), discrete mathematics (Section 1.3), and graph theory (Section 1.4). For readers familiar with these concepts, this chapter can be treated as a reference for notation and definitions.

1.1 Logic

Here, we review the basics of elementary logic. We pay special attention to the fundamental proof techniques used in this book.

In general, a *formal mathematical system* S consists of basic *symbols*, *formation rules*, *axioms*, and *inference rules*. Basic symbols, such as constants and operators, form components of *statements*, which are composed according to formation rules. Axioms are primitive statements whose validity is accepted without justification. By inference rules, some statements infer other statements. A *proof* of a statement s in S consists of a sequence of statements $s_1, \ldots, s_i, \ldots, s_n$ such that $s = s_n$ and each s_i is either an axiom of S or a statement inferred by some of the statements $s_1, \ldots, s_{i-1}$ according to the inference rules; s proved in this way represents a *theorem* of S.

Logical connectives join statements to create more complicated statements. The most common logical connectives are *not*, *and*, *or*, *implies*, and *if and only if*. In this list, *not* is unary, while the other connectives are binary. That is, if s is a statement, then *not* s is

Table 1.1. Truth table.

s_1	s_2	$s_1 \wedge s_2$	$s_1 \vee s_2$	$s_1 \implies s_2$	$s_1 \iff s_2$
0	0	0	0	1	1
0	1	0	1	1	0
1	0	0	1	0	0
1	1	1	1	1	1

a statement as well. Similarly, if s_1 and s_2 are statements, then s_1 *and* s_2, s_1 *or* s_2, s_1 *implies* s_2, and s_1 *if and only if* s_2 are also statements. We often write $\neg$, $\wedge$, $\vee$, $\implies$, and $\iff$ instead of *not*, *and*, *or*, *implies*, and *if and only if*, respectively. The following *truth table* presents the rules governing the *truth* or *falsity* concerning statements connected by the binary connectives. Regarding the unary connective $\neg$, if s is true, then $\neg s$ is false, and if s is false, then $\neg s$ is true.

According to Table 1.1, s_1 *and* s_2 is true if both statements are true; otherwise, s_1 *and* s_2 is false. Analogously, we can interpret the other rules governing the truth or falsity of a statement containing the other connectives from this table. A statement of equivalence, which has the form s_1 *if and only if* s_2, sometimes abbreviated to s_1 *iff* s_2, plays a crucial role in this book. A proof that such a statement is true usually consists of two parts.

The *only-if part* demonstrates that s_1 *implies* s_2 is true, while the *if part* proves that s_2 *implies* s_1 is true.

Example 1.1. There exists a useful way of representing ordinary *infix arithmetic expressions* without using parentheses. This notation is referred to as the *Polish notation*, which has two fundamental forms — *postfix* and *prefix notation*. The former is defined recursively as follows.

Let Ω be a set of binary operators, and let Σ be a set of operands:

(1) Every $a \in \Sigma$ is a postfix representation of a.
(2) Let AoB be an infix expression, where $o \in \Omega$, and A, B are infix expressions. Then, CDo is the postfix representation of AoB, where C and D are the postfix representations of A and B, respectively.
(3) Let C be the postfix representation of an infix expression A. Then, C is the postfix representation of (A).

Consider the infix logical expression $(\mathbf{1} \vee \mathbf{0}) \wedge \mathbf{0}$. The postfix expressions for $\mathbf{1}$ and $\mathbf{0}$ are $\mathbf{1}$ and $\mathbf{0}$, respectively. The postfix expression for $\mathbf{1} \vee \mathbf{0}$ is $\mathbf{1}\,\mathbf{0} \vee$, so the postfix expression for $(\mathbf{1} \vee \mathbf{0})$ is also $\mathbf{1}\,\mathbf{0} \vee$. Thus the postfix expression for $(\mathbf{1} \vee \mathbf{0}) \wedge \mathbf{0}$ is $\mathbf{1}\,\mathbf{0} \vee \mathbf{0} \wedge$.

Prefix notation is defined analogically except that in the second part of the definition, o is placed in front of AB; the details are left as an exercise. ■

There exist many logic laws useful to demonstrate that an implication is true. Specifically, the *contrapositive law* says (s_1 *implies* s_2) *if and only if* (($\neg s_2$) *implies* ($\neg s_1$)), so we can prove s_1 *implies* s_2 by demonstrating that ($\neg s_2$) *implies* ($\neg s_1$) holds true. We also often use a *proof by contradiction* or *reductio ad absurdum* so that we establish the validity of a statement by assuming it is false and inferring an obvious contradiction. A *proof by induction* demonstrates that a statement s_i is true for all integers $i \geq b$, where b is a non-negative integer. In general, a proof of this kind is made in this way:

Basis. Prove that s_b is true.

Induction hypothesis. Suppose that there exists an integer $n \geq b$ such that s_m is true for all $b \leq m \leq n$.

Induction step. Prove that s_{n+1} is true under the assumption that the inductive hypothesis holds.

A proof by contradiction and a proof by induction are illustrated at the beginning of the next section (see Example 1.2).

1.2 Sets and Languages

This section reviews the basic concepts of set theory. Most importantly, it defines languages as sets whose elements are finite sequences of symbols. Many language operations are covered, too.

In what follows, we suppose that there are certain *elements* taken from some prespecified *universe*. A *set* Σ is a collection of elements taken from this universe. To express that an element a is a *member* of Σ, we write $a \in \Sigma$. On the other hand, if a is not in Σ, we write $a \notin \Sigma$. We automatically assume that these utterly rudimentary notions exist, but they are left undefined, so they are considered *naive notions* of set theory.

If Σ has a finite number of members, then Σ is a *finite set*; otherwise, Σ is an *infinite set.* A finite set that has no members is an *empty set*, denoted by $\emptyset$. The *cardinality* of a finite set Σ, denoted by $|\Sigma|$, is the number of members in Σ; note that $|\emptyset| = 0$.

Convention 1.1. Throughout this book, $\mathbb{N}$ denotes the set of *natural numbers*, i.e., $\mathbb{N} = \{1, 2, \ldots\}$ and ${}_0\mathbb{N} = \{0\} \cup \mathbb{N}$. ■

Example 1.2. The purpose of this example is twofold. First, we give a few examples of sets. Second, as pointed out in the conclusion of the previous section, we illustrate how to make proofs by contradiction and by induction.

Let P be the set of all primes (a natural number n, $n \geq 2$, is prime if its only positive divisors are 1 and n).

A proof by contradiction. By contradiction, we next prove that P is infinite. That is, assume that P is finite. Set $k = |P|$. Thus, P contains k numbers, $p_1, p_2, \ldots, p_k$. Set $n = p_1 p_2 \cdots p_k + 1$. Observe that n is not divisible by any p_i, $1 \leq i \leq k$. As a result, either n is a new prime or n equals a product of new primes. In either case, there exists a prime out of P which contradicts the assumption that P contains all primes. Thus, P is infinite. Another proof by contradiction is given in Example 1.5.

A proof by induction. As already stated, by induction, we prove that the statement s_i holds for all $i \geq b$, where $b \in \mathbb{N}$. To illustrate, consider $\{i^2 \mid i \in \mathbb{N}\}$, and let s_i state

$$1 + 3 + 5 + \cdots + 2i - 1 = i^2$$

for all $i \in \mathbb{N}$; in other words, s_i says that the sum of the first i odd integers is a perfect square. An inductive proof of this statement follows next.

Basis. As $1 = 1^2$, s_1 is true.

Induction hypothesis. Assume that s_m is true for all $1 \leq m \leq n$, where n is a natural number.

Induction step. Consider $s_{n+1} = 1+3+5+\cdots+(2n-1)+(2(n+1) - 1) = (n+1)^2$. By the inductive hypothesis, $s_n = 1+3+5+\cdots+ (2n-1) = n^2$. Hence, $1 + 3 + 5 + \cdots + (2n-1) + (2(n+1)-1)$

$= n^2+2n+1 = (n+1)^2$. Consequently, s_{n+1} holds, and the inductive proof is completed. ■

A finite set, Σ, is customarily specified by listing its members, that is, $\Sigma = \{a_1, a_2, \ldots, a_n\}$, where a_1–a_n are all members of Σ; as a special case, we have $\{\} = \emptyset$. An infinite set, Ω, is usually specified by a property, p, so that Ω contains all elements satisfying p; in symbols, this specification has the general format $\Omega = \{a \mid p(a)\}$. Sets whose members are other sets are usually called families of sets rather than sets of sets.

Let Σ and Ω be two sets. Σ is a *subset* of Ω, symbolically written as $\Sigma \subseteq \Omega$, if each member of Σ also belongs to Ω. Σ is a *proper subset* of Ω, written as $\Sigma \subset \Omega$, if $\Sigma \subseteq \Omega$ and Ω contains an element that is not in Σ. If $\Sigma \subseteq \Omega$ and $\Omega \subseteq \Sigma$, Σ *equals* Ω, denoted by $\Sigma = \Omega$. The *power set* of Σ, denoted by $power(\Sigma)$, is the set of all subsets of Σ.

For two sets, Σ and Ω, their *union*, *intersection*, and *difference* are denoted by $\Sigma \cup \Omega$, $\Sigma \cap \Omega$, and $\Sigma - \Omega$, respectively, and defined as $\Sigma \cup \Omega = \{a \mid a \in \Sigma \text{ or } a \in \Omega\}$, $\Sigma \cap \Omega = \{a \mid a \in \Sigma \text{ and } a \in \Omega\}$, and $\Sigma - \Omega = \{a \mid a \in \Sigma \text{ and } a \notin \Omega\}$. If Σ is a set over a universe U, the complement of Σ is denoted by ${\sim}\Sigma$ and defined as ${\sim}\Sigma = U - \Sigma$.

The operations of union, intersection, and complement are related by *De Morgan's laws*, which state that ${\sim}({\sim}\Sigma \cup {\sim}\Omega) = \Sigma \cap \Omega$ and ${\sim}({\sim}\Sigma \cap {\sim}\Omega) = \Sigma \cup \Omega$, for any two sets Σ and Ω. If $\Sigma \cap \Omega = \emptyset$, Σ and Ω are *disjoint*. More generally, the n sets $\Sigma_1, \Sigma_2, \ldots, \Sigma_n$, where $n \geq 2$, are *pairwise disjoint* if $\Sigma_i \cap \Sigma_j = \emptyset$, for all $1 \leq i, j \leq n$, such that $i \neq j$.

A *sequence* is a list of elements from some universe. A sequence is *finite* if it consists of finitely many elements; otherwise, it is *infinite*. The *length* of a finite sequence x, denoted by $|x|$, is the number of elements in x. An *empty sequence*, denoted by ε, is a sequence that contains no elements, that is, $|\varepsilon| = 0$. For brevity, finite sequences are specified by listing their elements throughout. For instance, $(0, 1, 0, 0)$ is shortened to 0100; note that $|0100| = 4$.

Languages

Just like set theory has its naive notions, so does formal language theory. Indeed, it automatically assumes that there exists a prespecified infinite set, referred to as a *universal alphabet*, whose members are called *symbols*. An *alphabet*, Σ, is any finite non-empty subset of this universal alphabet. Any non-empty subset of Σ is a *subalphabet* of Σ. A finite sequence of symbols from Σ is a *string* over Σ; specifically, ε is referred to as an *empty string*, i.e., the string consisting of zero symbols. By Σ^*, we denote the set of all strings over Σ; $\Sigma^+ = \Sigma^* - \{\varepsilon\}$. Let $x \in \Sigma^*$. Like for any sequence, $|x|$ denotes the length of x, i.e., the number of symbols in x. For any $a \in \Sigma$, $occur(x, a)$ denotes the number of occurrences of as in x, so $occur(x, a)$ always satisfies $0 \leq occur(x, a) \leq |x|$. Furthermore, if $x \neq \varepsilon$, $symbol(x, i)$ denotes the ith symbol in x, where $1 \leq i \leq |x|$. Any subset $L \subseteq \Sigma^*$ is a *formal language* or, briefly, a *language* over Σ. Any subset of L is a *sublanguage* of L. If L represents a finite set of strings, L is a *finite language*; otherwise, L is an *infinite language*. For instance, Σ^*, which is called the *universal language* over Σ, is an infinite language while $\emptyset$ and $\{\varepsilon\}$ are finite; notably, $\emptyset \neq \{\varepsilon\}$ as $|\emptyset| = 0$ and $|\{\varepsilon\}| = 1$. Sets whose members are languages are called *families of languages*.

Example 1.3. The English alphabet, consisting of its twenty six letters, illustrates the definition of an alphabet as stated above, except that we refer to its members as symbols in this book. Our definition of a language includes all common artificial and natural languages. For instance, programming languages represent formal languages in terms of this definition, as do English, Navaho, and Japanese. All families of natural languages, including Indo-European, Sino-Tibetan, Niger-Congo, Afro-Asiatic, and Japonic families of languages, are language families according to the definition above. ■

Convention 1.2. In strings, for brevity, we simply juxtapose the symbols and omit the parentheses and all separating commas. That is, we write $a_1a_2 \cdots a_n$ instead of $(a_1, a_2, \ldots, a_n)$.

Let $\mathbf{L_{FIN}}$ and $\mathbf{L_{INFIN}}$ denote the families of finite and infinite languages, respectively. Let $\mathbf{L_{ALL}}$ denote the family of all languages; in other words, $\mathbf{L_{ALL}} = \mathbf{L_{FIN}} \cup \mathbf{L_{INFIN}}$. ■

Let $x, y \in \Sigma^*$ be two strings over an alphabet Σ, and let $L, K \subseteq \Sigma^*$ be two languages over Σ. As languages are defined as sets, all set operations apply to them. Specifically, $L \cup K$, $L \cap K$, and $L - K$

denote the union, intersection, and difference of languages L and K, respectively. Perhaps most importantly, the concatenation of x with y, denoted by xy, is the string obtained by appending y to x. Note that for every $w \in \Sigma^*$, $w\varepsilon = \varepsilon w = w$. The concatenation of L and K, denoted by LK, is defined as $LK = \{xy \mid x \in L, y \in K\}$.

Apart from binary operations, we also make some unary operations with strings and languages. Let $x \in \Sigma^*$ and $L \subseteq \Sigma^*$. The *complement* of L is denoted by $\sim L$ and defined as $\sim L = \Sigma^* - L$. The *reversal of* x, denoted by $reversal(x)$, is x written in the reverse order, and the *reversal of* L, $reversal(L)$, is defined as $reversal(L) = \{reversal(x) \mid x \in L\}$. For all $i \geq 0$, the *ith power of* x, denoted by x^i, is recursively defined as (1) $x^0 = \varepsilon$, and (2) $x^i = xx^{i-1}$, for $i \geq 1$. Observe that this definition is based on the *recursive definitional method.* To demonstrate the recursive aspect, consider, for instance, the ith power of x with $i = 3$. By the second part of the definition, $x^3 = xx^2$. By applying the second part to x^2 again, $x^2 = xx^1$. By another application of this part to x^1, $x^1 = xx^0$. By the first part of this definition, $x^0 = \varepsilon$. Thus, $x^1 = xx^0 = x\varepsilon = x$. Hence, $x^2 = xx^1 = xx$. Finally, $x^3 = xx^2 = xxx$. By using this recursive method, we frequently introduce new notions, including the *ith power of* L, L^i, which is defined as (1) $L^0 = \varepsilon$ and (2) $L^i = LL^{i-1}$, for $i \geq 1$. The *closure of* L, L^*, is defined as $L^* = L^0 \cup L^1 \cup L^2 \cup \cdots$, and the *positive closure of* L, L^+, is defined as $L^+ = L^1 \cup L^2 \cup \cdots$. Note that $L^+ = LL^* = L^*L$, and $L^* = L^+ \cup \{\varepsilon\}$. Let $w, x, y, z \in \Sigma^*$. If $xz = y$, then x is a *prefix* of y; if, in addition, $x \notin \{\varepsilon, y\}$, x is a *proper prefix* of y. By *prefixes*(y), we denote the set of all prefixes of y. Set *prefixes*$(L) = \{x \mid x \in$ *prefixes*(y) for some $y \in L\}$. For $i = 0, \ldots, |y|$, $prefix(y, i)$ denotes y's prefix of length i; note that $prefix(y, 0) = \varepsilon$ and $prefix(y, |y|) = y$. If $zx = y$, x is a *suffix* of y; if, in addition, $x \notin \{\varepsilon, y\}$, x is a *proper suffix* of y. By *suffixes*(y), we denote the set of all suffixes of y. Set *suffixes*$(L) = \{x \mid x \in$ *suffixes*(y) for some $y \in L\}$. For $i = 0, \ldots, |y|$, *suffix*(y, i) denotes y's suffix of length i. If $wxz = y$, x is a substring of y; if, in addition, $x \notin \{\varepsilon, y\}$, x is a *proper substring* of y. By *substrings*(y), we denote the set of all substrings of y. Observe that for all $v \in \Sigma^*$, *prefixes*$(v) \subseteq$ *substrings*(v), *suffixes*$(v) \subseteq$ *substrings*(v), and $\{\varepsilon, v\} \subseteq$ *prefixes*$(v) \cap$ *suffixes*$(v) \cap$ *substrings*(v). Set *symbols*$(y) = \{a \mid a \in$ *substrings*$(y), |a| = 1\}$. Furthermore, set $substrings(L) = \{x \mid x \in substrings(y)$ for some $y \in L\}$, and $symbols(L) = \{a \mid a \in symbols(y)$ for some $y \in L\}$.

Example 1.4. Consider the alphabet $\{0, 1\}$. For instance, ε, 1, and 010 are strings over $\{0, 1\}$. Note that $|\varepsilon| = 0$, $|1| = 1$, and $|010| = 3$. The concatenation of 1 and 010 is 1010. The third power of 1010 equals 101010101010. Observe that $reversal(1010) = 0101$. We have $prefixes(1010) = \{\varepsilon, 1, 10, 101, 1010\}$, where 1, 10, and 101 are proper prefixes of 1010 while ε and 1010 are not. We have $suffixes(1010) = \{\varepsilon, 0, 10, 010, 1010\}$, $substrings(1010) = \{\varepsilon, 0, 1, 01, 10, 010, 101, 1010\}$, and $symbols(1010) = \{0, 1\}$.

Set $K = \{0, 01\}$ and $L = \{1, 01\}$. Observe that $L \cup K$, $L \cap K$, and $L - K$ are equal to $\{0, 1, 01\}$, $\{01\}$, and $\{1\}$, respectively. The concatenation of K and L is $KL = \{01, 001, 011, 0101\}$. For L, $\sim L = \Sigma^* - L$, so $\sim L$ contains all strings in $\{0, 1\}^*$ but 1 and 01. Furthermore, $reversal(L) = \{1, 10\}$ and $L^2 = \{11, 101, 011, 0101\}$.

Next, L^* contains all strings from Σ^* such that every 0 is followed by at least one 1. To illustrate, the strings in L^* that consist of four or fewer symbols are ε, 1, 01, 11, 011, 101, 111, 0101, 0111, 1011, 1101, and 1111.

Since $\varepsilon \notin L$, $L^+ = L^* - \{\varepsilon\}$. Note that $prefixes(L) = \{\varepsilon, 1, 0, 01\}$, $suffixes(L) = \{\varepsilon, 1, 01\}$, $substrings(L) = \{\varepsilon, 0, 1, 01\}$, and $symbols(L) = \{0, 1\}$. ■

1.3 Relations and Translations

For the two objects a and b, (a, b) denotes the *ordered pair* consisting of a and b in this order. Let A and B be two sets. The *Cartesian product* of A and B, $A \times B$, is defined as $A \times B = \{(a, b) \mid a \in A$ and $b \in B\}$. A *binary relation* or, briefly, a *relation*, ρ, from A to B is any subset of $A \times B$, i.e., $\rho \subseteq A \times B$. If ρ represents a finite set, then it is a *finite relation*; otherwise, it is an *infinite relation*. For any $r \in \rho, r = (a, b)$, the left-hand side of r, $lhs(r)$, and the right-hand side of r, $rhs(r)$, are defined as $lhs(r) = a$ and $rhs(r) = b$, respectively. The *domain* of ρ, denoted by $domain(\rho)$, and the *range* of ρ, denoted by $range(\rho)$, are defined as $domain(\rho) = \{a \mid (a, b) \in \rho$ for some $b \in B\}$ and $range(\rho) = \{b \mid (a, b) \in \rho$ for some $a \in A\}$. If $A = B$, then ρ is a *relation on* A. A relation σ is a *subrelation* of ρ if $\sigma \subseteq \rho$. The *inverse of* ρ, denoted by *inverse*(ρ), is defined as *inverse*$(\rho) = \{(b, a) \mid (a, b) \in \rho\}$. Let $\chi \subseteq B \times C$ be a relation; the *composition of* ρ *with* χ is denoted by $\chi \circ \rho$ and defined as $\chi \circ \rho = \{(a, c) \mid (a, b) \in \rho, (b, c) \in \chi\}$. A *function from* A *to* B is a relation ϕ from A to B such that for every $a \in A$, $|\{b \mid b \in B \text{ and } (a, b) \in \phi\}| \leq 1$. If *domain*$(\phi) = A$, ϕ is *total.* If we want to emphasize that ϕ may not satisfy *domain*$(\phi) = A$, we say that ϕ is *partial.*

Furthermore, we say that a function ϕ is:

- an *injection* if for every $b \in B$, $|\{a \mid a \in A \text{ and } (a, b) \in \phi\}| \leq 1$;
- a *surjection* if for every $b \in B$, $|\{a \mid a \in A \text{ and } (a, b) \in \phi\}| \geq 1$;
- a *bijection* if ϕ is a total function that is both a surjection and an injection.

As relations and functions are defined as sets, the set operations apply to them too. For instance, if $\phi \subseteq A \times B$ is a function, its complement, $\sim\phi$, is defined as $(A \times B) - \phi$.

Convention 1.3. Let $\rho \subseteq A \times B$ be a relation. To express that $(a, b) \in \rho$, we usually write $a\rho b$. If ρ represents a function, we often write $\rho(a) = b$ instead of $a\rho b$. If $\rho(a) = b$, b is the *value* of ρ for *argument* a. ■

If there is a bijection from an infinite set Ψ to an infinite set Ξ, then Ψ and Ξ have the *same cardinality.* An infinite set, Ω, is *countable* or, synonymously, *enumerable*, if Ω and $\mathbb{N}$ have the same cardinality; otherwise, it is *uncountable* (according to Convention 1.1, $\mathbb{N}$ is the set of natural numbers).

Example 1.5. Consider the set of all even natural numbers, $\mathbb{E}$. Define the bijection $\phi(i) = 2i$, for all $i \in \mathbb{N}$. Observe that ϕ represents a bijection from $\mathbb{N}$ to $\mathbb{E}$, so they have the same cardinality. Thus, $\mathbb{E}$ is countable.

Consider the set ς of all functions mapping $\mathbb{N}$ to $\{\mathbf{1}, \mathbf{0}\}$. By contradiction, we prove that ς is uncountable. Suppose that ς is countable. Thus, there is a bijection from ς to $\mathbb{N}$. Let f_i be the function mapped to the ith positive integer, for all $i \geq 1$. Consider the total function g from $\mathbb{N}$ to $\{\mathbf{1}, \mathbf{0}\}$ defined as $g(j) = \mathbf{0}$ if and only if $f_j(j) = \mathbf{1}$, for all $i \geq 1$, so $g(j) = \mathbf{1}$ if and only if $f_j(j) = \mathbf{0}$. As ς contains g, $g = f_k$ for some $k \geq 1$. Specifically, $g(k) = f_k(k)$. However, $g(k) = \mathbf{0}$ if and

Table 1.2. Diagonalization.

	1	2	$\cdots$	k	$\cdots$
f_1	**0**	**1**	$\cdots$	**0**	$\cdots$
f_2	**1**	**1**	$\cdots$	**1**	$\cdots$
$\vdots$	$\vdots$	$\vdots$	$\ddots$	$\vdots$	
$g = f_k$	**0**	**0**	$\cdots$	**0** iff **1**	$\cdots$
$\vdots$	$\vdots$	$\vdots$		$\vdots$	$\ddots$

only if $f_k(k) = \mathbf{1}$, so $g(k) \neq f_k(k)$, which contradicts $g(k) = f_k(k)$. Thus, ς is uncountable.

The proof technique by which we have demonstrated that ς is uncountable is customarily called *diagonalization*. To see why, imagine an infinite table with $f_1, f_2, \ldots$ listed down the rows and $1, 2, \ldots$ listed across the columns (see Table 1.2). Each entry contains either **0** or **1**. Specifically, the entry in row f_i and column j contains **1** if and only if $f_i(j) = \mathbf{1}$, so this entry contains **0** if and only if $f_i(j) = \mathbf{0}$. A contradiction occurs at the diagonal entry in row f_k and column k because $g(k) = \mathbf{0}$ if and only if $f_k(k) = \mathbf{1}$ and $g(k) = f_k(k)$; in other words, this diagonal entry contains **0** if and only if it contains **1**, which is impossible. We make use of this proof technique several times in this book. ■

Let A be a set, ρ be a relation on A, and $a, b \in A$. For $k \geq 0$, the *k-fold product* of ρ, ρ^k, is recursively defined as: (1) $a\rho^0 b$ iff $a = b$, and (2) $a\rho^k b$ iff there exists $c \in A$ such that $a\rho c$ and $c\rho^{k-1}b$, for $k \geq 1$. The *transitive closure* of ρ, ρ^+, is defined as $a\rho^+ b$ if and only if $a\rho^k b$ for some $k \geq 1$, and the *reflexive and transitive closure* of ρ, ρ^*, is defined as $a\rho^* b$ if and only if $a\rho^k b$ for some $k \geq 0$.

Let T and U be two alphabets, $K \subseteq T^*$, and $L \subseteq U^*$. A relation τ from T^* to U^* with $domain(\tau) = K$ and $range(\tau) = L$ is a *translation* from K to L. Let σ be a translation from T^* to U^* such that σ is a total function from T^* to $power(U^*)$ satisfying $\sigma(uv) = \sigma(u)\sigma(v)$ for every $u, v \in T^*$; then, σ is a *substitution* from T^* to U^*. A total function φ from T^* to U^* such that $\varphi(uv) = \varphi(u)\varphi(v)$ for every $u, v \in T^*$ is a *homomorphism* from T^* to U^*; if $\varphi(a) \neq \varepsilon$ for all $a \in T$, φ is said to be an *ε-free homomorphism*. It is worth noting

Table 1.3. Morse code.

Letter	μ	Letter	μ	Letter	μ
A	$.-$	J	$.---$	S	$...$
B	$-...$	K	$-.-$	T	$-$
C	$-.-.$	L	$.-..$	U	$..-$
D	$-..$	M	$--$	V	$...-$
E	$.$	N	$-.$	W	$.--$
F	$..-.$	O	$---$	X	$-..-$
G	$--.$	P	$.--.$	Y	$-.--$
H	$....$	Q	$--.-$	Z	$--..$
I	$..$	R	$.-.$		

that a homomorphism from T^* to U^* does not necessarily represent an injection from T^* to U^*, as illustrated in Example 1.6.

Let T, U, σ, and φ have the same meaning as above. Observe that σ from T^* to $power(U^*)$ can be completely and properly specified by defining $\sigma(a)$ for every individual symbol $a \in T$. Indeed, since $\sigma(\varepsilon) = \{\varepsilon\}$ and $\sigma(a_1a_2 \cdots a_n) = \sigma(a_1)\sigma(a_2) \cdots \sigma(a_n)$, where $a_i \in T$, $1 \leq i \leq n$, for some $n \geq 1$, we can straightforwardly obtain $\sigma(w)$ for all $w \in T^*$. As any homomorphism is obviously a special case of a substitution, we can specify φ analogously. In this elegant way, using this natural *extension*, we always introduce every new notion of a substitution and a homomorphism throughout this book. In the following example, which illustrates this kind of introduction, we make it in a tabular way (see Table 1.3).

Example 1.6. Let Δ denote the English alphabet. The *Morse code*, denoted by μ, can be seen as a homomorphism from Δ^* to $\{.,-\}^*$ (see Figure 1.3). For instance,

$$\mu(\text{SOS}) = \cdots - - - \cdots$$

Note that μ is not an injection; for instance, $\mu(\text{SOS}) = \mu(\text{IJS})$.

Next, L^* contains all strings from Σ^* such that every 0 is followed by at least one 1. To illustrate, observe that this seemingly minor change implies that μ' from Δ^* to $\{.,-,\#\}^*$ is an injection. For instance, $\mu(A) = \mu(ET)$ but $\mu'(A) \neq \mu'(ET)$. ■

We conclude this section with the following example, which demonstrates how to represent non-negative integers as strings in a very simple way. More specifically, it introduces the function unary, which represents all non-negative integers as strings consisting of as.

Example 1.7. Let a be a symbol. To represent non-negative integers by strings over $\{a\}$, define the total function unary from ${}_0\mathbb{N}$ to $\{a\}^*$ as unary$(i) = a^i$, for all $i \geq 0$. For instance, unary$(0) = \varepsilon$, unary$(2) = aa$, and unary$(1000000) = a^{1000000}$. ■

1.4 Graphs

Let A be a set. A *directed graph* or, briefly, a *graph* is a pair $G = (A, \rho)$, where ρ is a relation on A. The members of A are called *nodes*, and the ordered pairs in ρ are called *edges*. If $(a, b) \in \rho$, then edge (a, b) *leaves* a and *enters* b. Let $a \in A$; then, the *in-degree* and *out-degree* of a are $|\{b \mid (b, a) \in \rho\}|$ and $|\{c \mid (a, c) \in \rho\}|$, respectively. A sequence of nodes, $(a_0, a_1, \ldots, a_n)$, where $n \geq 1$, is a *path of length* n from a_0 to a_n if $(a_{i-1}, a_i) \in \rho$ for all $1 \leq i \leq n$; if, in addition, $a_0 = a_n$, then $(a_0, a_1, \ldots, a_n)$ is a *cycle of length* n. In this book, we frequently label the edges of G with some attached information. Pictorially, we represent $G = (A, \rho)$ by drawing each edge $(a, b) \in \rho$ as an arrow from a to b, possibly along with its label, as illustrated in the following example.

Example 1.8. Consider a program p and its *call graph* $G = (P, \rho)$, where P represents the set of subprograms in p, and $(x, y) \in \rho$ iff subprogram x calls subprogram y. Specifically, let $P = \{a, b, c, d\}$ and $\rho = \{(a, b), (a, c), (b, d), (c, d)\}$, which says that a calls b and c, b calls d, and c calls d as well (see Figure 1.1). The in-degree of a is 0, and its out-degree is 2. Note that (a, b, d) is a path of length 2 in G.

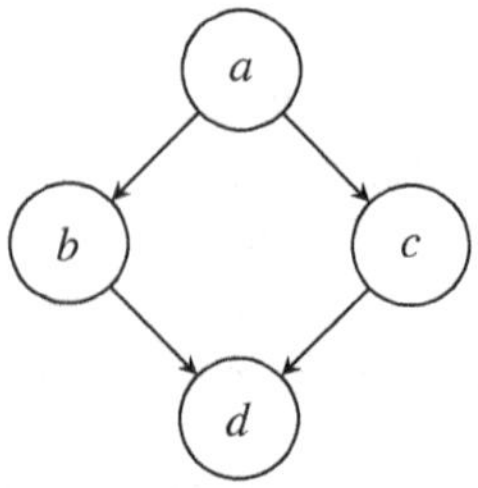

Figure 1.1. Graph.

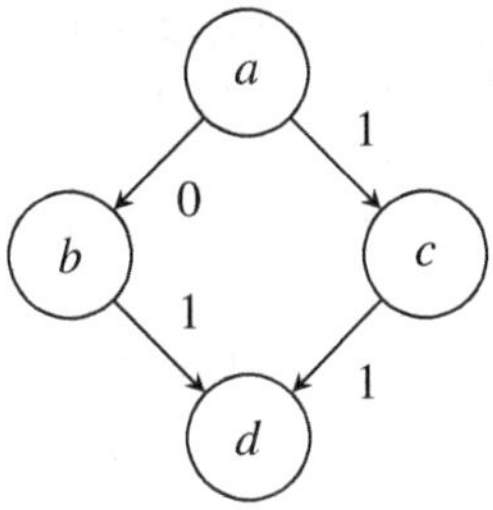

Figure 1.2. Labeled graph.

G contains no cycles because none of its paths start and end in the same node.

Suppose we use G to study the value of a global variable during the four calls. Specifically, we want to express that this value is zero when call (a, b) occurs; otherwise, it is one. We express this by labeling the edges of G, as given in Figure 1.2. ■

Let $G = (A, \rho)$ be a graph. G is an *acyclic graph* if it contains no cycles. If $(a_0, a_1, \ldots, a_n)$ is a path in G, then a_0 is an *ancestor* of a_n and a_n is a *descendant* of a_0; if, in addition, $n = 1$, then a_0 is a *direct ancestor* of a_n and a_n a *direct descendant* of a_0. A *tree* is an acyclic graph $T = (A, \rho)$ such that: (1) A contains a specified node called the *root* of T and denoted by $root(T)$; and (2) for every $a \in A - \{root(T)\}$, a is a descendant of $root(T)$ and the in-degree of a is one. If $a \in A$ is a node whose out-degree is 0, a is a *leaf*; otherwise, it is an *interior node*. In this book, a tree T is always considered an *ordered tree* in which each interior node $a \in A$ has all its direct descendants, b_1–b_n, where $n \geq 1$, ordered from left to right so that b_1 is the leftmost direct descendant of a and b_n is the rightmost direct descendant of a. At this point, a is the *parent* of its *children* b_1–b_n, and all these nodes, together with the edges connecting them, (a, b_1)–(a, b_n), are called a *parent–children portion* of T. The *frontier* of T, denoted by $frontier(T)$, is a sequence of T's leaves ordered from left to right. The *depth* of T, $depth(T)$, is the length of the longest path in T. A tree $S = (B, \nu)$ is a subtree of T if $\emptyset \subset B \subseteq A$, $\nu \subseteq \rho \cap (B \times B)$, and in T, no node in $A - B$ is a descendant of a node in B; S is an *elementary subtree* of T if $depth(S) = 1$.

Like any graph, a tree T can be described as a two-dimensional structure. To simplify this description, however, we draw a tree T with its root at the top and all edges directed down. Each parent has its children drawn from the left to the right according to its ordering. Drawing T in this way, we always omit all arrowheads.

Apart from this two-dimensional representation, however, it is frequently convenient to specify T by a one-dimensional representation, denoted by $odr(T)$, in which each subtree of T is represented by the expression appearing inside a balanced pair of $\langle$ and $\rangle$ with the node which is the root of that subtree appearing immediately to the left of $\langle$. More precisely, $odr(T)$ is defined by the following recursive rules:

(1) If T consists of a single node a, then $odr(T) = a$.
(2) Let (a, b_1)–(a, b_n), where $n \geq 1$, be the parent–children portion of T, $root(T) = a$, and T_k be the subtree rooted at b_k, $1 \leq k \leq n$, then $odr(T) = a\langle odr(T_1)\, odr(T_2) \cdots odr(T_n)\rangle$.

The following example illustrates both the one-dimensional $odr()$-representation and the two-dimensional pictorial representation of a tree. For brevity, we prefer the former throughout the rest of this book.

Example 1.9. Graph G discussed in Example 1.8 is acyclic. However, it is not a tree because the in-degree of node d is two. By removing the edge (b, d), we obtain a tree $T = (P, \nu)$, where $P = \{a, b, c, d\}$ and $\nu = \{(a, b), (a, c), (c, d)\}$. The nodes a and c are interior nodes, while b and d are leaves. The root of T is a. We define b and c as the first and second child of a, respectively. A parent–children portion

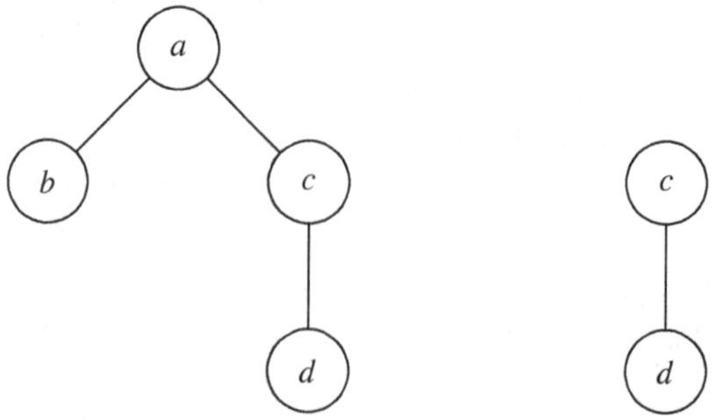

Figure 1.3. A tree and a subtree.

of T is, for instance, (a, b) and (a, c). Note that $frontier(T) = bd$ and $depth(T) = 2$. Following (1.4) and (1.4) above, we obtain the one-dimensional representation of T as $odr(T) = a\langle bc\langle d\rangle\rangle$. Its subtrees are $a\langle bc\langle d\rangle\rangle$, $c\langle d\rangle$, b, and d. In Figure 1.3, we pictorially describe $a\langle bc\langle d\rangle\rangle$ and $c\langle d\rangle$. ∎

Chapter 2

Automata

In this chapter, to give an insight into the subject of this book, we explain the meaning and significance of automata in computer science. Primarily, we describe all the notions from a theoretical viewpoint. Apart from this theoretical standpoint, we also point out their real practical applications in such computer science areas as language processing. All the notions informally described in this chapter are covered rigorously later in the text.

First and foremost, we consider automata as models of computation (Section 2.1). Then, we view them as language-defining models (Section 2.2). Crucially, we explain that under both of these inseparable considerations, we aim at the same fundamental goal: a formalization of computing that allows us to answer the very basic questions about it, such as what is computable and what is not.

2.1 Automata as Models of Computation

Indisputably, the intuitive notion of an effective *procedure* is central to computation as a whole. It consists of finitely many discrete steps, each of which can be executed mechanically in a fixed amount of time, and this finite specification represents its crucially important feature. When executed, a procedure reads input data, executes its instructions, and produces output data. As opposed to the procedure itself, both the input and output data may be infinite. An *algorithm* is a special case of a procedure that halts on all inputs. For instance, every computer program represents a procedure, and if it never enters an endless loop, it is an algorithm.

In essence, the notion of an *automaton* represents a formalization of procedures underlain by a mathematical notion of a finite size in order to reflect the very essential property of procedures — their finitary specifications. On the one hand, this notion should always reflect the formalized procedures properly and adequately from a real practical standpoint (for instance, as most syntax analyzers are based upon stacks, their automata should be based upon a formalization of this well-known data structure). On the other hand, as a mathematical abstraction, it allows us to explore the formalized procedures by using the invaluable apparatus of mathematics as a systematized body of unshakable knowledge obtained by precise and infallible reasoning.

Example 2.1. Let $Postfix\ Expressions$ denote the language consisting of all well-written postfix Polish expressions over $\{\mathbf{0}, \mathbf{1}, \vee, \wedge\}$. The following algorithm takes any string from $\{\mathbf{0}, \mathbf{1}, \vee, \wedge\}^*$, decides whether it belongs to $Postfix\ Expressions$, and if so, it determines its logical value, **0** or **1**.

We formalize this procedure by the finite eight-member binary function $\rightarrow$ from $\{\mathbf{0}, \mathbf{1}\}\{\mathbf{0}, \mathbf{1}\}\{\vee, \wedge\}$ to $\{\mathbf{0}, \mathbf{1}\}$ defined as

$$\mathbf{11}\vee \rightarrow \mathbf{1}, \mathbf{10}\vee \rightarrow \mathbf{1}, \mathbf{01}\vee \rightarrow \mathbf{1}, \mathbf{00}\vee \rightarrow \mathbf{0},$$

$$\mathbf{11}\wedge \rightarrow \mathbf{1}, \mathbf{10}\wedge \rightarrow \mathbf{0}, \mathbf{01}\wedge \rightarrow \mathbf{0}, \mathbf{00}\wedge \rightarrow \mathbf{0}.$$

Based on $\rightarrow$, we define the infinite relation $\Rightarrow$ over $\{\mathbf{0}, \mathbf{1}, \vee, \wedge\}^*$ as $uXYov \Rightarrow uUv$, for all $u, v \in \{\mathbf{0}, \mathbf{1}, \vee, \wedge\}^*, XYo \rightarrow U, X, Y, U \in \{\mathbf{0}, \mathbf{1}\}$, and $o \in \{\vee, \wedge\}$. Take its transitive-reflexive closure $\Rightarrow^*$. Observe that for all $x \in \{\mathbf{0}, \mathbf{1}, \vee, \wedge\}^*$ and $i \in \{\mathbf{0}, \mathbf{1}\}$,

$x \Rightarrow^* i$ if and only if x is a postfix expression whose logical value is i.

For instance, $\mathbf{10} \vee \mathbf{0}\wedge \Rightarrow \mathbf{10}\wedge \Rightarrow \mathbf{0}$, so $\mathbf{10} \vee \mathbf{0}\wedge \Rightarrow^* \mathbf{0}$. Thus, $\mathbf{10} \vee \mathbf{0}\wedge$ is valid, and its value is **0**. On the other hand, $\mathbf{101}\wedge \Rightarrow \mathbf{10}$, and from **10**, no further rewriting step can be made, so $\mathbf{101}\wedge$ is not a postfix Polish expression.

So far, this book has not defined any notion of an automaton strictly mathematically. Nevertheless, one should easily and naturally see that $\rightarrow$ might underlain an automaton for postfix expressions, denoted by PEA, which formalizes the above algorithm. The reader is encouraged to fill in all the missing details and define PEA based on $\rightarrow$ quite rigorously while keeping in mind the only essential requirement: each specific instance according to the resulting definition has a finitary description. ■

From a general standpoint, by formalizing procedures and, thereby, fulfilling the crucially important role of mathematical models of computation, automata lead us to build up a certain *metaphysics of computation* based on them. Indeed, they should help us answer the following fundamental questions, which are of great interest in computer science, mathematics, and philosophy: Are computers omnipotent? What can they compute? Does there exist a procedure that computes any mathematical formula? Is any yes–no problem solvable mechanically by an algorithm?

Answering questions of this kind requires an introduction to the notion of an automaton as an utterly *general model of computation* which covers all possible computers as well as procedures. Without going into any details, let us simply suppose we have somehow defined this model of computation, denoted by M, which is general enough to satisfy this requirement. Concerning the questions above, let us modestly narrow our attention only to the functions and ask whether there is a procedure that computes any function. The answer is no. To justify this answer, we should realize that no matter how we define M, each of its instances is finitely describable, for instance, in terms of a mathematical formula of a finite size. The set of all these instances is countable because we can make a list of all their finite descriptions, for instance, according to length and alphabetic order, so the set of these descriptions is equal in cardinality to $\mathbb{N}$. However, the set theory tells us that the set of all functions is uncountable (see Aho and Ullman, 1972a, p. 1), so there necessarily exist functions that cannot be computed by any procedure. Simply and briefly put, there are more functions than procedures.

More surprisingly, even if we further narrow our attention only to the set ϕ that contains all total functions over the set of non-negative integers, we can prove the existence of a specific function $g \in \phi$ that cannot be computed by any M-formalized procedure. We sketch a proof of this important result by using the diagonalization proof technique (see Meduna, 2014, p. 6). Since each function $h \in \phi$ is total, it has to be computed by a procedure that represents an algorithm, which always halts and produces $h(j)$ for all $j \in \mathbb{N}$. For the sake of contradiction, suppose that all functions in ϕ are computed by an algorithm formalized by an instance of M. Consider all the descriptions of the M instances that compute the functions in ϕ. Let

$$M_1, M_2, \ldots, M_i, \ldots$$

be an enumeration of these finite descriptions. By f_i, we denote the function computed by the algorithm formalized by the model described as M_i in the enumeration. Define the function g as $g(k) = f_k(k) + 1$, for all $k \in \mathbb{N}$. As $g \in \phi$, the above enumeration $M_1, M_2, \ldots$ contains M_j such that f_j coincides with g for some $j \geq 1$. Then, $f_j(j) = g(j) = f_j(j) + 1$, which is a contradiction. Thus, no M-formalized algorithm computes g.

Uncomputable functions, illustrated by g given above, are of particular interest in the theory of *computability*, which definitely belongs to the very heart of the *theory of computation* as a whole. Apart from uncomputable functions, however, there also exist undecidable problems which cannot be decided by any algorithm. More regretfully and surprisingly, the theory of *decidability* has even proved that there will never exist algorithms that decide some simple and natural yes–no problems whose genuine significance is more than evident. To illustrate, it is undecidable whether a program always halts; in other words, the existence of a general algorithm that decides this problem is ruled out once and for all.

As a matter of fact, even if we restrict our attention only to decidable problems and take a closer look at them, we find out that they significantly differ in terms of their *time and space computational complexity*. Indeed, two decidable problems may differ so that the computation of one problem takes a reasonable amount of time while the computation of the other does not; that is, compared to the first problem, the other problem is considered *intractable* because its solution requires an unmanageable amount of time. Thus, apart from theoretically oriented investigation, this study of computational complexity is obviously crucially important to most application-oriented areas of computer science as well.

2.2 Automata as Language Models

The previous section concludes its discussion by pointing out several problems and results, many of which have strongly restrictive implications in terms of computers and their power. As is obvious, all these results make automata the subject of an even greater investigative interest. It thus comes as no surprise that their automata theory also approaches automata from a slightly different angle, namely, as models for languages. Just as procedures have their input data, automata have their input strings, from which they accept the strings that belong to the languages they define. This viewpoint is particularly appreciated when the languages accepted in this way represent infinite sets, which obviously cannot be specified by an exhaustive enumeration of all their strings. For instance, the automaton PEA discussed in Example 2.1 can also be straightforwardly rephrased as a language model. Indeed, as input, it has any string $x \in \{\mathbf{0}, \mathbf{1}, \vee, \wedge\}^*$, and if it reduces x to $\mathbf{0}$ or $\mathbf{1}$, it accepts x as a member of the infinite language $Postfix\ Expressions$, consisting of all well-formed postfix expressions over $\{\mathbf{0}, \mathbf{1}, \vee, \wedge\}$.

Reformulated as a general model for languages, M consists of a *finite state control* and a *workspace* represented by a string, which can be limitlessly extended to the right. On its workspace, M has a read–write head, and the symbol under this head is the *current input symbol.* The finite control consists of finitely many *states* and computational *rules*, and in essence, it can be considered a program that dictates the behavior of M. During this behavior, a picture of the current situation of M is captured with the notion of a *configuration* that describes the current state together with the contents of the workspace, including the location of the head. To describe this behavior in greater detail, M works by making *moves* between its configurations. Every move is made according to a computational rule that describes how the current state and symbol are changed and where the head is shifted. M has one state defined as the start state and some states designated as final states. Starting from its start state with an input string on its workspace, M performs a sequence of moves, and if it enters a final state, the input string is *accepted.* The language of M consists of all strings it accepts. M conceptualized in this way is called a *Turing machine* after its originator, Alan Turing. It constitutes, by far, the most general model for languages and, perhaps more importantly, the most widely used general model of computation due to the bold *Church–Turing thesis*, proclaiming that any procedure can be realized as a Turing machine.

Over its history, the automata theory has placed several restrictions upon Turing machines and, thereby, introduced other important types of automata as their special cases. Most restrictively, this theory defines the notion of a *finite automaton* (FA) as a strictly finitary language device whose initial input string is unchangeable, its workspace is unextendable, and its head can only be shifted to the right.

The automata theory has relaxed this strict FA restriction by introducing *pushdown automata* as finite automata with potentially infinite stacks, customarily referred to as pushdowns. To express their concept in terms of the Turing-machine workspace, they use one part of this space as a strictly unchangeable input string, just like finite automata do. In addition, however, they make use of the other workspace part, which can be extended in a potentially infinite manner, but this extension is always based on the well-known last-in-first-out principle. In practice, these automata underlie many real software components, most of which deal with language processing, such as syntax analyzers. To illustrate this use, reconsider $Postfix\ Expressions$ from Example 2.1. To accept this infinite language, design a pushdown automaton, denoted by PDA, that works in the following way. If an operand $Z \in \{\mathbf{0}, \mathbf{1}\}$ occurs as the current input symbol, PDA pushes Z onto the pushdown. If an operator $o \in \{\vee, \wedge\}$ appears as the input symbol and, in addition, the topmost two pushdown symbols, X and Y, are operands from $\{\mathbf{0}, \mathbf{1}\}, PDA$ replaces them with the logical value resulting from XoY. If, in this way, PDA reads all the input and, simultaneously, its pushdown contains only $\mathbf{0}$ or $\mathbf{1}$, PDA accepts; otherwise, it rejects.

Let us make use of PDA and PEA (see Example 2.1) to illustrate two important topics concerning automata — *determinism* and *equivalence*.

Determinism. PDA represents an example of an automaton that works deterministically. Indeed, PDA works in this way because, under any configuration it occurs in, it makes no more than one next move. On the other hand, PEA from Example 2.1 defines the same language in a non-deterministic way because, in some computational situations, it can continue its work in several different ways (for instance, from $\mathbf{10} \vee \mathbf{10} \wedge \vee$, PEA makes both $\mathbf{10} \vee \mathbf{10} \wedge \vee \Rightarrow \mathbf{110} \wedge \vee$ and $\mathbf{10} \vee \mathbf{10} \wedge \vee \Rightarrow \mathbf{10} \vee \mathbf{0} \vee$). As obvious, non-deterministic versions of automata are mathematically convenient, but they are difficult to implement; therefore, they are mainly used in theory as opposed to their deterministic versions, which are primarily used in practice. That is also why automata theory usually proceeds in the following way when introducing the notion of a new automaton. First, it defines its basic model non-deterministically. Then, it places a restriction on their computational rules so that the resulting model that is restricted in this way necessarily works deterministically. Finally, it studies whether any of its non-deterministic versions can be converted to a deterministic version so that both define the same language, and if so, it struggles to construct the best possible algorithm that performs this conversion.

Equivalence. Language models that define the same language are said to be *equivalent.* As both PDA and PEA define $Prefix$ $Expressions$, PDA and PEA are equivalent. More generally, if two types of language models, such as non-deterministic and deterministic versions of finite automata, define the same language family, they are equivalent or, synonymously, *equally powerful.* If one type of language model defines a proper superfamily of the language family defined by another type of language model, we say that the former is stronger than the latter. Concerning the three types of automata sketched above, Turing machines are stronger than pushdown automata, and pushdown automata are stronger than finite automata.

Apart from automata, formal languages are often defined by their grammars in a similar way that natural languages, such as English, are defined. The *theory of formal languages* always conceptualizes its grammars based on finitely many grammatical rules, by which the grammars generate their languages. Grammars that make this generation contextually independent, in the sense that they make each step of this generation process regardless of the context in which the applied rule is used, are of particular interest both in theory and, perhaps more significantly, in practice. Apart from this significance, these grammars, referred to as *context-free grammars,* define the same language family as pushdown automata do, and in this sense, they actually represent their grammatical counterparts. That is also why this book pays special attention to them.

By now, we should see that the two fundamental approaches to automata — automata as language models and automata as models of computation — are nothing but two different, but literally inseparable, viewpoints of the same thing. Consequently, we approach automata in this twofold way throughout the upcoming text. More specifically, we always introduce a new type of automata as language models and establish fundamental results about them while paying special attention to the consequences of these results in terms of computation.

Part 2

Theory

The primary goal of this part is to give an account of the classical mathematical theory of automata. We proceed from the weakest to the strongest automata. Specifically, Chapters 3, 4, and 5 cover finite automata, pushdown automata, and Turing machines, respectively. Chapter 6 gives an account of the grammatical counterparts to these automata. Most importantly, based on Turing machines, which are viewed as the universal model of computation, Chapter 7 builds up a metaphysics of computation, which establishes the limits of computation from an utterly general standpoint.

Chapter 3

Finite Automata

The notion of a Finite Automaton (FA) is defined in an extremely restrictive computational way. Not only is it a strictly finitary model of computation but also its workspace, represented by a string of input symbols, is neither extendible nor modifiable in any way. Over the workspace, the automaton always works from the left to the right within the workspace while reading the symbols with a read head.

Conceptually, the automaton contains finitely many states and input symbols. Its heart consists in a program, represented by finitely many rules, which controls its computation, consisting of a sequence of moves. Every move is made from the current state in which the automaton occurs, with the current input symbol under the read head. The move is performed according to a computational rule that specifies the next state entered from the current state (possibly back to the state itself) and whether the head remains stationary over the current input symbol or is shifted to the next right-neighboring symbol. In the set of states, one state is defined as the start state, and some states are designated as final states. The automaton starts every computation from the start state with its read head over the leftmost symbol of an input string placed on the workspace. If the computation eventually brings it to a final state after the entire input string is read, then the input string is accepted. All the strings accepted in this way represent the language accepted by the automaton.

An Introductory Example

To illustrate finite automata with a real-world example, consider a common ticket machine, such as a public transportation ticket machine, which dispenses a ticket b after it has received a coin a. The machine cyclically repeats this receiving–dispensing loop indefinitely; in other words, its repetition goes to infinity.

A FA that formalizes this machine can be designed in the following way: It has two states — **OFF** and **ON**, where **OFF** is the start state as well as its only final state, so **ON** is a non-final state. An insertion of a coin brings the machine from state **OFF** to state **ON**, from which it disposes b while returning to **OFF**. Formally, the automaton thus works according to the following two rules:

$$\text{Rule 1: from } \mathbf{OFF} \text{ with } a, \text{ move to } \mathbf{ON};$$

$$\text{Rule 2: from } \mathbf{ON} \text{ with } b, \text{ move to } \mathbf{OFF}.$$

Consequently, the automaton accepts the strings

$$\varepsilon, ab, abab, ababab, \ldots.$$

Therefore, its language equals $\{ab\}^*$, which properly reflects the behavior of the ticket machine.

Historically, finite automata were first used to model neural networks as a circuit of biological neurons. Following these original steps, modern artificial intelligence and cognitive modeling make use of them to create artificial neural networks, composed of artificial neurons, for solving their problems. In fact, computer science makes use of finite automata in a large variety of its research fields, ranging from logical circuits, autonomous robots, speech and image recognition, software agents in computers and video games, to text editors and lexical analysis, as explored in Section 11.1. From a broader perspective, look around to see that finite automata can formalize numerous common everyday machines, including elevators, traffic semaphores, telephones, and vending machines.

3.1 Definitions

Definition 3.1. A *finite automaton* (FA) is a quintuple:

$$M = \big(Q, \Delta, R, s, F\big),$$

where Q and Δ are two finite sets such that $Q \cap \Delta = \emptyset$,

$$R \subseteq Q \times (\Delta \cup \{\varepsilon\}) \times Q,$$

$s \in Q$, and $F \subseteq Q$. In M, the components Q, Δ, s, and F are referred to as the *set of states*, the *input alphabet*, the *start state*, and the *set of final states*, respectively. R is called the *set of rules* of M. Any rule $(q, a, p) \in R$ is written as $qa \to p \in R$ in what follows, where $q, p \in Q, a \in \Delta \cup \varepsilon$. For brevity, we often denote $qa \to p$ by a unique label m as $m : qa \to p$, and we briefly use m instead of $qa \to p$ under this denotation.

A *configuration* of M is any string of the form qv, where $q \in Q$ and $v \in \Delta^*$; X_M denotes the set of all configurations of M. Over X_M, we define the *computation step* $\Rightarrow$ as $\beta \Rightarrow \chi$ in M, for all $\beta, \chi \in X_M, \beta = qax, \chi = px, qa \to p \in R, q, p \in Q, a \in \Delta \cup \varepsilon$. As usual, $\Rightarrow^*$ and $\Rightarrow^+$ denote the transitive and transitive-reflexive closure of $\Rightarrow$, respectively. If $\beta \Rightarrow \chi$ in M, where $\beta, \chi \in X_M$, we say that M makes a *move* from β to χ. M makes a computation from β to χ if $\beta \Rightarrow^* \chi$ in M, where $\beta, \chi \in X_M$.

M accepts $w \in \Delta^*$ if $sw \Rightarrow^* f$ in M with $f \in F$, and the set of all strings M accepts is the *language accepted by* M, denoted by $L(M)$; mathematically,

$$L(M) = \big\{w \mid w \in \Delta^*, sw \Rightarrow^* f, f \in F\big\}.$$

We denote the set of all finite automata by **FA** and set $\mathbf{L_{FA}} = \{L(M) \mid M \in \mathbf{FA}\}$. ∎

To rephrase this definition less formally, every configuration is of the form qav, where $q \in Q, a \in \Delta \cup \varepsilon$, and $v \in \Delta^*$. By using a rule of the form $qa \to p \in R$, where $q, p \in Q, a \in \Delta \cup \varepsilon$, M makes a move that reads a and changes q to p. By repeatedly performing moves like this, M reads the input string $w \in \Delta^*$ from the left to the right.

M accepts an input string $w \in \Delta^*$ if it starts from sw, reads all the symbols of w, and ends up in a final state.

Convention 3.1. For every $M \in \mathbf{FA}$, we automatically assume that Δ, Q, s, F, and R denote its alphabet of input symbols, the set of states, the start state, the set of final states, and the set of rules of M, respectively. If there exists any danger of confusion, we mark Q, Δ, s, F, and R with M as Q_M, Δ_M, s_M, F_M, and R_M, respectively, in order to explicitly relate these components to M (in particular, we make these marks when several automata are simultaneously discussed). ■

Representations of finite automata

Throughout this book, we represent any $M \in \mathbf{FA}$ in one of the following five ways:

(I) The *formal description* of all M spells out the states, symbols, and rules of M strictly according to Definition 3.1.

(II) M is defined by simply listing its rules together with specifying the start state and final states of M.

(III) M is specified by its *state table* (Table 3.1), whose columns and rows are denoted by the members of $\Delta \cup \{\varepsilon\}$ and the states of Q, respectively. The start state denotes the first row. The final states are specified by underlining. For each $q \in Q$ and each $a \in \Delta \cup \{\varepsilon\}$, the entry in row q and column a contains $\{p \mid qa \to p \in R\}$. For brevity, we omit the braces in the sets of these entries; a blank entry means $\emptyset$.

(IV) M is specified by its *state diagram* in a pictorial way. That is, this diagram is a labeled directed graph such that each node is labeled with a state $q \in Q$, and for two nodes $q, p \in Q$, there is an edge (q, p) labeled with $\{a \mid qa \to p \in R, a \in \Delta \cup \{\varepsilon\}\}$. For simplicity, we entirely omit every edge labeled with $\emptyset$ in the diagram. Furthermore, in the specification of edge-labeling non-empty sets, we omit the braces; for instance, instead of $\{a, b\}$ as an edge label, we just write a, b. To symbolically state that a state s is the start state, we point to it with a short arrow, like in Figure 3.1. The final states are doubly circled.

(V) We give an *informal description* of M. That is, we describe it as a procedure, omitting various details concerning its components. Describing M in this way, we always make sure that the translation from this description to the corresponding formal description represents a straightforward task.

Table 3.1. State table.

State	a	b	ε
1			2, 4
2	2, 3		
3		3	
4		4, 5	
5			

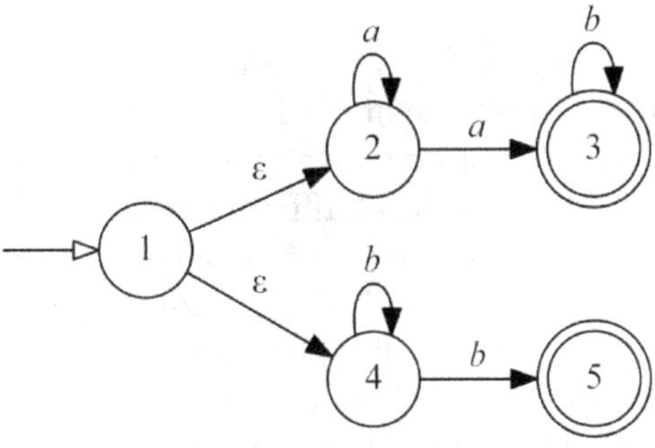

Figure 3.1. State diagram.

Example 3.1. In this example, we design an FA M that accepts $\{a\}^+\{b\}^* \cup \{b\}^+$, symbolically written as

$$L(M) = \{a\}^+\{b\}^* \cup \{b\}^+.$$

We describe M in all five ways sketched above.

Informal description. We design M so it accepts either w_1w_2 or w_3, where $w_1 \in \{a\}^+, w_2 \in \{b\}^*, w_3 \in \{b\}^+$; therefore, $L(M) = \{a\}^+\{b\}^* \cup \{b\}^+$. From its start state 1, without reading any input symbol, M moves either to state 2 to accept w_1w_2 or to state 4 to accept w_3. Looping in 2, M can read any number of as. In addition, M can make a move from state 2 to state 3 while reading a. In state 3, M reads any number of bs. As 3 is a final state, M completes the acceptance of w_1w_2 in this way. To accept w_3, in 4, M reads any number of bs. In addition, M can make a move from state 4 to state 5 with b. Since 5 is a final state, M completes the acceptance of w_3.

List of rules. M is defined by the following seven rules:

$$1 \to 2,\ 1 \to 4,\ 2a \to 2,\ 2a \to 3,\ 3b \to 3,\ 4b \to 4,\ 4b \to 5,$$

where 1 is the start state, while 3 and 5 are final states.

Formal description. Let $M = (Q, \Delta, R, s, F)$, where $Q = \{1, 2, 3, 4, 5\}$, $\Delta = \{a, b\}, F = \{3, 5\}$, and $s = 1$. Furthermore, R contains the seven rules listed above.

For instance, M accepts aab in the following way

$$1aab \quad \Rightarrow \quad 2aab \quad \Rightarrow \quad 2ab \quad \Rightarrow \quad 3b \quad \Rightarrow \quad 3.$$

Considering the design of M in terms of its implementation, we see that it deserves an improvement. Indeed, M can be redesigned so it contains fewer rules and states. Even more importantly, the current version of M can perform several different computations on the same input, and this non-determinism obviously complicates its use in practice. To put it more generally, we obviously prefer implementing FAs that work deterministically. Therefore, mostly from this pragmatically oriented perspective, we give several useful algorithms that convert FAs into their equivalent versions that are easy to implement and apply in practice. ■

3.2 Restrictions

FAs introduced in the previous section represent a mathematically convenient formal model. On the other hand, they are so general that they are difficult to implement and apply in practice. Therefore, in this section, we simplify FAs so that they are more convenient to implement than their general versions. That is, we introduce several restricted versions of these automata and explain how to transform FAs into them, so they accept the same languages. Consequently, all the upcoming restricted versions are as powerful as FAs, so they also accept $\mathbf{L_{FA}}$ (see Definition 3.1).

Removal of ε-rules

FAs may contain *ε-rules*, by which they make *ε-moves*, during which they change states without reading any symbols. Consequently, on some input strings, they may loop endlessly. As is obvious, in practice, we prefer using FAs without any ε-rules because, during every move, they read a symbol, so they cannot loop endlessly. Therefore, we next explain how to remove ε-rules from any FA without changing its language.

Definition 3.2. Let $M \in \mathbf{FA}$. M is an *ε-free finite automaton* (${}_{\varepsilon\text{-free}}$FA for short) if every rule in R_M is of the form $qa \to p$, where $q, p \in Q_M$ and $a \in \Delta_M$ (see Definition 3.1). ■

Before transforming any FA to an equivalent ${}_{\varepsilon\text{-free}}$FA, we explain how to construct the set of states that an FA can reach by *ε-moves* from a given set of states E because this construction plays an important role in the transformation.

Convention 3.2. Let $M \in \mathbf{FA}$ and $E \subseteq Q_M$. By *ε-moves*(E), we denote the set of states that M can reach by sequences of *ε-moves* from states in E; formally,

$$\varepsilon\text{-}moves(E) = \{q \mid o \Rightarrow^* q, \text{ where } o \in E, q \in Q_M\}.$$

We often omit the braces in E if E contains a single state; in other words, *ε-moves*$(\{p\})$ is shortened to *ε-moves*(p), where $p \in Q_M$. ■

Basic idea. Given $M \in \mathbf{FA}$ and $E \subseteq Q_M$, we determine *ε-moves*(E) in the following way. Initially, set *ε-moves*(E) to E because M can reach any state in E by performing zero ε-moves. If Q_M contains a state p such that $q \to p \in R_M$ with $q \in$ *ε-moves*(E), add p to *ε-moves*(E), and in this way, keep extending *ε-moves*(E) until no more states can be included in *ε-moves*(E). The resulting set *ε-moves*(E) satisfies *ε-moves*$(E) = \{q \mid o \Rightarrow^* q$, where $o \in E, q \in Q_M\}$ as desired.

Algorithm 3.1. States reachable without reading.

Input: An FA $M = (Q, \Delta, R, s, F)$ and $E \subseteq Q$.
Output: $\varepsilon\text{-}moves(E) = \{q \mid o \Rightarrow^* q$, where $o \in E, q \in Q\}$.
Method:

begin
  set $\varepsilon\text{-}moves(E)$ to E
  repeat
    $\varepsilon\text{-}moves(E) = \varepsilon\text{-}moves(E) \cup \{p \mid q \to p$ **and**
    $q \in \varepsilon\text{-}moves(E)\}$
  until $\{\varepsilon\text{-}moves(E)$ cannot be further expanded$\}$
end

Lemma 3.1. *Algorithm 3.1 is correct.*

Proof. To establish this lemma, we prove Claims A and B, in which $\varepsilon\text{-}moves(E_i)$ denotes the set of states that $\varepsilon\text{-}moves(E)$ contains after the ith iteration of the **repeat** loop, where $i = 0, 1, \ldots, h$, for some $h \leq |Q_M|$.

Claim A. For every $j \geq 0$ and every $p \in Q_M$, if $p \in \varepsilon\text{-}moves(E_j)$, then there exists $q \in \varepsilon\text{-}moves(E)$ such that $q \Rightarrow^* p$ in M.

Proof of Claim A by induction on $j \geq 0$.

Basis. Let $j = 0$. Observe that $p \in \varepsilon\text{-}moves(E_0)$ implies $p \in \varepsilon\text{-}moves(E)$. As $p \Rightarrow^0 p$, the basis holds.

Induction hypothesis. Assume that Claim A holds for all $j = 0, \ldots, i$, where i is a non-negative integer.

Induction step. Let $p \in \varepsilon\text{-}moves(E_{i+1})$. Next, we distinguish the following two cases, $p \in \varepsilon\text{-}moves(E_j)$, for some $j \leq i$ and $p \in (\varepsilon\text{-}moves(E_{i+1}) - \varepsilon\text{-}moves(E_i))$:

(I) Let $p \in \varepsilon\text{-}moves(E_j)$, for some $j \leq i$. By the induction hypothesis, there exists $q \in Q_M$ such that $q \Rightarrow^* p$ in M, so the inductive step holds in this case.
(II) Let $p \in (\varepsilon\text{-}moves(E_{i+1}) - \varepsilon\text{-}moves(E_i))$. Examine the **repeat** loop. Observe that there exists $o \to p \in R_M$ for some $o \in \varepsilon\text{-}moves(E_i)$. By the induction hypothesis, $q \Rightarrow^* o$ in M for some $q \in E$, so $q \Rightarrow^* o \Rightarrow p$ in M. Thus, $q \Rightarrow^* p$ in M, and the inductive step holds in this case as well.

Claim B. For all $j \geq 0$, if $q \Rightarrow^j p$ in M with $q \in E$, then $p \in$ $\varepsilon\text{-}moves(E)$.

Proof of Claim B by induction on $j \geq 0$.

Basis. Let $j = 0$. That is, $q \Rightarrow^0 p$ in M with $q \in E$, so $q = p$. Then, the algorithm includes p in $\varepsilon\text{-}moves(E)$ even before the first iteration of the **repeat** loop; formally, $p \in \varepsilon\text{-}moves(E_0)$. Thus, the basis holds.

Induction hypothesis. Assume that Claim B holds for all $j = 0, \ldots, i$, where i is a non-negative integer.

Induction step. Let $q \Rightarrow^{i+1} p$ in M with $q \in E$. Next, we first consider $p \in \varepsilon\text{-}moves(E_j)$, for some $j \leq i$; then, we study the case when $p \notin \varepsilon\text{-}moves(E_j)$, for any $j \leq i$:

(I) Let $p \in \varepsilon\text{-}moves(E_j)$, for some $j \leq i$. Recall that no iteration of the **repeat** loop removes any states from $\varepsilon\text{-}moves(E)$. Therefore, $p \in \varepsilon\text{-}moves(E)$.

(II) Let $p \notin \varepsilon\text{-}moves(E_j)$, for any $j \leq i$. As $i+1 \geq 1$, we can express $q \Rightarrow^{i+1} p$ in M as $q \Rightarrow^i o \Rightarrow p$, and the induction hypothesis implies $o \in \varepsilon\text{-}moves(E)$. R_M contains $o \rightarrow p$ because $o \Rightarrow p$ in M. As $o \in \varepsilon\text{-}moves(E)$ and $o \rightarrow p \in R_M$, the **repeat** loop adds p to $\varepsilon\text{-}moves(E)$ during iteration $i+1$, so $p \in \varepsilon\text{-}moves(E)$.

By Claims A and B, $q \Rightarrow^* p$ in M, with $q \in E$ if and only if $p \in \varepsilon\text{-}moves(E)$. Hence, Algorithm 3.1 correctly determines $\varepsilon\text{-}moves(E) = \{q \mid o \Rightarrow^* q, \text{ where } o \in E, q \in Q_M\}$, so the lemma holds. □

We are now ready to convert any FA I to an equivalent ${}_{\varepsilon\text{-free}}$FA O.

Basic idea. By using Algorithm 3.1, we find out whether $s_I \Rightarrow^* f$, with $f \in F_I(s_I$ denoting the start state of I according to Definition 3.1), which implies that I accepts ε, and if this is the case, we include s_I into F_O in order that O accepts ε as well. Furthermore, set

$$R_O = \{qa \rightarrow p \mid q \in Q_I, a \in \Delta_I, oa \rightarrow p \in R_I \\ \text{for some } o \in \varepsilon\text{-}moves(\{q\})\},$$

where ε-$moves(\{q\})$ is constructed by Algorithm 3.1. In this way, by $qa \to p \in R_O, O$ simulates $qa \Rightarrow^* oa \Rightarrow p$ in I.

Algorithm 3.2. Removal of ε-rules.

Input: An FA $I = (Q_I, \Delta_I, R_I, s_I, F_I)$.
Output: An $_{\varepsilon\text{-free}}$FA $O = (Q_O, \Delta_O, R_O, s_O, F_O)$ such that $L(O) = L(I)$.
Method:

begin
  set $Q_O = Q_I, \Delta_O = \Delta_I$ and $s_O = s_I$
  set $F_O = \{q \mid q \in Q_I, \varepsilon\text{-}moves(q) \cap F_I \neq \emptyset\}$
  set R_O to $\{qa \mid q \in Q_I, a \in \Delta_I, oa \to p$ for some
    $o \in \varepsilon\text{-}moves$ in $I\}$
end

Theorem 3.1. *Algorithm 3.2 is correct. Therefore, for every FA I, there exists an equivalent $_{\varepsilon\text{-free}}$FA O.*

Proof. Note that Algorithm 3.2 produces $\Delta_O = \Delta_I, Q_O = Q_I$, and $s_O = s_I$. Therefore, for simplicity, we just write Δ, Q, and s, respectively, throughout this proof because there exists no danger of confusion. However, we carefully distinguish F_I from F_O because they may differ from each other.

As is obvious, R_O contains no ε-rules. To establish $L(O) = L(I)$, we first prove the following claim.

Claim. For every $q, p \in Q$ and $w \in \Delta^+, qw \Rightarrow^+ p$ in O iff $qw \Rightarrow^+ p$ in I.

Proof of Claim.

Only if. By induction on $|w| \geq 1$, we next prove that for every $q, p \in Q$ and $w \in \Delta^+, qw \Rightarrow^+ p$ in O implies $qw \Rightarrow^+ p$ in I.

Basis. Let $|w| = 1$ and $qw \Rightarrow^+ p$ in O. Since $|w| = 1$, w is a symbol in Δ. As R_O contains no ε-rules, $qw \Rightarrow^+ p$ is, in effect, a one-move computation of the form $qw \Rightarrow p$ in O. Thus, $qw \to p$ in R_O. By the definition of $R_O, ow \to p \in R_I$ for some $o \in \varepsilon\text{-}moves(\{q\})$ in I, so $q \Rightarrow^+ o$ and $ow \Rightarrow p$ in I. Consequently, $qw \Rightarrow^+ p$ in I.

Induction hypothesis. Assume that the only-if part of this claim holds for all $w \in \Delta^+$, with $|w| \leq n$, where n is a non-negative integer.

Induction step. Let $qw \Rightarrow^+ p$ in O, where $q, p \in Q$ and $w \in \Delta^+$, with $|w| = n + 1$, so $|w| \geq 2$. Because $|w| \geq 1$ and R_O contains no ε-rules, we can express $qw \Rightarrow^+ p$ as $qva \Rightarrow^+ ha \Rightarrow p$, where $h \in Q, a \in \Delta, v$ is a prefix of w, and $|v| = n$. As $qv \Rightarrow^+ h$ in O and $|v| = n, qv \Rightarrow^+ h$ in I by the inductive hypothesis. Since $ha \Rightarrow p$ in $O, ha \to p \in R_O$. By the definition of $R_O, ha \Rightarrow^+ oa \Rightarrow p$ in I, where $o \in Q$. Putting $qv \Rightarrow^+ h$ and $ha \Rightarrow^+ oa \Rightarrow p$ in I together, $qva \Rightarrow^+ ha \Rightarrow^+ p$ in I. Because $va = w, qw \Rightarrow^+ p$ in I. Thus, the only-if part of the claim holds.

If. The if part of the claim is left as an exercise.

To prove that $L(O) = L(I)$, consider the above claim for $q = s$. That is, $sw \Rightarrow^+ p$ in O iff $sw \Rightarrow^+ p$ in I for all $w \in \Delta^+$, so $L(O)$-$\{\varepsilon\} = L(I)$-$\{\varepsilon\}$. As $F_O = \{q \mid q \in Q_I, \varepsilon\text{-}moves(q) \cap F_I \neq \emptyset\}, s \Rightarrow^* p$ in O with $p \in F_O$ iff $s \Rightarrow^* p \Rightarrow^* f$ in I with $f \in F_I$. Therefore, $L(O) = L(I)$, and the lemma holds. □

Example 3.2. Return to the FA in Figure 3.1. By Algorithm 3.1, we next determine ε-*moves*(1), which denotes the set of states reachable from state 1 by ε-moves. The first iteration of the **repeat** loop adds 2 and 4 to ε-*moves*(1) because $1 \to 2$ and $1 \to 4$ are ε-rules in this FA, so this iteration produces $\varepsilon\text{-}moves(1) = \{1, 2, 4\}$. During the second iteration of the **repeat** loop, this iteration adds no new state to ε-*moves*(1), so the **repeat** loop exits. As $1 \to 2$ and $1 \to 4$ are the only ε-rules in the FA, $\varepsilon\text{-}moves(i) = \{i\}$ for all the other states $i = 2, \ldots, 5$. Having constructed these sets, we use Algorithm 3.2 to transform the FA into an equivalent ${}_{\varepsilon\text{-free}}$FA. For instance, as $\varepsilon\text{-}moves(1) = \{1, 2, 4\}$ and $2b \to 2$ is a rule of the input automaton, Algorithm 3.2 introduces a rule of the form $1b \to 2$ into the output automaton. The complete list of the resulting output ${}_{\varepsilon\text{-free}}$FA, displayed in Figure 3.2, is as follows:

$1a \to 2, 1a \to 3, 1b \to 4, 1b \to 5, 2a \to 2, 2a \to 3, 3b \to 3, 4b \to 4,$

and $4b \to 5$.

■

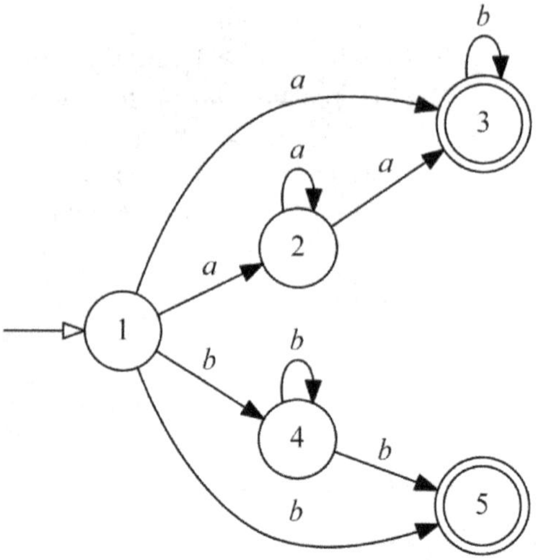

Figure 3.2. $_{\varepsilon\text{-free}}$FA.

3.3 Determinism

An $_{\varepsilon\text{-free}}$FA can make several different moves on the same symbol from the same state. For instance, take the $_{\varepsilon\text{-free}}$FA in Figure 3.2; from state 1, it can enter state 2 or 3 on a. As is obvious, this non-determinism represents an undesirable phenomenon in practice. Therefore, in this section, we explain how to convert any $_{\varepsilon\text{-free}}$FA to its deterministic version that makes no more than one move on every symbol from the same state. In other words, a deterministic FA makes a unique sequence of moves on every input string, and this property clearly simplifies its use in practice.

Definition 3.3. Let $M = (Q, \Delta, R, s, F)$ be an $_{\varepsilon\text{-free}}$FA. M is a *deterministic FA* (DFA for short) if for every $q \in Q$ and every $a \in \Delta$, there exists no more than one $p \in Q$ such that $qa \to p \in R$. ■

Next, we explain how to convert any $_{\varepsilon\text{-free}}$FA into an equivalent DFA.

Basic idea. Consider any $_{\varepsilon\text{-free}}$FA $I = (Q_I, \Delta_I, R_I, s_I, F_I)$. To turn I into an equivalent DFA $O = (Q_O, \Delta_O, R_O, s_O, F_O)$, set $Q_O = \{\langle W \rangle \mid W \subseteq Q_I\}$, that is, any set $W \subseteq Q_I$ is represented by a unique symbol $\langle W \rangle$. Furthermore, if $W \cap F_I \neq \emptyset$, include $\langle W \rangle$ in F_O. For every $\langle X \rangle \in Q_O$ and every $a \in \Delta$, add $\langle X \rangle a \to \langle Y \rangle$ to R_O with $Y = \{q \mid xa \to q \in R_I$, for some $x \in X$ and $a \in \Delta_I\}$. Consequently,

$\langle\{s_I\}\rangle w \Rightarrow^* \langle Z\rangle$ in O iff $s_I w \Rightarrow^* p$ in I, for all $p \in Z$ and $w \in \Delta^*$. Specifically, for every $\langle E\rangle \in F_O$,$\langle\{s_I\}\rangle w \Rightarrow^* \langle E\rangle$ in O iff $s_I w \Rightarrow^* f$ in I, for some $f \in E \cap F_I$. In other words, O accepts w iff I accepts w, so $L(O) = L(I)$.

Algorithm 3.3. ${}_{\varepsilon\text{-free}}$FA-to-DFA conversion.

Input: An ε-*free* FA $I = (Q_I, \Delta_I, R_I, s_I, F_I)$.
Output: A DFA $O = (Q_O, \Delta_O, R_O, s_O, F_O)$ such that $L(O) = L(I)$.
Method:

begin
  set $\Delta_O = \Delta_I$
  set $Q_O = \{\langle W\rangle \mid W \subseteq Q_I\}$
  set $F_O = \{\langle U\rangle \mid \langle U\rangle \in Q_O$ and $U \cap F_I \neq \emptyset\}$
  set $s_O = \langle\{s_I\}\rangle$
  for every $\langle X\rangle \subseteq Q_O$ and every $a \in \Delta_O$
    add $\langle X\rangle a \to \langle Y\rangle$ to R_O with $Y = \{q \mid pa \to q \in R_I, p \in X\}$
end

Theorem 3.2. *Algorithm 3.3 is correct. Therefore, for every* ${}_{\varepsilon\text{-free}}$*FA, there exists an equivalent DFA.*

Proof. Clearly, for any subset $O \subseteq Q_I$ and any input symbol $a \in \Delta_I$, there exists a single set, $P \subseteq Q_I$, satisfying the following equivalence:

$$p \in P \text{ iff } oa \to p \in R_I \text{ for some } o \in O.$$

Note that $\Delta_I = \Delta_O$. Consequently, for any $\langle O\rangle \in Q_O$ and $a \in \Delta_O$, there exists a unique state $\langle P\rangle \in Q_O$ such that $\langle O\rangle a \to \langle P\rangle \in R_O$, so O is deterministic. A rigorous proof that $L(I) = L(O)$ is left as an exercise. □

Algorithm 3.3 works simply, but this simplicity is perhaps its only advantage. Observe that it produces $O = (Q_O, \Delta_O, R_O, s_O, F_O)$, with $|Q_O| = |power(Q_I)|$ and $|R_O| = |power(Q_I) \times \Delta|$. This exponential increase in the number of states and rules represents an indisputable drawback, as illustrated by the following example.

Convention 3.3. For simplicity, we omit braces in the states of the form $\langle\{\ldots\}\rangle$ and write $\langle\ldots\rangle$ instead, for example, $\langle\{2, 3\}\rangle$ is simplified to $\langle 2, 3\rangle$. ■

Table 3.2. DFA state table.

State	a	b	State	a	b
$\langle\emptyset\rangle$	$\langle\emptyset\rangle$	$\langle\emptyset\rangle$	$\langle 1,2,3\rangle$	$\langle 2,3\rangle$	$\langle 3,4,5\rangle$
$\langle 1\rangle$	$\langle 2,3\rangle$	$\langle 4,5\rangle$	$\langle 1,2,4\rangle$	$\langle 2,3\rangle$	$\langle 4,5\rangle$
$\langle 2\rangle$	$\langle 2,3\rangle$	$\langle\emptyset\rangle$	$\langle 1,2,5\rangle$	$\langle 2,3\rangle$	$\langle 4,5\rangle$
$\langle 3\rangle$	$\langle\emptyset\rangle$	$\langle 3\rangle$	$\langle 1,3,4\rangle$	$\langle 2,3\rangle$	$\langle 3,4,5\rangle$
$\langle 4\rangle$	$\langle\emptyset\rangle$	$\langle 4,5\rangle$	$\langle 1,3,5\rangle$	$\langle 2,3\rangle$	$\langle 3,4,5\rangle$
$\langle 5\rangle$	$\langle\emptyset\rangle$	$\langle\emptyset\rangle$	$\langle 1,4,5\rangle$	$\langle 2,3\rangle$	$\langle 4,5\rangle$
$\langle 1,2\rangle$	$\langle 2,3\rangle$	$\langle 4,5\rangle$	$\langle 2,3,4\rangle$	$\langle 2,3\rangle$	$\langle 3,4,5\rangle$
$\langle 1,3\rangle$	$\langle 2,3\rangle$	$\langle 3,4,5\rangle$	$\langle 2,3,5\rangle$	$\langle 2,3\rangle$	$\langle 3\rangle$
$\langle 1,4\rangle$	$\langle 2,3\rangle$	$\langle 4,5\rangle$	$\langle 2,4,5\rangle$	$\langle 2,3\rangle$	$\langle 4,5\rangle$
$\langle 1,5\rangle$	$\langle 2,3\rangle$	$\langle 4,5\rangle$	$\langle 3,4,5\rangle$	$\langle\emptyset\rangle$	$\langle 3,4,5\rangle$
$\langle 2,3\rangle$	$\langle 2,3\rangle$	$\langle 3\rangle$	$\langle 1,2,3,4\rangle$	$\langle 2,3\rangle$	$\langle 3,4,5\rangle$
$\langle 2,4\rangle$	$\langle 2,3\rangle$	$\langle 4,5\rangle$	$\langle 1,2,3,5\rangle$	$\langle 2,3\rangle$	$\langle 3,4,5\rangle$
$\langle 2,5\rangle$	$\langle 2,3\rangle$	$\langle\emptyset\rangle$	$\langle 1,2,4,5\rangle$	$\langle 2,3\rangle$	$\langle 4,5\rangle$
$\langle 3,4\rangle$	$\langle\emptyset\rangle$	$\langle 3,4,5\rangle$	$\langle 1,3,4,5\rangle$	$\langle 2,3\rangle$	$\langle 3,4,5\rangle$
$\langle 3,5\rangle$	$\langle\emptyset\rangle$	$\langle 3\rangle$	$\langle 2,3,4,5\rangle$	$\langle 2,3\rangle$	$\langle 3,4,5\rangle$
$\langle 4,5\rangle$	$\langle\emptyset\rangle$	$\langle 4,5\rangle$	$\langle 1,2,3,4,5\rangle$	$\langle 2,3\rangle$	$\langle 3,4,5\rangle$

Example 3.3. Reconsider ${}_{\varepsilon\text{-free}}$FA in Figure 3.2. Recall its rules:

$$1a \to 2, 1a \to 3, 1b \to 4, 1b \to 5, 2a \to 2, 2a \to 3, 3b \to 3, 4b \to 4,$$
$$\text{and } 4b \to 5,$$

in which states 3 and 5 are final. This automaton works nondeterministically. For instance, from 1 on b, it can make a move to 4 or 5. With this automaton as its input, Algorithm 3.3 produces an equivalent DFA in which every state that contains 3 or 5 is final. The automaton has 32 states because $2^5 = 32$.

Note that Algorithm 3.3 may introduce some states completely uselessly. Specifically, it may introduce states that the output DFA can never reach from its start state. For instance, in Table 3.2, $\langle 1, 2\rangle$ is obviously unreachable as follows from their absence in the table columns denoted by either of the two input symbols. ■

Definition 3.4. In a DFA, $M = (Q, \Delta, R, s, F)$, a state $q \in Q$ is *reachable* if there exists $w \in \Delta^*$ such that $sw \Rightarrow^* q$ in M; otherwise, q is *unreachable* in M. ■

The following algorithm converts any ${}_{\varepsilon\text{-free}}$FA to its deterministic version that contains only reachable states.

Basic idea. Take any ${}_{\varepsilon\text{-free}}$FA, $I = (Q_I, \Delta_I, R_I, s_I, F_I)$. The following algorithm transforms I into an equivalent DFA, $O = (Q_O, \Delta_O, R_O, s_O, F_O)$, so it parallels the previous algorithm except that a new state is introduced into Q_O only if it is reachable. Initially, Q_O contains only the start state $\langle s_I \rangle$. Then, if Q_O already contains $\langle W \rangle$, where $W \subseteq Q_I$, and $P = \{p \mid oa \Rightarrow p, \text{ for some } o \in W\}$ is non-empty, where $a \in \Delta$, then we add $\langle P \rangle$ to Q_O and include $\langle W \rangle a \rightarrow \langle P \rangle$ into R_O.

Algorithm 3.4. ${}_{\varepsilon\text{-free}}$FA-to-DFA conversion without unreachable states.

Input: ${}_{\varepsilon\text{-free}}$FA $I = (Q_I, \Delta_I, R_I, s_I, F_I)$.
Output: A DFA $O = (Q_O, \Delta_O, R_O, s_O, F_O)$ such that $L(O) = L(I)$ and Q_O contains only reachable states.
Method:

begin
  set $\Delta_O = \Delta_I$
  set $Q_O = \{\langle s_I \rangle\}$
  set $s_O = \langle s_I \rangle$
  set $R_O = \emptyset$
  repeat
    if $a \in \Delta_I, X \subseteq Q_I, \langle X \rangle \in Q_O, Y = \{p \mid qa \Rightarrow p \text{ in } I$
      with $q \in X\}, Y \neq \emptyset$
      add $\langle Y \rangle$ to Q_O
      add $\langle X \rangle a \rightarrow \langle Y \rangle$ to R_O
  until no change
  set $F_O = \{\langle Y \rangle \mid \langle Y \rangle \in Q_O, Y \cap F_I \neq \emptyset\}$
end

Example 3.4. Convert ${}_{\varepsilon\text{-free}}$FA from Figure 3.2 to an equivalent DFA by using Algorithm 3.4, which initializes the start state of the DFA to $\langle 1 \rangle$. From 1 on a, the ${}_{\varepsilon\text{-free}}$FA enters 2 or 3, so introduce $\langle 2, 3 \rangle$ as a new state and $\langle 1 \rangle a \rightarrow \langle 2, 3 \rangle$ as a new rule in the output DFA. From 2 on a, the ${}_{\varepsilon\text{-free}}$FA enters 2 or 3, and from 3 on a, it does not enter any state; therefore, add $\langle 2, 3 \rangle a \rightarrow \langle 2, 3 \rangle$ to the set of rules in the DFA. Complete the construction of the DFA as an exercise.

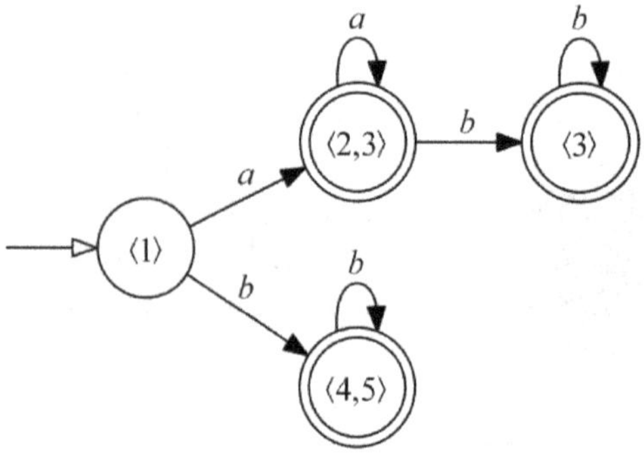

Figure 3.3. State diagram of DFA.

The resulting DFA is defined by its rules:

$$\langle 1\rangle a \to \langle 2,3\rangle, \langle 1\rangle b \to \langle 4,5\rangle, \langle 2,3\rangle a \to \langle 2,3\rangle,$$
$$\langle 2,3\rangle b \to \langle 3\rangle, \langle 3\rangle b \to \langle 3\rangle, \text{ and } \langle 4,5\rangle b \to \langle 4,5\rangle,$$

where states $\langle 2,3\rangle$, $\langle 3\rangle$, and $\langle 4,5\rangle$ are final. Figure 3.3 gives the state diagram of the resulting output DFA. ■

Besides unreachable states, a DFA may contain useless states, from which there is no computation that terminates in a final state. In the rest of this section, we explain how to detect and remove states like these. We only give a gist of this removal.

Definition 3.5. In a DFA, $M = (Q, \Delta, R, s, F)$, a state $q \in Q$ is *terminating* if there exists $w \in \Delta^*$ such that $qw \Rightarrow^* f$ in M with $f \in F$; otherwise, q is *non-terminating.* ■

Next, we sketch a two-phase transformation of any DFA into an equivalent DFA in which each state is both reachable and terminating.

Basic idea. Consider any DFA, $I = (Q_I, \Delta_I, R_I, s_I, F_I)$. For I, we construct a DFA, $O = (Q_O, \Delta_O, R_O, s_O, F_O)$, such that $L(O) = L(I)$ and each state in Q_O is both reachable and terminating by performing (1) and (2), given as follows:

(1) Determine the set of all terminating states ${}_{\text{term}}Q_I \subseteq Q_I$. Observe that all final states are automatically terminating because ε takes I from any final state to the same state by making zero moves.

Therefore, initialize ${}_{\text{term}}Q_I$ to F. If there exists $a \in \Delta$ and $p \in {}_{\text{term}}Q_I$ such that $oa \to p \in R_I$, with $o \in Q_I - {}_{\text{term}}Q_I$, then o is also terminating, so add o to ${}_{\text{term}}Q_I$. Repeat this extension of ${}_{\text{term}}Q_I$ until no more states can be added to ${}_{\text{term}}Q_I$. The resulting set ${}_{\text{term}}Q_I$ contains all terminating states in I.

(2) Without any loss of generality, suppose that Q_I only contains reachable states. Having determined all terminating states ${}_{\text{term}}Q_I$ in (1), we are now ready to construct O from I. If $s_I \notin {}_{\text{term}}Q_I$, $L(I) = \emptyset$, and the construction of O is trivial. If $s_I \in {}_{\text{term}}Q_I$, remove all non-terminating states from Q_I and eliminate all rules containing non-terminating states in R_I. Take the resulting automaton as O.

Theorem 3.3. *For every DFA I, there exists an equivalent DFA O in which all states are both reachable and terminating.* ■

Complete specification

In a general case, a DFA, $M = (Q, \Delta, R, s, F)$, may get stuck on some input strings. Indeed, for a state $q \in Q_M$ and an input symbol $a \in \Delta_M, R_M$ may have no rule with qa on its left-hand side. As a result, when occurring in q and reading a somewhere in the middle of an input string, M gets stuck, so it never completes reading the string. As is obvious, this phenomenon is ruled out if M contains a rule for every state and every input symbol because it always completely reads every single input string, which is a property often highly appreciated in theory as well as in practice. Indeed, from a theoretical viewpoint, this property simplifies achieving some significant results concerning DFAs, such as their minimization, which is described later in this section. From a practical viewpoint, it makes their implementation easier (see Section 11.1). Therefore, we explain how to adapt any DFA so it satisfies this property.

Definition 3.6. A DFA, $M = (Q, \Delta, R, s, F)$, is a *completely specified deterministic finite automaton* (CSDFA for short) if for every $q \in Q$ and every $a \in \Delta$, there exists precisely one $p \in Q$ such that $qa \to p \in R$. ■

We next sketch a transformation of any DFA, $I = (Q_I, \Delta_I, R_I, s_I, F_I)$, to a completely specified DFA, $O = (Q_O, \Delta_O, R_O, s_O, F_O)$, satisfying $L(O) = L(I)$.

Basic idea. If I is completely specified, we take O as I, and we are done. Suppose that I is not completely specified. To obtain an equivalent CSDFA O, set Q_O to $Q_I \cup \{o\}$ and R_O to $R_I \cup \{qa \to o \mid q \in Q_O, a \in \Delta_M$, and $qa \to p \notin R_M$ for any $p \in Q_I\}$, where o is a new state.

It is worth noting that $Q_O = Q_I \cup \{o\}$, where o represents a new non-terminating state. By Theorems 3.3 and 3.4, for any FA, there is an equivalent DFA that has only reachable and terminating states. As a result, we obtain the following theorem, whose rigorous proof is left as an exercise.

Theorem 3.4. *For every DFA, $I = (Q_I, \Delta_I, R_I, s_I, F_I)$, there exists a CSDFA, $O = (Q_O, \Delta_O, R_O, s_O, F_O)$, such that*

(I) $L(O) = L(I)$;
(II) $|Q_I| \leq |Q_O| \leq |Q_I| + 1$;
(III) *in Q_O, all states are reachable and no more than one state is non-terminating.* ■

Example 3.5. Return to the DFA from Example 3.4 (see Figure 3.3). Recall its rules:

$$\langle 1\rangle a \to \langle 2,3\rangle, \langle 1\rangle b \to \langle 4,5\rangle, \langle 2,3\rangle a \to \langle 2,3\rangle,$$
$$\langle 2,3\rangle b \to \langle 3\rangle, \langle 3\rangle b \to \langle 3\rangle, \text{ and } \langle 4,5\rangle b \to \langle 4,5\rangle.$$

To convert this DFA into an equivalent CSDFA, follow the basic idea above. That is, by using a new non-final state o, change this DFA to a CSDFA defined as

$$\langle 1\rangle a \to \langle 2,3\rangle, \langle 1\rangle b \to \langle 4,5\rangle, \langle 2,3\rangle a \to \langle 2,3\rangle,$$
$$\langle 2,3\rangle b \to \langle 3\rangle, \langle 3\rangle a \to o, \langle 3\rangle b \to \langle 3\rangle, \langle 4,5\rangle a \to o,$$
$$\langle 4,5\rangle b \to \langle 4,5\rangle, oa \to o, \text{ and } ob \to o.$$

■

Minimization

Even a DFA, M, may contain some states that can be merged together without affecting $L(M)$. By merging these states, we can minimize the size of M with respect to the number of states; as is obvious, a minimization like this is frequently highly appreciated in practice. Before explaining how to achieve it, we need some terminology.

Definition 3.7. Let $M = (Q, \Delta, R, s, F)$ be a DFA:

(I) A string $w \in \Delta^*$ *distinguishes* p from q, where $p, q \in Q$, if $pw \Rightarrow^* u$ and $qw \Rightarrow^* v$ in M, for some $u, v \in Q$ satisfying either ($u \in F$ and $v \in Q - F$) or ($v \in F$ and $u \in Q - F$).

(II) State p is *distinguishable* from q if there exists an input string that distinguishes p from q; otherwise, p and q are *indistinguishable*.

(III) M is a *minimum-state deterministic finite automaton* (MDFA) if each $q \in Q$ is reachable, terminating, and distinguishable from all $p \in Q - \{q\}$. ■

Next, we explain how to transform any DFA, $I = (Q_I, \Delta_I, R_I, s_I, F_I)$, into an equivalent MDFA, $O = (Q_O, \Delta_O, R_O, s_O, F_O)$. Without any loss of generality, suppose that I represents a CSDFA in which all states are reachable and no more than one state is non-terminating (see Theorem 3.4). Clearly, if $Q_I = F_I$, then $L(I) = \Delta_I^*$ because I is a CSDFA, and if $F_I = \emptyset$, then $L(I) = \emptyset$. In either case, the MDFA that accepts $L(I)$ has a single state; a trivial proof of this result is left as an exercise. Therefore, we assume that $Q_I \neq F_I$ and $F_I \neq \emptyset$ in the following transformation.

Basic idea. First, the transformation of I into O constructs the set of states Q_O so that $Q_O \subseteq \{\langle W \rangle \mid W \subseteq Q_I\}$. Initially, set Q_O to $\{\langle F_I \rangle, \langle Q_I - F_I \rangle\}$ because ε obviously distinguishes $\langle F_I \rangle$ from $\langle Q_I - F_I \rangle$, so $\langle F_I \rangle$ and $\langle Q_I - F_I \rangle$ are necessarily distinguishable. Let $a \in \Delta_I, \emptyset \subset U \subset X \subset Q_I$, and $\langle X \rangle \in Q_O$. If $\{y \mid ua \Rightarrow y \text{ in } I, u \in U\}$ and $\{z \mid va \Rightarrow z \text{ in } I, v \in X - U\}$ are two non-empty disjoint sets, then any state $u \in U$ is distinguishable from any state $v \in X - U$, so replace $\langle X \rangle$ with the two states $\langle U \rangle$ and $\langle X - U \rangle$ in Q_O and, thereby, increase $|Q_O|$ by one. Keep extending Q_O in this way until no further extension is possible. As I may contain one non-terminating state (see Definition 3.6 and Theorem 3.4), the resulting output automaton may have a non-terminating state as well; if this is the case, remove it, as well as all rules containing it. To complete the construction of O, set $s_O = \langle Y \rangle$, where $\langle Y \rangle \in Q_O$, with $s_I \in Y, F_O = \{\langle Y \rangle \mid Y \subseteq F_I\}$, and $R_O = \{\langle X \rangle a \rightarrow \langle Y \rangle \mid qa \Rightarrow p, q \in X, p \in Y, a \in \Delta_I\}$.

Algorithm 3.5. DFA minimization.

Input: A CSDFA $I = (Q_I, \Delta_I, R_I, s_I, F_I)$ such that all states are reachable, no more than one state is non-terminating, $Q_I \neq F_I$, and $F_I \neq \emptyset$.

Output: An MDFA $O = (Q_O, \Delta_O, R_O, s_O, F_O)$ such that $L(O) = L(I)$.

Method:

begin
  set $\Delta_O = \Delta_I$
  set $Q_O = \{\langle F_I \rangle, \langle Q_I - F_I \rangle\}$
  while there exist $a \in \Delta_I, U, X \subset Q_I$ such that
    $\emptyset \subset U \subset X$,
    $\langle X \rangle \in Q_O$, and
    $\{y \mid ua \Rightarrow y \text{ in } I, u \in U\}$ and $\{z \mid va \Rightarrow z \text{ in } I, v \in X - U\}$
    are two non-empty
    disjoint sets **do**
  begin
    set $Q_O = (Q_O - \{\langle X \rangle\}) \cup \{\langle U \rangle, \langle XU \rangle\}$
  end
  set $s_O = \langle Y \rangle$, where $\langle Y \rangle \in Q_O$ with $s_I \in Y$
  set $R_O = \{\langle X \rangle a \rightarrow \langle Y \rangle \mid qa \Rightarrow p, q \in X, p \in Y, a \in \Delta\}$
  set $F_O = \{\langle Y \rangle \mid Y \subseteq F_I\}$
  if $\langle Z \rangle \in Q_O - F_O$ **and** $\langle Z \rangle$ is non-terminating **then**
    set $R_O = R_O - \{\langle X \rangle a \rightarrow \langle Y \rangle \mid \{\langle X \rangle, \langle Y \rangle\} \cap \{\langle Z \rangle\} \neq \emptyset,$
    $a \in \Delta\}$
    remove $\langle Z \rangle$ from Q_O
end

As an exercise, based on the basic idea that precedes Algorithm 3.5, prove the following theorem.

Theorem 3.5. *Algorithm 3.5 is correct. Therefore, for every CSDFA, there exists an equivalent MDFA.* ■

There exist many alternative algorithms that also minimize a DFA. Nevertheless, from any DFA, they all construct an essentially unique

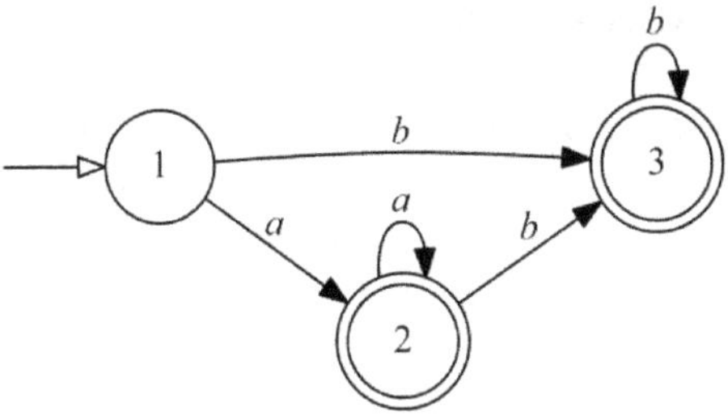

Figure 3.4. Minimum-state finite automaton.

MDFA O equivalent to I, that is, any other MDFA equivalent to I completely coincides with O except that its names of states may differ. A proof of this crucial result is omitted because it is beyond the scope of this introductory text.

Example 3.6. Return to the CSDFA from Example 3.5. By Algorithm 3.5, convert this automaton into an equivalent MDFA O. This transformation first starts with two states — $\langle\{\langle 1\rangle, o\}\rangle$ and $\langle\{\langle 2,3\rangle,\langle 3\rangle,\langle 4,5\rangle\}\rangle$, which correspond to the sets of final and non-final states in the CSDFA. After this start, the **while** loop replaces $\langle\{\langle 2,3\rangle,\langle 3\rangle,\langle 4,5\rangle\}\rangle$ with $\langle\{\langle 2,3\rangle\}\rangle$ and $\langle\{\langle 3\rangle,\langle 4,5\rangle\}\rangle$, because of these rules:

$$\langle 2,3\rangle a \to \langle 2,3\rangle, \langle 2,3\rangle b \to \langle 3\rangle, \langle 3\rangle a \to o, \langle 3\rangle b \to \langle 3\rangle, \langle 4,5\rangle a \to o,$$

$$\langle 4,5\rangle b \to \langle 4,5\rangle.$$

Then, this loop replaces $\langle\{\langle 1\rangle, o\}\rangle$ with $\langle\{\langle 1\rangle\}\rangle$ and $\langle\{o\}\rangle$. After this replacement, the set of states can no longer be extended by the algorithm. For simplicity and readability, rename states $\langle\{\langle 1\rangle\}\rangle$,$\langle\{\langle 2,3\rangle\}\rangle$,$\langle\{\langle 3\rangle,\langle 4,5\rangle\}\rangle$, and $\langle\{\text{ o }\}\rangle$ to 1, 2, 3, and 4, respectively. States 2 and 3 are final. Construct the set of rules

$$1 \to 2, 1b \to 3, 2 \to 2, 2b \to 3, 3 \to 4, 3b \to 3, 4 \to 4, \text{ and } 4b \to 4.$$

State 4 is non-terminating. Remove it, as well as all the rules that contain 4. The resulting MDFA O is given in Figure 3.4. ■

To sum up the previous theorems of this chapter (see Theorems 3.1–3.5), all the variants of FAs are equally powerful, as stated in the following corollary.

Corollary 3.1. *FAs, $_{\varepsilon\text{-free}}$FAs, DFAs, CSDFAs, and MDFAs are equally powerful. They all characterize $\mathbf{L}_{\mathbf{FA}}$.* ■

3.4 Regular Expressions

In this section, we introduce regular expressions and define regular languages with them. Then, we demonstrate that these expressions and FAs are equivalent — more precisely, they both characterize the family of regular languages.

Regular expressions

Regular expressions represent simple language-denoting formulas based on the operations of concatenation, union, and closure.

Definition 3.8. Let Δ be an alphabet. The *regular expressions* (REs for short) over Δ and the *languages that these expressions denote* are defined recursively as follows:

(I) $\emptyset$ is an RE denoting the empty set;
(II) ε is an RE denoting $\{\varepsilon\}$;
(III) a, where $a \in \Delta$, is an RE denoting $\{a\}$;
(IV) if r and s are REs denoting the languages R and S, respectively, then

  (i) $(r|s)$ is an RE denoting $R \cup S$,
  (ii) (rs) is an RE denoting RS,
  (iii) $(r)^*$ is an RE denoting R^*.

The language denoted by an RE, r, is symbolically written as $L(r)$. A language $L \subseteq \Delta^*$ is *regular* iff there exists an RE, r, satisfying $L = L(r)$. The family of all regular languages is denoted by $\mathbf{L_{REG}}$. ■

In practice as well as in theory, REs sometimes use + instead of |. This book uses the latter throughout.

REs are said to be *equivalent* if they define the same language. For instance, as a simple exercise, prove that $(a)^*$ and $((a)^*\varepsilon)$ are equivalent, but $(a)^*$ and $((a)^*\emptyset)$ are not. In practice, we frequently simplify the *fully parenthesized REs* created strictly by this definition by reducing the number of parentheses in them. To do so, we introduce the following convention concerning the priority of the three operators.

Convention 3.4. We automatically assume that * has higher precedence than the operation concatenation, which has higher precedence than |. In addition, the expressions rr^* and r^*r are usually written as r^+. ■

Based on Convention 3.4, we can make REs more succinct and readable. For instance, instead of $((a^*)(b(b^*))$, we can write $(a^*)(b^+)$ or even a^*b^+ according to Convention 3.4. From a broader point of view, Convention 3.4 allows us to establish general equivalences between REs. To illustrate, all the following REs are equivalent:

$$(r)^*, (r^*), ((r|\emptyset)^*), (r|\emptyset)^*, r^*|\emptyset^*, \text{ and } r^*.$$

Specifically, instead of (r^*), we can thus equivalently write $(r)^*$ or r^*, and we frequently make use of these equivalent expressions later in this section (see, for instance, the proof of Lemma 3.2). There exist a number of other equivalences as well as algebraic laws obeyed by REs, and many of them allow us to further simplify REs, as demonstrated in Exercises.

Next, this section gives an example of how REs are applied to English vocabulary.

Example 3.7. We open this example in terms of formal English. Let Δ be the English alphabet, together with a hyphen (-). Let $W \subseteq \Delta^*$ be the set of all possible well-formed English words, including compound words, such as *made-up*. Note that W is infinite. For instance, the subset $V \subseteq W$ contains infinitely many words of the form

$$(great\text{-})^i grandparents,$$

for all $i \geq 0$. Purely theoretically, they all represent well-formed English words, although most of them, such as

great-great-great-great-great-great-great-great-great-great-grandparents,

are rather uncommon words, which most people never utter during their lifetimes. Although V represents an infinite set of rather bizarre

English words, it is easily and rigorously defined by a short regular expression of the form

$$(great\text{-})^*grandparents.$$

To illustrate the use of REs in terms of informal English, *blah* is a word that is repeatedly used to describe words or feelings where the specifics are considered unimportant or boring to the speaker or writer. An example of this use might be a sentence like the following:

He keeps making excuses that he is tired and sleepy blah-blah-blah.

As the repetition of blah is unlimited, the set of all these compound words contains

$$blah, blah\text{-}blah, blah\text{-}blah\text{-}blah, blah\text{-}blah\text{-}blah\text{-}blah, \ldots.$$

Simply put, this set represents the infinite language $\{blah\text{-}\}^*\{blah\}$. Observe that this language is denoted by an RE of the form $(blah\text{-})^*blah$. ■

Equivalence with FAs

First, we explain how to transform any DFA into an equivalent RE. Then, we describe the transformation of any RE into an equivalent FA and, thereby, establish the equivalence between REs and FAs.

From FAs to REs

Next, we explain how to convert any DFA into an equivalent RE.

Basic idea. Let $M = (Q, \Delta, R, s, F)$ be a DFA and $|Q| = n$, for some $n \geq 1$. Rename states in M so that $Q = \{q_1, \ldots, q_n\}$, with $q_1 = s$. For all $i, j = 1, \ldots, n$ and all $k = 0, \ldots, n$, let $[i, k, j]$ denote the language consisting of all input strings on which M makes a computation from q_i to q_j, during which every state q_l that M enters and leaves satisfies $l \leq k$. That is, for all $x \in \Delta^*$, $x \in [i, k, j]$ iff $q_i x \Rightarrow^+ q_j$ and for every $y \in prefix(x) - \{\varepsilon, x\}, q_i y \Rightarrow^+ q_l$, with $l \leq k$.

Definition 3.9. For all $k = 0, \ldots, n$, we recursively define $[i, k, j]$ as

(1) $[i, 0, j] = \{a \mid a \in \Delta \cup \{\varepsilon\}, q_i a \Rightarrow^m q_j$, with $m \leq 1\}$;
(2) for $k \geq 1$, $[i, k, j] = [i, k-1, j] \cup [i, k-1, k][k, k-1, k]^* [k, k-1, j]$. ■

According to (1), $a \in [i, 0, j]$ iff $q_i a \to q_j \in R$, and $\varepsilon \in [i, 0, j]$ iff $i = j$. According to (2), $[i, k-1, j] \subseteq [i, k, j]$, so $[i, k, j]$ contains all input strings that take M from q_i to q_j, during which it never passes through q_l, with $l \geq k-1$. Furthermore, $[i, k, j]$ includes $[i, k-1, k][k, k-1, k]^*[k, k-1, j]$, which contain all input strings $x = uvw$ such that (a) u takes M from q_i to q_k, (b) v takes M from q_k back to q_k zero or more times, and (c) w takes M from q_k to q_j. Observe that during (a)–(c), M never enters and leaves q_l, with $l \geq k-1$. Furthermore, note that M performs (b) zero times iff $v = \varepsilon$. Based on these properties, the following lemma demonstrates that for every $[i, k, j]$, there is an RE ${}_{i,k,j}E$ such that $L({}_{i,k,j}E) = [i, k, j]$. From this lemma, Theorem 3.6 derives that $L(M)$ is denoted by the RE ${}_{1,n,j_1}E|{}_{1,n,j_2}E| \cdots |{}_{1,n,j_h}E$, with $F = \{q_{j_1}, q_{j_2}, \ldots, q_{j_h}\}$.

Lemma 3.2. *For all $i, j = 1, \ldots, n$ and all $k = 0, \ldots, n$, there exists an RE ${}_{i,k,j}E$ such that $L({}_{i,k,j}E) = [i, k, j]$, where $[i, k, j]$ has the meaning given in Definition 3.9.*

Proof by induction on $k \geq 0$

Basis. Let $i, j \in \{1, \ldots, n\}$ and $k = 0$. By (1) in Definition 3.9, $[i, 0, j] \subseteq \Delta \cup \{\varepsilon\}$, so there surely exists an RE ${}_{i,0,j}E$ such that $L({}_{i,0,j}E) = [i, 0, j]$.

Induction hypothesis. Suppose that there exist $l \in \{1, \ldots, n\}$ such that for each $i, j \in \{1, \ldots, n\}$ and $k \leq l-1$, there exists an RE ${}_{i,k,j}E$ such that $L({}_{i,k,j}E) = [i, k, j]$.

Induction step. Consider any $[i, k, j]$, where $i, j \in \{1, \ldots, n\}$ and $k = l$. By the recursive formula (2) above, $[i, k, j] = [i, k-1, j] \cup [i, k-1, k][k, k-1, k]^*[k, k-1, j]$. By the induction hypothesis, there exist REs ${}_{i,k-1,j}E$, ${}_{i,k-1,k}E$, ${}_{k,k-1,k}E$, and ${}_{k,k-1,j}E$ such that $L({}_{i,k-1,j}E) = [i, k-1, j]$, $L({}_{i,k-1,k}E) = [i, k-1, k]$, $L({}_{k,k-1,k}E) = [k, k-1, k]$, and $L({}_{k,k-1,j}E) = [k, k-1, j]$. Thus, $[i, k, j]$ is denoted by the RE ${}_{i,k-1,j}E|{}_{i,k-1,k}E({}_{k,k-1,k}E)^*{}_{k,k-1,j}E$. □

Theorem 3.6. *For every DFA M, there exists an equivalent RE E.*

Proof. By the definition of $[i, k, j]$ for $k = n$ (see Definition 3.9), where $i, j \in \{1, \ldots, n\}$, $[i, n, j] = \{x \mid x \in \Delta^*, q_i x \Rightarrow^* q_j\}$. Thus, $L(M) = \{x \mid x \in \Delta^*, x \in [1, n, j], q_j \in F\}$. Thus,

$$L(M) = {}_{1,n,j_1}E|{}_{1,n,j_2}E|\cdots|{}_{1,n,j_h}E,$$

where $F = \{q_{j_1}, q_{j_2}, \ldots, q_{j_h}\}$. Consequently, the theorem holds true. □

Example 3.8. Consider the DFA M defined by its rules:

$$q_1 a \to q_1, \quad q_1 b \to q_2, \quad \text{and} \quad q_2 b \to q_2,$$

where q_1 is the start state and q_2 is the only final state. Note that $L(M) = \{a\}^*\{b\}^+$. Following the idea described above, construct

$${}_{1,0,1}E = a|\varepsilon, {}_{1,0,2}E = b, {}_{2,0,2}E = b|\varepsilon, {}_{2,0,1}E = \emptyset.$$

Furthermore, we obtain

$$\begin{aligned}
{}_{1,1,1}E &= {}_{1,0,1}E|{}_{1,0,1}E({}_{1,0,1}E)^*{}_{1,0,1}E = (a|\varepsilon)|(a|\varepsilon)(a|\varepsilon)^*(a|\varepsilon) = a^*,\\
{}_{1,1,2}E &= {}_{1,0,2}E|{}_{1,0,1}E({}_{1,0,1}E)^*{}_{1,0,2}E = b|(a|\varepsilon)(a|\varepsilon)^*(b) = a^*b,\\
{}_{2,1,1}E &= {}_{2,0,1}E|{}_{2,0,1}E({}_{1,0,1}E)^*{}_{1,0,1}E = \emptyset|\emptyset(a|\varepsilon)^*(a|\varepsilon) = \emptyset,\\
{}_{2,1,2}E &= {}_{2,0,2}E|{}_{2,0,1}E({}_{1,0,1}E)^*{}_{1,0,2}E = (b|\varepsilon) \mid \emptyset(a|\varepsilon)^*b = b|\varepsilon,\\
{}_{1,2,1}E &= {}_{1,1,1}E|{}_{1,1,2}E({}_{2,1,2}E)^*{}_{2,1,1}E = a^*|a^*b(b|\varepsilon)^*\emptyset = a^*,\\
{}_{1,2,2}E &= {}_{1,1,2}E|{}_{1,1,2}E({}_{2,1,2}E)^*{}_{2,1,2}E = a^*b|a^*b(b|\varepsilon)^*(b|\varepsilon) = a^*b^+,\\
{}_{2,2,1}E &= {}_{2,1,1}E|{}_{2,1,2}E({}_{2,1,2}E)^*{}_{2,1,1}E = \emptyset|(b|\varepsilon)(b|\varepsilon)^*\emptyset = \emptyset,\\
{}_{2,2,2}E &= {}_{2,1,2}E|{}_{2,1,2}E({}_{2,1,2}E)^*{}_{2,1,2}E = (b|\varepsilon)|(b|\varepsilon)(b|\varepsilon)^*(b|\varepsilon) = b^*.
\end{aligned}$$

M has two states — q_1 and q_2, where q_1 is the start state and q_2 is the only final state. Therefore, ${}_{1,2,2}E$ denotes $L(M)$. Indeed, $L(M) = \{a\}^*\{b\}^+ = L({}_{1,2,2}E) = L(a^*b^+)$. ■

From REs to FAs

Consider fully parenthesized REs over an alphabet Δ, that is, REs constructed strictly according to Definition 3.8 without involving any simplification introduced in Convention 3.4. Next, we transform these expressions into equivalent FAs. To achieve this transformation, we first prove the following three statements:

(I) There exist FAs equivalent to the trivial REs $\emptyset$,ε, and $a \in \Delta$.
(II) For any pair of REs, I and J, there exists an FA that accepts $L(I) \cup L(J)$ and $L(I)L(J)$.
(III) For any RE, I, there exists an FA that accepts $L(I)^*$.

By induction on the number of operators occurring in REs, we then make use of these statements to obtain the desired transformation that turns any RE into an equivalent FA.

Lemma 3.3. *There exist FAs that accept the empty set,* $\{\varepsilon\}$, *and* $\{a\}$, *with* $a \in \Delta$.

Proof. First, any FA with no final state accepts the empty set. Second, $\{\varepsilon\}$ is accepted by any one-state FA that contains no rule. Its only state is the start state; simultaneously, this state is final. As is obvious, this automaton accepts $\{\varepsilon\}$. Third, let $a \in \Delta$. Consider a one-rule FA defined by one rule, $sa \to f$, where s is the start non-final state and f is the only final state. Clearly, $L(M) = \{a\}$. □

Next, we convert any two FAs, I and J, into an FA O satisfying $L(O) = L(I) \cup L(J)$.

Basic idea (see Figure 3.5). Let us consider any two FAs, I and J. Without any loss of generality, we assume that I and J have disjoint sets of states (if I and J contain some states in common, we rename states in either I or J so that they have no state in common). Construct O so that from its start state s_O, it enters s_I or s_J by an ε-move. From s_I, O simulates I, and from s_J, it simulates J. Whenever occurring in a final state of I or J, O can enter its only final state f_O by an ε-move and stop. Thus, $L(O) = L(I) \cup L(J)$.

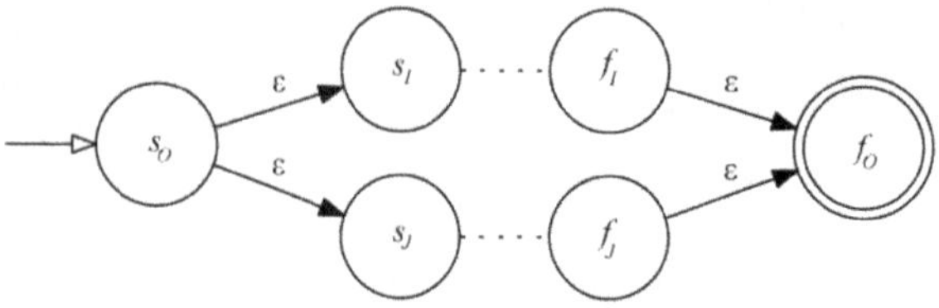

Figure 3.5. FA for union.

Algorithm 3.6. FA for union.

Input: Two FAs, $I = (Q_I, \Delta_I, R_I, s_I, F_I)$ and $J = (Q_J, \Delta_J, R_J, s_J, F_J)$, such that $Q_I \cap Q_J = \emptyset$.

Output: An FA $O = (Q_O, \Delta_O, R_O, s_O, F_O)$ such that $L(O) = L(I) \cup L(J)$.

Method:

begin
set $\Delta_O = \Delta_I \cup \Delta_J$
set $Q_O = Q_I \cup Q_J$
introduce two new states, s_O and f_O, into Q_O
set $F_O = \{f_O\}$
set $R_O = R_I \cup R_J \cup \{s_O \to s_I, s_O \to s_J\} \cup \{p \to f_O \mid p \in F_I \cup F_J\}$
end

Lemma 3.4. *Algorithm 3.6 is correct.*

Proof. To establish $L(O) = L(I) \cup L(J)$, we first prove the following claim.

Claim. For every $w \in \Delta_O^*, qw \Rightarrow^* p$ in O iff $qw \Rightarrow^* p$ in K, with $K \in \{I, J\}, q, p \in Q_K$.

Proof of Claim.

Only if. By induction on $i \geq 0$, we prove for every $w \in \Delta_O^*$ and every $i \geq 0$,

$$qw \Rightarrow^i p \text{ in } O \text{ implies } qw \Rightarrow^i p \text{ in } K,$$

where $K \in \{I, J\}$ and $q, p \in Q_K$. As is obvious, the only-if part follows from this implication.

Basis. Let $i = 0$, so $qw \Rightarrow^0 p$ in O, where $q, p \in Q_K$, and $K \in \{I, J\}$. Then, $q = p$ and $w = \varepsilon$. Clearly, $q \Rightarrow^0 q$ in K, and the basis holds.

Induction hypothesis. Assume that the implication holds for all $0 \leq i \leq n$, where $n \in {}_0\mathbb{N}$.

Induction step. Let $qw \Rightarrow^{n+1} p$ in O, where $q, p \in Q_K$ for some $K \in \{I, J\}$, and $w \in \Delta_O^*$. Because $n + 1 \geq 1$, express $qw \Rightarrow^{n+1} p$ as $qva \Rightarrow^n oa \Rightarrow p$ in O, where $o \in Q_K, w = va$, with $a \in \Delta_O \cup \{\varepsilon\}$ and $v \in \Delta_O^*$. Since $oa \Rightarrow p$ in $O, oa \rightarrow p \in R_O$. Recall that $o, p \in Q_K$. Thus, by the definition of R_O (see Algorithm 3.6), R_K contains $oa \rightarrow p$, so $oa \Rightarrow p$ in K. By the inductive hypothesis, $qv \Rightarrow^n o$ in K. Putting $qv \Rightarrow^n o$ and $oa \Rightarrow p$ in K together, $qva \Rightarrow^n oa \Rightarrow p$ in K. Because $va = w, qw \Rightarrow^{n+1} p$ in K, which completes the induction step.

If. By analogy with the only-if part of the claim, prove its if part as an exercise.

Consequently, the claim holds true.

Observe that O makes every accepting computation (see Definition 3.1) in this way. It starts by applying a rule from $\{s_O \rightarrow s_I, s_O \rightarrow s_J\}$ and ends by applying a rule from $\{p \rightarrow f_O \mid p \in F_I \cup F_J\}$. More formally, O accepts every $w \in L(O)$ by a computation of the form

$$s_O w \quad \Rightarrow \quad s_K w \quad \Rightarrow^* \quad f_K \Rightarrow f_O,$$

where $K \in \{I, J\}$ and $f_K \in K_F$. Consider the previous claim for $q = s_K$ and $p = K_F$ to obtain

$$s_K w \quad \Rightarrow^* \quad f_K \text{ in } O \text{ iff } s_K w \quad \Rightarrow^* \quad f_K \text{ in } K.$$

Thus,

$$s_O w \quad \Rightarrow \quad s_K w \quad \Rightarrow^* \quad f_K \quad \Rightarrow \quad F_O \text{ iff } s_K w \quad \Rightarrow^* \quad f_K \text{ in } K.$$

Hence, $w \in L(O)$ iff $w \in L(K)$. In other words, $L(O) = \{w \in L(K) \mid K \in \{I, J\}\}$. That is, $L(O) = L(I) \cup L(J)$, and the lemma holds. □

Before going any further, we need the notion of a stop state, used in Algorithm 3.7.

Definition 3.10. Let M be an FA. In Q_M, a *stop state* is a state that does not occur on the left-hand side of any rule in R_M. ■

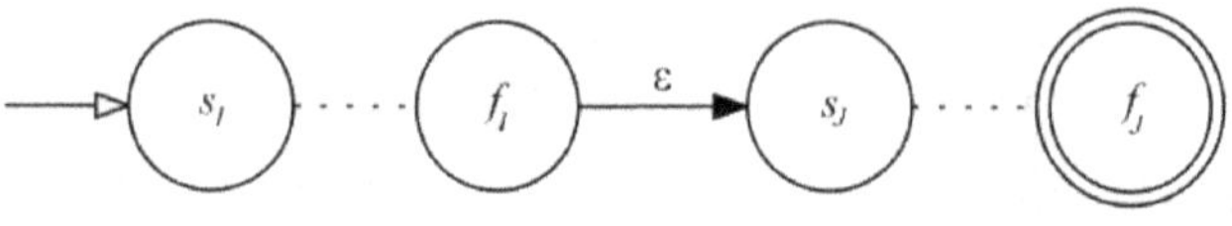

Figure 3.6. FA for concatenation.

By this definition, any FA can never leave any of its stop states. By the following lemma, without any loss of generality, we can always assume that an FA has precisely one final state, which is also a stop state.

Lemma 3.5. *For every FA I, there exists an equivalent FA O such that $F_O = \{f_O\}$ and f_O is a stop state.*

Proof. Let I be an FA. Take any FA J such that $L(J) = \emptyset$ (see Lemma 3.3). By using Algorithm 3.6, convert I and J to an FA O satisfying $L(O) = L(I) \cup L(J) = L(I) \cup \emptyset = L(I)$. Observe that O constructed in this way has a single final state F_O, which is also a stop state. □

We are now ready to convert any pair of FAs, I and J, into an FA O that accepts $L(O) = L(I)L(J)$.

Basic idea (see Figure 3.6). Consider any two FAs, I and J, such that $Q_I \cap Q_J = \emptyset$. Without any loss of generality, suppose that I has a single final state, f_I, such that f_I is also a stop state, and J has also only one final state, f_J, and this state is a stop state (see Lemma 3.5). Construct O as follows. Starting from s_I, O simulates I until it enters f_I, from which O makes an ε-move to s_J. From s_J, O simulates J until O enters f_J, which is also the only final state in O. Thus, $L(O) = L(I)L(J)$.

Lemma 3.6. *Algorithm 3.7 is correct.*

Proof. Note that O accepts every $w \in L(O)$ by a computation of the form

$$s_I uv \quad \Rightarrow^* \quad f_I v \quad \Rightarrow \quad s_J v \quad \Rightarrow^* \quad f_J,$$

where $w = uv$. Thus, $s_I u \Rightarrow^* f_I$ in I, and $s_J v \Rightarrow^* f_J$ in J. Therefore, $u \in L(I)$ and $v \in L(J)$, so $L(O) \subseteq L(I)L(J)$. In a similar way, demonstrate $L(I)L(J) \subseteq L(O)$. Hence, $L(O) = L(I)L(J)$. A rigorous version of this proof is left as an exercise. □

Algorithm 3.7. FA for concatenation.

Input: Two FAs, $I = (Q_I, \Delta_I, R_I, s_I, F_I)$ and $J = (Q_J, \Delta_J, R_J, s_J, F_J)$, such that $Q_I \cap Q_J = \emptyset$, $F_I = \{f_I\}, F_J = \{f_J\}, f_I$ and f_J are both stop states (see Lemma 3.5).
Output: An FA $O = (Q_O, \Delta_O, R_O, s_O, F_O)$ such that $L(O) = L(I)L(J)$.
Method:

begin
set $\Delta_O = \Delta_I \cup \Delta_J$
set $Q_O = Q_I \cup Q_J$
set $s_O = s_I$
set $F_O = \{f_J\}$
set $R_O = R_I \cup R_J \cup \{f_I \to s_J\}$
end

We now convert any FA I to an FA O satisfying $L(I) = L(O)^*$.

Basic idea (Figure 3.7). Consider an FA, I. Without any loss of generality, suppose that I has a single final state f_I such that f_I is also a stop state (see Lemma 3.5). Apart from all states from I, O has two new states, s_O and f_O, where s_O is its start state and f_O is its only final state. From s_O, O can enter f_O without reading any symbol, thus accepting ε. In addition, it can make an ε-move to s_I to simulate I. Occurring in f_I, O can enter s_I or f_O by making an ε-move. If O enters s_I, it starts simulating another sequence of moves in I. If O enters f_O, it successfully completes its computation. Therefore, $L(O) = L(I)^*$.

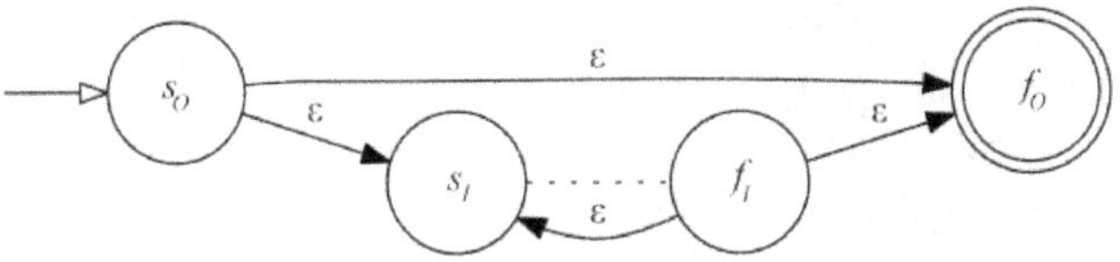

Figure 3.7. FA for iteration.

Algorithm 3.8. FA for iteration.

Input: An FAs $I = (Q_I, \Delta_I, R_I, s_I, F_I)$ such that $F_I = \{f_I\}$ and f_I is a stop state (see Lemma 3.5).
Output: An FA $O = (Q_O, \Delta_O, R_O, s_O, F_O)$ such that $L(O) = L(I)^*$.
Method:

begin
  set $\Delta_O = \Delta_I$
  set $Q_O = Q_I$
  introduce two new states, s_O and f_O, into Q_O
  set $F_O = \{f_O\}$
  set $R_O = R_I \cup \{s_O \rightarrow f_O, s_O \rightarrow s_I, f_I \rightarrow s_I, f_I \rightarrow f_O\}$
end

Lemma 3.7. *Algorithm 3.8 is correct.*

Proof. To see the reason why $L(O) = L(I)^*$, observe that O accepts every $w \in L(O) - \{\varepsilon\}$ in this way

$$\begin{aligned} s_O v_1 v_2 \cdots v_n &\Rightarrow^* s_I v_1 v_2 \cdots v_n \\ &\Rightarrow^* f_I v_2 \cdots v_n \\ &\Rightarrow\ s_I v_2 \cdots v_n \\ &\Rightarrow^* f_I v_3 \cdots v_n \\ &\ \ \vdots \\ &\Rightarrow\ s_I v_n \\ &\Rightarrow^* f_I \\ &\Rightarrow\ f_O, \end{aligned}$$

where $n \in \mathbb{N}, w = v_1 v_2 \cdots v_n$, with $v_i \in L(I)$ for all $i = 1, \ldots, n$. Furthermore, O accepts ε as $s_O\varepsilon \Rightarrow f_O$. Therefore, $L(I)^* \subseteq L(O)$. Similarly, prove $L(O) \subseteq L(I)^*$. As $L(I)^* \subseteq L(O)$ and $L(O) \subseteq L(I)^*, L(O) = L(I)^*$. A rigorous version of this proof is left as an exercise. □

We are now ready to prove that for any RE, there is an equivalent FA. In the proof, we consider fully parenthesized REs defined strictly according to Definition 3.8 (that is, we do not consider their simplified versions introduced in Convention 3.4).

Theorem 3.7. *Let r be an RE; then, there exists an FA M such that $L(r) = L(M)$.*

Proof by induction on the number of operators in r. Let r be a fully parenthesized RE over an alphabet Δ (see Definition 3.8).

Basis. Let r be a fully parenthesized RE that contains no operator. By Definition 3.8, r is of the form $\emptyset, \varepsilon$, or a, with $a \in \Delta$. Then, this basis follows from Lemma 3.3.

Induction hypothesis. Suppose that Theorem 3.7 holds for every RE containing n or fewer operators, for some $n \in {}_0\mathbb{N}$.

Induction step. Let e be any fully parenthesized RE containing $n+1$ operators. Thus, e is of the form $(r|s)$, (rs), or $(r)^*$, where r and s are REs with no more than n operators, so for r and s, Theorem 3.7 holds by the induction hypothesis. These three forms of e are considered in the following:

(I) Let $e = (r|s)$, so $L(e) = L(r) \cup L(s)$. As Theorem 3.7 holds for r and s, there are FAs I and J such that $L(r) = L(I)$ and $L(s) = L(J)$. By using Algorithm 3.6, verified by Lemma 3.4, convert I and J into an FA O satisfying $L(O) = L(I) \cup L(J)$, so $L(O) = L(e)$.

(II) Let $e = (rs)$, so $L(e) = L(r)L(s)$. Let I and J be FAs such that $L(r) = L(I)$ and $L(s) = L(J)$. By using Algorithm 3.7, turn I and J into an FA O satisfying $L(O) = L(I)L(J)$, so $L(O) = L(e)$.

(III) Let $e = (r)^*$, so $L(e) = L((r)^*)$. Let I be an FA such that $L(r) = L(I)$. By using Algorithm 3.8, convert I into an FA O satisfying $L(O) = L(I)^* = L(r)^* = L(e)$. □

Taking a closer look at the previous proof, we see that it actually presents a method of converting any fully parenthesized RE into an equivalent FA, as described in the following.

Table 3.3. REs and FAs.

	RE	FA
1	a	M_1
2	b	M_2
3	c	M_3
4	$(b\|c)$	M_4
5	$((b\|c))^*$	M_5
6	$(a((b\|c))^*)$	M_6

Basic idea. Consider any RE. Processing from the innermost parentheses out, determine how the expression is constructed by Definition 3.8. Follow the conversion described in the previous proof and simultaneously create the equivalent FAs by the algorithms referenced in this proof. More specifically, let r and s be two REs obtained during the construction:

(1) Let r be an RE of the form ε or a, where $a \in \Delta$. Turn r into an equivalent FA by the method given in the proof of Lemma 3.3.
(2) Let r and s be two REs, and let I and J be FAs equivalent to r and s, respectively. Then,
  (i) $(r|s)$ is equivalent to the FA constructed from I and J by Algorithm 3.6;
  (ii) (rs) is equivalent to the FA constructed from I and J by Algorithm 3.7;
  (iii) $(r)^*$ is equivalent to the FA constructed from I by Algorithm 3.8.

Leaving an algorithm that rigorously reformulates this method as an exercise, we next illustrate it by an example.

Example 3.9. Consider $(a((b|c))^*)$ as a fully parenthesized RE. By Definition 3.8, we see that this expression is constructed step by step by numbering expressions $a, b, c, (b|c), ((b|c))^*$, and $(a((b|c))^*)$ as 1, 2, 3, 4, 5, and 6, respectively, for brevity; in this example, we construct the six FAs, M_1–M_6, so M_i is equivalent to RE i, where $i = 1, \ldots, 6$ (see Table 3.3). During the construction of $M_i, 1 \leq i \leq 5$, we always introduce two new states, denoted by s_I and f_I (during the construction of M_6, we need no new state). ■

Consider the first three elementary subexpressions a, b, and c that denote the languages $\{a\}$,$\{b\}$, and $\{c\}$, respectively. Based on the construction given in the proof of Lemma 3.3, we construct M_1, M_2, and M_3 that accept $\{a\}$,$\{b\}$, and $\{c\}$, respectively. From expressions b and c, we make expression $(b|c)$ that denotes $\{b\} \cup \{c\}$. Recall that $_2$ and M_3 are equivalent to expressions b and c, respectively, that is, $L(M_2) = \{b\}$ and $L(M_3) = \{c\}$. Thus, with M_2 and M_3 as the input of Algorithm 3.6, we construct M_4 that accepts $L(M_2) \cup L(M_3) = \{b\} \cup \{c\} = \{b, c\}$. From $(b|c)$, we can make $((b|c))^*$ that denotes $(\{b\} \cup \{c\})^* = \{b, c\}^*$. Recall that M_4 is equivalent to RE 4. Therefore, with M_4 as the input of Algorithm 3.8, we construct M_5 equivalent to RE 5. From RE 1 and RE 5, we make RE 6 that denotes $\{a\}(\{b\}\cup\{c\})^* = \{a\}\{b, c\}^*$. M_1 and M_5 are equivalent to expressions RE 1 and RE 5, respectively. Therefore, with M_1 and M_5 as the input of Algorithm 3.7, we construct an FA M_6 equivalent to RE 6 as desired. Figures 3.8–3.11 summarize this construction.

Using algorithms given earlier in this chapter, we can now turn M_6 into an equivalent MDFA M depicted in Figure 3.12.

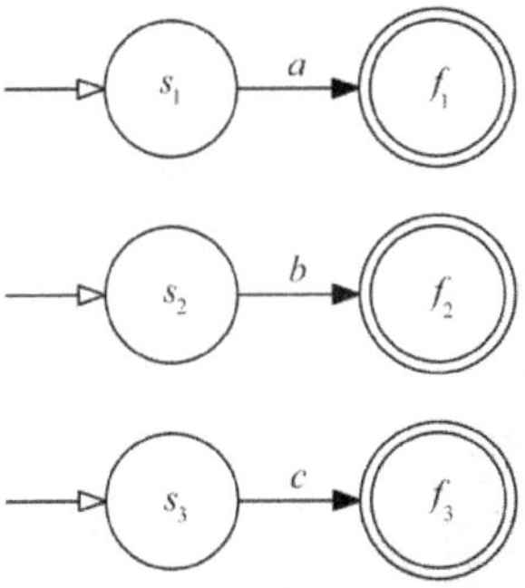

Figure 3.8. M_1, M_2, and M_3.

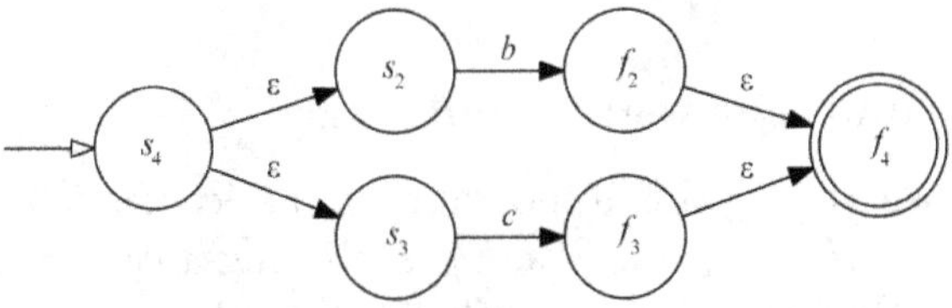

Figure 3.9. M_4.

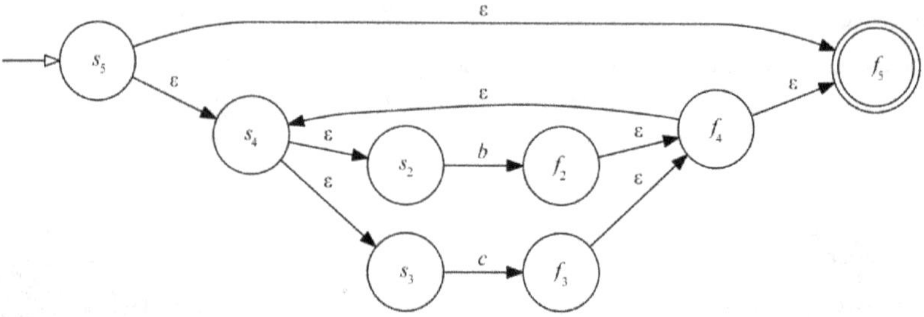

Figure 3.10. M_5.

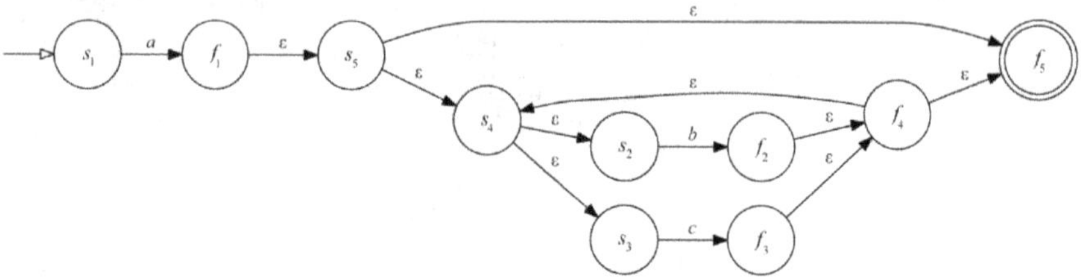

Figure 3.11. M_6.

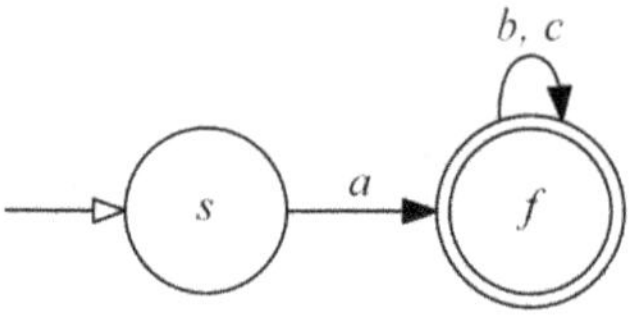

Figure 3.12. M_{min}.

We close this chapter with the following crucially important theorem, which summarizes Theorems 3.6, and 3.7, and Corollary 3.1.

Theorem 3.8. *The REs and the FAs are equivalent, that is, they both characterize* $\mathbf{L}_{\mathbf{REG}}$. *Consequently, all the restricted versions of FAs mentioned in Corollary 3.1 characterize* $\mathbf{L}_{\mathbf{REG}}$ *too.*

Pumping lemma for regular languages

To give an insight into this important lemma, consider any $L \in \mathbf{L}_{\mathbf{FA}}$ and any DFA, $M = (Q, \Delta, R, s, F)$, satisfying $L(M) = L$. Set $k = |Q|$. Suppose that $z \in L(M)$, with $|z| \geq k$. As M reads a symbol during every move, M accepts z by making a sequence of $|z|$ moves;

therefore, there necessarily exists at least one state that M visits two or more times when accepting z. Take the first two visits of the same state, q, in this sequence of moves and decompose z according to them as follows. Set $z = uvw$ so that u, v, and w satisfy the following properties (I)–(IV):

(I) u takes M from the start state to the first visit of q;
(II) v takes M from the first visit of q to the second visit of q;
(III) apart from the two visits of q in (I) and (II), M never visits any state twice on uv;
(IV) w takes M from the second visit of q to a final state f.

As M is deterministic and, therefore, makes no ε-moves, $0 < |v|$. By (I)–(III), $|uv| \leq k$. Most importantly, M can obviously iterate the sequence of moves between the two visits of q described in (I) and (II) m times, for any $m \geq 0$, and during each of these iterations, M reads v. Consequently, M accepts every string of the form $uv^m w$. Next, we state and prove this lemma rigorously.

Lemma 3.8. *Pumping lemma for regular languages. Let $L \in \boldsymbol{L_{FA}}$. Then, there exists a positive integer $k \in \mathbb{N}$ such that every string $z \in L$ satisfying $|z| \geq k$ can be expressed as $z = uvw$, where $0 < |v| \leq |uv| \leq k$, and $uv^m w \in L$, for all $m \geq 0$.*

Proof. Let $L \in \mathbf{L_{FA}}$. Let L be finite. Set k to any positive integer satisfying $|z| < k$, for all $z \in L$, so the lemma trivially holds simply because L contains no string whose length is greater than or equal to k. Therefore, suppose that L is infinite. Let $M = (Q, \Delta, R, s, F)$ be a DFA such that $L(M) = L$. Set $k = |Q|$. Consider any string $z \in L$, with $|z| \geq k$. As $z \in L, sz \Rightarrow^* f\ [\rho]$, where s is the start state of $M, f \in F$, and $\rho \in R^*$ is a sequence of rules from R. Because M makes no ε-moves and, thus, reads a symbol on every step, $|\rho| = |z|$. As $|z| \geq k$ and $k = |Q|, |\rho| \geq |Q|$. Therefore, during $sz \Rightarrow^* f\ [\rho]$, M enters some states more than once. Consider the shortest initial part of this sequence of moves, during which M visits the same state, $q \in Q$. More formally, express $sz \Rightarrow^* f\ [\rho]$ as

$$\begin{aligned} suvw &\Rightarrow^* qvw \quad [\sigma] \\ &\Rightarrow^+ qw \quad [\tau] \\ &\Rightarrow^* f \quad [\upsilon], \end{aligned}$$

where $\rho = \sigma\tau\upsilon, q \in Q, |\tau| \geq 1$, and during $suvw \Rightarrow^* qw$, M visits q precisely twice and any other state no more than once.

The following claims verify that uvw satisfies the conditions stated in the lemma.

Claim A. $0 < |v|$ and $|uv| \leq k$.

Proof of Claim A. As $|v| = |\tau|$ and $|\tau| \geq 1, 0 < |v|$. Because during $suvw \Rightarrow^* qw\ [\sigma\tau]$, M visits q twice and any other state no more than once, $|\sigma\tau| \leq |Q|$. As $|uv| = |\sigma\tau|$ and $|Q| = k, |uv| \leq k$. Thus, $|uv| \leq k$.

Claim B. For all $m \geq 0, uv^m w \in L$.

Proof of Claim B. Let $m \geq 0$. M accepts $uv^m w$ by repeating the sequence of moves according to τ m times; formally,

$$\begin{aligned} suv^m w &\Rightarrow^* qv^m w & [\sigma] \\ &\Rightarrow^* qw & [\tau^m] \\ &\Rightarrow^* f & [\upsilon]. \end{aligned}$$

Thus, for all $m \geq 0, uv^m w \in L$; note that for $m = 0$, M actually accepts uw, so it completely omits the sequence of moves made according to τ and, therefore, computes

$$\begin{aligned} suw &\Rightarrow^* qw & [\sigma] \\ &\Rightarrow^* f & [\upsilon]. \end{aligned}$$

□

Example 3.10. To illustrate the technique used in the previous proof, take the following regular language:

$$L = \{a\}\{bc\}^*\{b\}.$$

Consider the DFA M defined by the following three rules:

$$1 : sa \to p, \quad 2 : pb \to f, \quad 3 : fc \to p,$$

where s is the start state of M and $f \in F$. As is obvious, $L(M) = L$. Since M has three states, set $k = 3$. Consider $z = abcb \in L$. As $|z| = 4$

and $k = 3$, z satisfies $|z| \geq k$. M accepts z as $sabcb \Rightarrow^* f$ [1232], which can be expressed in a move-by-move way as follows:

$$\begin{aligned} sabcb &\Rightarrow pbcb \quad [1] \\ &\Rightarrow fcb \quad [2] \\ &\Rightarrow pb \quad [3] \\ &\Rightarrow f \quad [2]. \end{aligned}$$

The shortest initial part of this computation that contains two occurrences of the same state is

$$\begin{aligned} sabcb &\Rightarrow pbcb \quad [1] \\ &\Rightarrow fcb \quad [2] \\ &\Rightarrow pb \quad [3], \end{aligned}$$

where p is the state M visits twice. Following the above proof, we express $sabcb \Rightarrow^* f$ [1232] as

$$\begin{aligned} suvw &\Rightarrow^* qvw \quad [1] \\ &\Rightarrow^* qw \quad [23] \\ &\Rightarrow^* f \quad [2], \end{aligned}$$

with $u = a, v = bc$, and $w = b$. Observe that $0 < |v| = 2$ and $|uv| = 3 \leq k = 3$. As $a(bc)^m b = u(v)^m w \in L$, for all $m \geq 0$, all the conditions stated in Lemma 3.8 hold. Specifically, take $m = 2$ to obtain $a(bc)^m b = a(bc)^2 b = abcbcb$. M accepts $abcbcb$ by iterating the partial computation according to 23 twice as follows:

$$\begin{aligned} sabcbcb &\Rightarrow^* pbcbcb \quad [1] \\ &\Rightarrow^* pbcb \quad [23] \\ &\Rightarrow^* pb \quad [23] \\ &\Rightarrow^* f \quad [2]. \end{aligned}$$

■

Applications of the pumping lemma for regular languages

As already pointed out, we primarily apply Lemma 3.8 to prove that a given language L is out of $\mathbf{L_{FA}}$. Of course, a proof like this is made by contradiction, and its typical structure is as follows:

(1) Assume that $L \in \mathbf{L_{FA}}$.
(2) Consider the constant k from the pumping lemma, and select a string $z \in L$ whose length depends on k so that $|z| \geq k$ is surely true.
(3) Consider all possible decompositions of z into uvw satisfying $|uv| \leq k$ and $v \neq \varepsilon$, and for each of these decompositions, demonstrate that there exists $m \geq 0$ such that $uv^m w \notin L$, which contradicts the third condition in Lemma 3.8.
(4) The contradiction obtained in (3) implies that the assumption in (1) was incorrect, so $L \notin \mathbf{L_{FA}}$.

Example 3.11. Let $L = \{x \mid x \in \{0,1\}^*, \mathit{occur}(x,0) = \mathit{occur}(x,1)\}$. In other words, L is the language consisting of strings containing the same number of 0s and 1s. Following the proof scheme above almost literally, we easily demonstrate that $L \notin \mathbf{L_{FA}}$:

(1) Assume that $L \in \mathbf{L_{FA}}$.
(2) As $L \in \mathbf{L_{FA}}$, there exists a natural number k satisfying Lemma 3.8. Set $z = 0^k 1^k$. Note that $z \in L$. Note that $|z| \geq k$ because $|z| = 2k \geq k$.
(3) By Lemma 3.8, z can be expressed as $z = uvw$, so the conditions of the pumping lemma hold. As $0 < |v|$ and $|uv| \leq k, v \in \{0\}^+$. Consider $uv^0w = uw$. Then, $uw = 0^j 1^k$, with $j = k - |v|$; therefore, $uw \notin L$. However, by the pumping lemma, $uv^m w \in L$, for all $m \geq 0$, including the case when $m = 0$, so $uw \in L$ — a contradiction.
(4) By the contradiction in (3), $L \notin \mathbf{L_{FA}}$. ∎

The previous example followed the recommended four-step proof idea, consisting of (1)–(4), almost literally. The following example makes use of the pumping lemma in a more original and ingenious way.

Example 3.12. Clearly, $\{a\}^+ \in \mathbf{L_{FA}}$. Consider its sublanguage $P \subseteq \{a\}^+$ defined as $P = \{a^n \mid n \text{ is prime}\}$ (a positive integer n is prime if its only positive divisors are 1 and n). As demonstrated in the following, $P \notin \mathbf{L_{FA}}$. Thus, from a more general viewpoint, we see that a sublanguage of a regular language may be non-regular.

Assume that P is regular. Then, there exists a positive integer k satisfying the pumping lemma. As P is infinite, there surely exists a string $z \in P$ such that $|z| \geq k$. By Lemma 3.8, z can be written as $z = uvw$, so the conditions of the pumping lemma hold. Consider $uv^m w$, with $m = |uw| + 2|v| + 2$. As $|uv^m w| = |uw| + m|v| = |uw| + (|uw| + 2|v| + 2)|v| = (|uw| + 2|v|) + (|uw| + 2|v|)|v| = (|uw| + 2|v|)(1 + |v|)$, $|uv^m w|$ is no prime, and thus, $uv^m w \notin P$. By the pumping lemma, however, $uv^m w \in P$ — a contradiction. Thus, $P \notin \mathbf{L_{FA}}$.

Observe that this result has its significant practical consequences. Indeed, as P is non-regular, no FA accepts P (see Theorem 3.8). Consequently, informally, the FAs are not strong enough to handle the primes. ■

Lemma 3.8 is a powerful tool to disprove that certain languages are in $\mathbf{L_{FA}}$, but it cannot be applied in the positive sense. That is, it cannot be used to prove that certain languages are regular because some non-regular languages satisfy the pumping lemma conditions, as the following example illustrates. Consequently, proving that a language satisfies these conditions does not necessarily imply that the language is regular.

Example 3.13. Take

$$L = \{a^o b^n c^n \mid o \geq 1 \text{ and } n \geq 0\} \cup \{b^o c^n \mid o, n \geq 0\}.$$

Observe that L satisfies the conditions of the pumping lemma for regular languages, although $L \notin \mathbf{L_{FA}}$. To see that L satisfies the pumping lemma conditions, set $k = 1$. Consider any string $z \in L$ and $|z| = k$. As $z \in L$, either $z \in \{a^o b^n c^n \mid o \geq 1 \text{ and } n \geq 0\}$ or $z \in \{b^o c^n \mid o, n \geq 0\}$.

Let $z = a^o b^n c^n$, for some $o \geq 1$ and $n \geq 0$. Express z as $z = uvw$, with $u = \varepsilon, v = a$, and $w = a^{o-1} b^n c^n$. Clearly, $v \neq \varepsilon$ and $|uv| = k = 1$. Furthermore, note that $a^m b^n c^n$, for all $m \geq 1$, so $uv^m w \in L$, for all $m \geq 0$. Thus, the pumping lemma conditions are satisfied in this case.

Let $z = b^o c^n$, for some o, $n \geq 0$. Express z as $z = uvw$, with $u = \varepsilon$, v is the leftmost symbol of $b^o c^n$, and w is the remaining suffix of z. For $o = 0, v = c$ and $w = c^{n-1}$, and all three conditions hold. For $o \geq 1, v = b$ and $w = b^{o-1} c^n$, and all three conditions hold in this case as well.

Thus, L satisfies the conditions of the pumping lemma for regular languages.

Finally, prove that $L \notin \mathbf{L_{FA}}$ by contradiction. That is, assume that $L \in \mathbf{L_{FA}}$. Let $M = (Q, \Delta, R, s, F)$ be a DFA such that $L(M) = L$. As an exercise, show how to transform M into an FA N such that $L(N) = \{ab^nc^n \mid n \geq 0\}$, and by analogy with Example 3.11, prove that $\{ab^nc^n \mid n \geq 0\} \notin \mathbf{L_{FA}}$. From this contradiction, it follows that $L \notin \mathbf{L_{FA}}$. ■

Chapter 4

Pushdown Automata

In essence, the notion of a pushdown automaton is that of a finite automaton extended by a potentially infinite stack, referred to as a pushdown in automata theory. In other words, the automaton has its workspace divided into two parts — the input workspace and the pushdown workspace. The former is represented by a strictly unchangeable input string, just like in any finite automaton (FA). The latter is extendable, but this extension is always based on the well-known last-in-first-out principle, which requires any symbol to be entered or removed only at the top of the pushdown. That is, the topmost pushdown symbol is replaced with a string x, and thus, the symbol previously second from the top becomes the $|m|$th. Note that if $x = \varepsilon$, the symbol previously second actually becomes the topmost pushdown symbol, which corresponds to popping up the contents of the pushdown.

In greater detail, the notion of a pushdown automaton contains finitely many states, input symbols, and pushdown symbols. It has two heads: An input read head and a pushdown read–write head. Its heart consists of a program, represented by finitely many rules, which controls its computation that consists of a sequence of consecutive moves. Every move is made from the current state in which the automaton occurs, with the current input symbol under the read head and the topmost pushdown symbol under its read–write head. A move is performed according to a computational rule that specifies: (1) the next state entered from the current state (possibly back to the same state), (2) whether the read head remains stationary over the current input symbol or shifts to the next right-neighboring symbol, and (3) the string substituted for the current topmost pushdown symbol (thus, no move can be made if the pushdown is empty). In the set of states, a state is defined as the start state, and some states are designated as final states. The pushdown automaton starts every computation from the start state with its read head over the leftmost symbol of an input string and its pushdown containing only a special start pushdown symbol. If, by making a sequence of moves, the computation eventually brings the automaton into a final state after the entire input string is read and the pushdown is emptied, then the input string is accepted. All the strings accepted in this way represent the language accepted by the automaton.

An Introductory Example

Reconsider the ticket machine described in the introductory example of Chapter 3. Modify its behavior so that the machine dispenses n tickets b after it has received n coins a, for all $n \geq 0$, after which the machine is ready to repeat this receiving–dispensing loop. Consider just one repetition, formalized by the language $\{a^n b^n \mid n \geq 0\}$. Note that no FA accepts this language. Indeed, by using finitely many states, it cannot count the number of as before the first b is encountered because n goes to infinity. As opposed to this incapability of finite automata, pushdown automata are capable of accepting this language, as sketched in the following.

A pushdown automaton that accepts $\{a^n b^n \mid n \geq 0\}$ can operate, in principle, in the following way. It copies the initial prefix of all as onto its pushdown. Then, starting from the first appearance of b, it pops one a from the pushdown for each input b. If working this way, the machine removes all as from the pushdown and, simultaneously, exhausts all input bs, it accepts; otherwise, it rejects.

From a historical perspective, stacks or, to put it in terms of this book, pushdowns have been in use from the earliest days of computing in the 1950s. Pushdown automata were formally conceptualized in the 1960s. Compilers represent perhaps the most significant application area of these automata, as Sections 11.2–11.5 demonstrate in terms of their syntax analyzers. Apart from syntax analysis, they are also used for other compiler-related purposes, such as the organization of symbol tables, keeping track of the return addresses and environments invoked by recursive subprogram calls, keeping track of locally defined variables when block-structured languages are translated, and evaluating postfix notation (see Chapter 2), as well as the target code generation resulting from this notation. However, applications of pushdown automata go beyond techniques employed in compiler writing. To illustrate, hardware verification often makes use of them in virtual machines when moving short-lived temporary values to and from pushdowns. Interestingly enough, they are also used for solving problems within some mathematical games, such as the famous *Tower of Hanoi*. As a matter of fact, even in our quite common daily lives, many objects are stored and supplied based upon pushdown-based methods, including plates and soup bowls in cafeterias and restaurants, piles of books in bookstores and libraries, and ice cream cones in confectioner's stores.

4.1 Definitions

Definition 4.1. A *pushdown automaton* (PDA for short) is a septuple

$$M = \big(Q, \Delta, \Gamma, R, s, S, F\big),$$

where Q, Δ, and Γ are three finite sets such that $Q \cap (\Gamma \cup \Delta) = \emptyset$,

$$R \subseteq \Gamma \times Q \times (\Delta \cup \{\varepsilon\}) \times \Gamma^* \times Q$$

is finite, $s \in Q, S \in \Gamma$, and $F \subseteq Q$. In M, the components Q, Δ, Γ, S, s, and F are referred to as the *set of states*, the *input alphabet*, the *pushdown alphabet*, the *start pushdown symbol*, the *start state*, and the *set of final states*, respectively. R is called the *set of rules of* M. Every rule $(A, q, a, x, p) \in R$ is written as $Aqa \rightarrow xp \in R$ in what follows, where $A \in \Gamma, q, p \in Q, a \in \Delta \cup \{\varepsilon\}, x \in \Gamma^*$. For brevity, we often denote $Aqa \rightarrow xp$ by a unique label m as $m : Aqa \rightarrow xp$, and we briefly use m instead of $Aqa \rightarrow xp$ under this denotation.

A *configuration* of M is any string of the form uqv, where $u \in \Gamma^*, q \in Q$, and $v \in \Delta^*$; X_M denotes the set of all configurations of M. Over X_M, we define the binary relation $\Rightarrow$ as $\beta \Rightarrow \chi$ in M, for all $\beta, \chi \in X_M$, such that $\beta = uXqav, \chi = uxqv, Xqa \rightarrow xp \in R$,

$u \in \Gamma^*, X \in \Gamma, q, p \in Q, a \in \Delta \cup \{\varepsilon\}, v \in \Delta^*$. As usual, $\Rightarrow^*$ denotes the transitive-reflexive closure of $\Rightarrow$. If $\beta \Rightarrow \chi$ in M, where $\beta, \chi \in X_M$, we say that M makes a *move* from β to χ. M makes a *computation* from β to χ if $\beta \Rightarrow^* \chi$ in M, where $\beta, \chi \in X_M$.

M accepts $w \in \Delta^*$ if $Ssw \Rightarrow^* uf$ in M with $u \in \Gamma^*$ and $f \in F$, and the set of all strings M accepts is the *language accepted by M*, denoted by $L(M)$; mathematically,

$$L(M) = \{w \mid w \in \Delta^*, Ssw \Rightarrow^* f, f \in F\}.$$

We denote the set of all pushdown automata by **PDA**. Set $\mathbf{L_{PDA}} = \{L(M) \mid M \in \mathbf{PDA}\}$; in other words, $\mathbf{L_{PDA}}$ is the family of languages accepted by all PDAs. ■

To rephrase this definition less formally, every configuration is of the form uqv, where $u \in \Gamma^*, q \in Q$, and $v \in \Delta^*$. By using a rule of the form $Xqa \to xp \in R$, M makes a move that changes the topmost pushdown symbol X to x, reads a, and changes the current state q to p, where $X \in \Gamma, q, p \in Q, a \in \Delta \cup \{\varepsilon\}, x \in \Gamma^*$. M accepts an input string $w \in \Delta^*$ if starting from Ssw, it performs a sequence of moves so it reads all the symbols of w, the pushdown is empty, and it ends up in a final state.

Convention 4.1. For every $M \in \mathbf{PDA}$, we automatically assume that $Q, \Delta, \Gamma, R, s, S$, and F denote its total alphabet, the input alphabet, the pushdown alphabet, the set of rules, the start state, and the set of final states, respectively. If there exists any danger of confusion, we mark $Q, \Delta, \Gamma, R, s, S$, and F, with M as $Q_M, \Delta_M, \Gamma_M, R_M, s_M, S_M$, and F_M, respectively, in order to explicitly relate these components to M. ■

Example 4.1. Consider the language $L = \{a^k b^k \mid k \geq 1\}$. In this example, we construct $M \in \mathbf{PDA}$ that accepts L.

Informally, M works in the following two-phase way. First, it pushes down as. Then, when the first occurrence of b appears, M pops up as and pairs them off with bs. If the number of as and bs coincides, M accepts the input string; otherwise, it rejects. Formally,

$M = (Q, \Delta, \Gamma, R, s, S, F)$, where $\Gamma = \{S, a\}, \Delta = \{a, b\}, Q = \{s, f\}, F = \{f\}$, and

$$R = \{Ssa \rightarrow as, asa \rightarrow aas, asb \rightarrow f, afb \rightarrow f\}.$$

Under Definition 4.1, we can succinctly specify M by listing its four rules, labeled by 1–4, as

$$1 : Ssa \rightarrow as, \quad 2 : asa \rightarrow aas, \quad 3 : asb \rightarrow f, \quad 4 : afb \rightarrow f.$$

For instance, on $aaabbb$, M performs this sequence of moves:

$$\begin{aligned} Ssaaabbb &\Rightarrow asaabbb \ [1] \\ &\Rightarrow aasabbb \ [2] \\ &\Rightarrow aaasbbb \ [2] \\ &\Rightarrow aafbb \quad [3] \\ &\Rightarrow afb \quad\ \ [4] \\ &\Rightarrow f \qquad\ \ [4]. \end{aligned}$$

As f is final, M accepts $aaabbb$.

In general, observe that M makes every acceptance according to a sequence of rules 12^n34^n, for some $n \geq 0$; more specifically, by 12^n34^n5, M accepts $a^{n+1}b^{n+1}$. Consequently, $L(M) = \{a^k b^k \mid k \geq 1\}$; a rigorous proof of this identity is left as an exercise. ■

Example 4.2. Consider

$$L = \{vw \mid v, w \in \{a, b\}^*, v = \text{reversal}(w)\}$$

Next, we construct a PDA M such that $L(M) = L$. In many respects, M works similarly to the PDA designed in Example 4.1. Indeed, let $z \in \{a, b\}^*$ be an input string. M first pushes down a prefix of z. However, it can take a guess any time that it has pushed down precisely $|z|/2$ initial symbols of z, so it starts to pop up the symbols from the pushdown and pair them off with the remaining symbols of w. When and if M simultaneously empties the pushdown and completes reading z, it accepts; otherwise, it rejects.

Formally, $M = (Q, \Delta, \Gamma, R, s, S, F)$, where $\Gamma = \{S, a, b\}, \Delta = \{a, b\}, Q = \{s, q, f\}, F = \{f\}$, and

$$\begin{aligned} R = &\{Ssc \to Scs \mid c \in \{a, b\}\} \\ &\cup \{dsc \to dcs \mid c, d \in \{a, b\}\} \\ &\cup \{s \to q\} \\ &\cup \{cqc \to q \mid c \in \{a, b\}\} \\ &\cup \{Sq \to f \mid c \in \{a, b\}\}. \end{aligned}$$

A proof that $L(M) = L$ is left as an exercise. ■

Equivalent types of acceptance

Consider any M in **PDA**. According to Definition 4.1, $L(M) = \{w \mid w \in \Delta^*, Ssw \Rightarrow^* f, f \in F\}$. In other words, M accepts w by empty pushdown and a final state because, after reading w, M has to: (i) empty its pushdown and (ii) end up in a final state to accept $w \in \Delta_M^*$. Next, we relax this way of acceptance by requiring that only one of conditions (i) or (ii) is satisfied.

Definition 4.2. Let $M = (Q, \Delta, \Gamma, R, s, S, F)$ be a PDA:

(I) If $Ssw \Rightarrow^* q$ in M with $q \in Q, M$ *accepts* w. *by empty pushdown.* The set of all strings that M accepts in this way is the *language accepted by M by empty pushdown*, denoted by $L_\varepsilon(M)$.
(II) If $Ssw \Rightarrow^* vf$ in M with $v \in \Gamma^*$ and $f \in F, M$ *accepts* w *by final state.* The set of all strings that M accepts in this way is the *language accepted by M by final state*, denoted by $L_f(M)$. ■

Example 4.3. Consider a PDA M defined by its five rules:

$$\begin{gathered} 1: Ssa \to Sas, \quad 2: asa \to aas, \quad 3: asb \to q, \\ 4: aqb \to q, \quad 5: Sq \to f, \end{gathered}$$

where $\Gamma_M = \{S, a\}, \Delta_M = \{a, b\}, Q_M = \{s, q, f\}$, and $F_M = \{s, f\}$. To describe how M works, observe that M reads a and pushes down Sa by using rule 1. Then, M pushes down as by using rule 2 while remaining in the final state s. When and if the first occurrence of b

appears, M begins to pop up as and pair them off with bs while remaining in the only non-final state q. For instance, on aa, M works as

$$\begin{aligned} Ssaa &\Rightarrow Sasa \quad [1] \\ &\Rightarrow Saas \quad [2]. \end{aligned}$$

On $aabb$, M works in the following way:

$$\begin{aligned} Ssaabb &\Rightarrow Sasabb \quad [1] \\ &\Rightarrow Saasbb \quad [2] \\ &\Rightarrow Saqb \quad [3] \\ &\Rightarrow Sq \quad [4] \\ &\Rightarrow f \quad [5]. \end{aligned}$$

As s and f are final, M accepts aa and $aabb$ by final states.

Consider the three types of acceptance and the corresponding three languages that M accepts — $L_f(M), L_\varepsilon(M)$, and $L(M)$ — in terms of Definitions 4.1 and 4.2. Observe that

$$L_f(M) = \{a^k b^k \mid k \geq 1\} \cup \{a\}^+ \text{ and } L_\varepsilon(M) = L(M) = \{a^k b^k \mid k \geq 1\}.$$

■

Convention 4.2. Set $\mathbf{L_{PDA\text{-}\varepsilon}} = \{L_\varepsilon(M) \mid M \in \mathbf{PDA}\}$ and $\mathbf{L_{PDA\text{-}f}} = \{L_f(M) \mid M \in \mathbf{PDA}\}$. ■

We next establish Lemmas 4.1–4.4, whose proofs describe how to convert the three types of acceptance into each other. As a consequence, these lemmas imply $\mathbf{L_{PDA}} = \mathbf{L_{PDA\text{-}\varepsilon}} = \mathbf{L_{PDA\text{-}f}}$ (see Definition 4.1 for $\mathbf{L_{PDA}}$).

Lemma 4.1. *For any $I \in \boldsymbol{PDA}$, there is $O \in \boldsymbol{PDA}$ such that $L_f(I) = L_\varepsilon(O)$. Therefore, $\boldsymbol{L_{PDA\text{-}f}} \subseteq \boldsymbol{L_{PDA\text{-}\varepsilon}}$.*

Proof. Let $I \in \mathbf{PDA}$. O keeps its start symbol S_O on the pushdown bottom; otherwise, it simulates I move by move. If I enters a final state and accepts, O completely empties its pushdown list, including S_O, and accepts too. If I empties its pushdown while occurring in a non-final state and, therefore, does not accept its input by final state, then the pushdown of O contains S_O, so O does not accept its input either. A fully rigorous proof is left as an exercise. □

Lemma 4.2. *For any $I \in$ **PDA**, there is $O \in$ **PDA** such that $L(I) = L_f(O)$. Therefore,* $\mathbf{L_{PDA}} \subseteq \mathbf{L_{PDA\text{-}f}}$.

Proof. Let $I \in$ **PDA**. Construct $O \in$ **PDA** with $F_O = \{f\}$, where f is a new final state, by analogy with the proof of Lemma 4.1. That is, O keeps S_O on the pushdown bottom while simulating I. If I empties its pushdown, enters a final state, and accepts, then after simulating this computation in I, O occurs in the same state, with the pushdown containing S_O; from this configuration, O enters f in order to accept the input by final state. □

Lemma 4.3. *For any $I \in$ **PDA**, there is $O \in$ **PDA** such that $L_\varepsilon(I) = L_f(O) = L(O)$. Therefore,* $\mathbf{L_{PDA\text{-}\varepsilon}} \subseteq \mathbf{L_{PDA\text{-}f}}$ *and* $\mathbf{L_{PDA\text{-}\varepsilon}} \subseteq \mathbf{L_{PDA}}$.

Proof. Let $I \in$ **PDA**. Construct $O \in$ **PDA** so it has the same final states like I. O keeps its start symbol S_O on the pushdown bottom while simulating I. If I empties its pushdown and accepts its input, then O has its pushdown containing only S_O, and from this configuration, it enters a final state while removing S_O from the pushdown. Observe that O constructed in this way satisfies $L_\varepsilon(I) = L_f(O) = L(O)$. □

Lemma 4.4. *For any $I \in$ **PDA**, there is $O \in$ **PDA** such that $L(I) = L_\varepsilon(O)$. Therefore,* $\mathbf{L_{PDA}} \subseteq \mathbf{L_{PDA\text{-}\varepsilon}}$.

Proof. Let $I \in$ **PDA**. Construct $O \in$ **PDA** so that it keeps its start symbol S_O on the pushdown bottom; otherwise, it simulates I move by move. If I empties its pushdown and enters a final state, then O has its pushdown containing only S_O, so it removes S_O from the pushdown to accept the input by empty pushdown. □

The previous three lemmas imply that $\mathbf{L_{PDA\text{-}f}} \subseteq \mathbf{L_{PDA\text{-}\varepsilon}} \subseteq \mathbf{L_{PDA}} \subseteq \mathbf{L_{PDA\text{-}f}}$, so we obtain the following theorem.

Theorem 4.1. $\mathbf{L_{PDA}} = \mathbf{L_{PDA\text{-}\varepsilon}} = \mathbf{L_{PDA\text{-}f}}$.

4.2 Determinism

In general, PDAs work in a non-deterministic way; that is, they can make many different sequences of moves on the same input. Since this non-determinism obviously complicates their implementation and application in practice, we study their deterministic versions in this section. First, we demonstrate that the equivalence of the three types of acceptance discussed in the previous section does not hold in terms of these deterministic versions, although it does hold in terms of PDAs. Then, we show that the deterministic versions of PDAs are less powerful than PDAs, although they are stronger than FAs.

Definition 4.3. Let $M \in \mathbf{PDA}$. M is a *deterministic* PDA (dPDA for short) if for every $xqa \to yp \in R_M$, where $x, y \in \Gamma^*, q, p \in Q$, and $a \in \Delta \cup \{\varepsilon\}$, it holds that $lhs(r) \notin \Gamma^*\{xq\}\{a, \varepsilon\}$, for all $r \in R_M - \{xqa \to yp\}$. ■

In the following convention, we introduce three language families: $\mathbf{L_{dPDA}}, \mathbf{L_{dPDA\text{-}\varepsilon}}$, and $\mathbf{L_{dPDA\text{-}f}}$, corresponding to the three types of acceptance discussed in Section 4.1 in terms of PDAs.

Convention 4.3. We denote the set of dPDAs by **dPDA** and set $\mathbf{L_{dPDA}} = \{L(M) \mid M \in \mathbf{dPDA}\}, \mathbf{L_{dPDA\text{-}\varepsilon}} = \{L_\varepsilon(M) \mid M \in \mathbf{dPDA}\}$, and $\mathbf{L_{dPDA\text{-}f}} = \{L_f(M) \mid M \in \mathbf{dPDA}\}$ (see Definition 4.2 for $L_\varepsilon(M)$ and $L_f(M)$). ■

Recall that $\mathbf{L_{PDA}} = \mathbf{L_{PDA\text{-}\varepsilon}} = \mathbf{L_{PDA\text{-}f}}$ (see Theorem 4.1). Surprisingly, this identity cannot be reformulated in terms of $\mathbf{L_{dPDA}}, \mathbf{L_{dPDA\text{-}\varepsilon}}$, and $\mathbf{L_{dPDA\text{-}f}}$. Indeed, $\mathbf{L_{dPDA\text{-}\varepsilon}} = \mathbf{L_{dPDA}} \subset \mathbf{L_{dPDA\text{-}f}}$, as demonstrated in the following.

Theorem 4.2. $\mathbf{L_{dPDA\text{-}\varepsilon}} = \mathbf{L_{dPDA}}$.

Proof. Establish this identity by analogy with the proofs of Lemmas 4.3 and 4.4. □

Lemma 4.5. $\{a^k b^k \mid k \geq 1\} \cup \{a\}^+ \notin \mathbf{L_{dPDA}}$.

Proof. Let $L = \{a^k b^k \mid k \geq 1\} \cup \{a\}^+$. We prove $L \notin \mathbf{L_{dPDA}}$ by contradiction. Suppose that $L = L(M)$ for some $M \in \mathbf{dPDA}$. As F_M is finite and M accepts any string in $\{a\}^+$, there surely exist $i, j \in \mathbb{N}$ such that $1 \leq i < j$, and M accepts both a^i and a^j so that after reading both a^i and a^j, M empties its pushdown and enters the same final state, $f \in F_M$. Consider $a^j b^j \in L(M)$. Since M is deterministic, during the acceptance of $a^j b^j$, M empties its pushdown and enters f after reading both the prefix a^i and the prefix a^j. Thus, M also accepts $a^i b^j$, which is out of L. Hence, $L \notin \mathbf{L_{dPDA}}$, and Lemma 4.5 holds. □

Theorem 4.3. ***$L_{dPDA} \subset L_{dPDA\text{-}f}$.***

Proof. Consider the conversion of any PDA I to a PDA O so that $L(I) = L_f(O)$ described in the proof of Lemma 4.2. Observe that if I is a dPDA, then O is also a dPDA. Thus, $\mathbf{L_{dPDA}} \subseteq \mathbf{L_{dPDA\text{-}f}}$. Return to the PDA M in Example 4.3. Recall that $L_f(M) = \{a^k b^k \mid k \geq 1\} \cup \{a\}^+$. Observe that M is deterministic. Thus, by Lemma 4.5, $\mathbf{L_{dPDA\text{-}f}} - \mathbf{L_{dPDA}} \neq \emptyset$, so $\mathbf{L_{dPDA}} \subset \mathbf{L_{dPDA\text{-}f}}$. □

According to Definition 4.3, any $M \in$ **dPDA** makes no more than one move from any configuration, so it makes a unique sequence of moves on any input — a property highly appreciated in practice, when M is implemented. Unfortunately, dPDAs also have a disadvantage: They are less powerful than their non-deterministic versions, as demonstrated in the following.

Lemma 4.6. *Let $L = \{vw \mid v, w \in \{a, b\}^*, v = reversal(w)\}$. Then, $L \notin \mathbf{L_{dPDA\text{-}f}}$.*

Proof. We only give a gist underlying this proof, whose rigorous version is beyond the scope of this introductory text. Any PDA M such that $L_f(M) = L$ necessarily works on its input $u \in \{a, b\}^*$ by performing the following four computational phases:

(1) M pushes down a prefix of u;
(2) M takes a guess that it is right in the middle of u, which means that the length of its current pushdown equals the length of the suffix that remains to be read;
(3) M pops up the symbols from the pushdown and pairs them off with the same input symbols;
(4) when and if M completes reading u, it accepts.

Note, however, that M ends (1) and begins (2) based on its guess, so M makes necessarily more than one move from the same configuration at this point. As a result, M cannot be deterministic, so $L \notin \mathbf{L_{dPDA\text{-}f}}$. □

Theorem 4.4. $\mathbf{L}_{dPDA\text{-}f} \subset \mathbf{L}_{PDA}$.

Proof. By Definitions 4.1 and 4.3, dPDAs are special cases of PDAs, so $\mathbf{L}_{\mathbf{dPDA\text{-}f}} \subseteq \mathbf{L}_{\mathbf{PDA\text{-}f}}$. By Theorem 4.1, $\mathbf{L}_{\mathbf{PDA\text{-}f}} = \mathbf{L}_{\mathbf{PDA}}$. Let $L = \{vw \mid v, w \in \{a,b\}^*, v = reversal(w)\}$. By Example 4.2 and Lemma 4.6, $L \in \mathbf{L}_{\mathbf{PDA}} - \mathbf{L}_{\mathbf{dPDA\text{-}f}}$, so $\mathbf{L}_{\mathbf{PDA}} - \mathbf{L}_{\mathbf{dPDA\text{-}f}} \neq \emptyset$. Thus, Theorem 4.4 holds true. □

Clearly, dPDAs are stronger than FAs.

Theorem 4.5. $\mathbf{L}_{FA} \subset \mathbf{L}_{dPDA}$.

Proof. By Definitions 3.3 and 4.3, any DFA can be seen as a dPDA that keeps its pushdown empty during any computation. As $\mathbf{L}_{\mathbf{FA}} = \mathbf{L}_{\mathbf{DFA}}$ (see Corollary 3.1), $\mathbf{L}_{\mathbf{FA}} \subseteq \mathbf{L}_{\mathbf{dPDA}}$. In Example 4.1, M is a dPDA such that $L(M) = \{a^k b^k \mid k \geq 1\}$. By analogy with Example 3.13, prove that $\{a^k b^k \mid k \geq 1\} \notin \mathbf{L}_{\mathbf{REG}}$. Since $\mathbf{L}_{\mathbf{REG}} = \mathbf{L}_{\mathbf{FA}}$ (see Theorem 3.8), $\mathbf{L}_{\mathbf{FA}} \subset \mathbf{L}_{\mathbf{dPDA}}$. □

Putting together Theorems 4.1–4.5, we obtain the following corollary, which is of fundamental importance.

Corollary 4.1. $\mathbf{L}_{FA} \subset \mathbf{L}_{dPDA\text{-}\varepsilon} = \mathbf{L}_{dPDA} \subset \mathbf{L}_{dPDA\text{-}f} \subset \mathbf{L}_{PDA} = \mathbf{L}_{PDA\text{-}\varepsilon} = \mathbf{L}_{PDA\text{-}f}$. ■

dPDAs fulfill a crucial role in the application of PDAs in practice. Perhaps most significantly, they are central to syntax analysis (see Chapter 11).

Chapter 5

Turing Machines

The notion of a Turing machine represents ultimately the general model of computation because of the conjecture that every procedure in the intuitive sense can be realized as a Turing machine — the *Church–Turing thesis*. Therefore, there exists a procedure for computing a mathematical formula, such as a function, if and only if a Turing machine computes it too. It follows from Definition 5.1 that the notion of a Turing machine represents a relatively simple language-defining model, so it obviously constitutes a procedure in the intuitive sense. Much more surprisingly, the Church–Turing thesis also asserts that every procedure in the intuitive sense can be completely formalized by this strictly mathematical notion of a Turing machine. Observe that it is really a thesis, not a theorem, because it cannot be proved. Indeed, any proof of this kind would necessitate a formalization of our intuitive notion of a procedure so that it can be rigorously compared with the notion of a Turing machine. At this point, however, there would be a problem of whether this newly formalized notion is equivalent to the intuitive notion of a procedure, which would give rise to another thesis similar to the Church–Turing thesis. Therefore, any attempt to prove this thesis inescapably ends in an infinite regression. However, the evidence supporting the Church–Turing thesis is hardly disputable because, throughout its history, computer science has formalized the notion of a procedure in the intuitive sense through other mathematical models, including Post systems, μ-recursive functions, and λ-calculus, and all of them have eventually turned out to be equivalent to Turing machines. Even more importantly, nobody has ever come up with a procedure in the intuitive sense and demonstrated that no Turing machine can formalize it. Thus, the strictly mathematical notion of a Turing machine has become the universally accepted formalization of the intuitive notion of a procedure. In other words, it is a general enough mathematical model for the notion of a procedure in the intuitive sense, so any possible procedure can be realized as a Turing machine. Playing such an utmost important role, the Turing machine model underlies the most fundamental ideas behind computation, as demonstrated in Chapter 7.

This chapter, however, discusses Turing machines as language models. In many respects, their basic definition resembles the notion of a finite automaton (see Chapter 3). Indeed, it is based on finitely many states and symbols, and over its workspace, it operates by making moves that depend on the current state and the symbol under its head. In addition, however, apart from reading, its head can also write and move both to the right and the left within the workspace. Finally, and most importantly, the workspace can be limitlessly extended to the right.

An Introductory Example

Reconsider the pushdown automaton sketched in the introductory example of Chapter 4. Recall that it accepts $\{a^n b^n \mid n \geq 0\}$, which formalizes the ticket machine behavior consisting in dispensing n tickets b after it has received n coins a, for all $n \geq 0$. Redesign this behavior so that after receiving n coins a, the machine dispenses n tickets b, followed by dispensing n ticket receipts c (for instance, for the sake of the subsequent refunding of the spent coins at the buyer's office). Formalize this behavior with the language $\{a^n b^n c^n \mid n \geq 0\}$. Note that no pushdown automaton accepts this language. Indeed, by a single pushdown, it can verify that an input string starts with $a^n b^n$ for some $n \geq 0$ in the way described in Chapter 4. After this verification, however, its pushdown is empty, so no information about n is maintained; thus, any verification that the remaining input suffix equals c^n is ruled out. Turing machines can accept this language, as sketched in the following.

A Turing machine that accepts $\{a^n b^n c^n \mid n \geq 0\}$ can operate as follows. With x as an input string initially placed on the workspace, the machine first ensures that $x \in \{a\}^*\{b\}^*\{c\}^*$. Then, it cyclically repeats erasing one a, one b, and one c until at least one of these three symbols is missing within the workspace. Finally, if the machine gets out of this cycle with its workspace completely cleared of all as, bs, and cs, it accepts; otherwise, it rejects.

5.1 Definitions

Definition 5.1. A *Turing machine* (TM for short) is a sextuple

$$M = (Q, \Delta, \Gamma, R, s, F),$$

where Q, Δ, and Γ are three finite sets such that $Q \cap (\Gamma \cup \Delta) = \emptyset, s \in Q, F \subseteq Q, \Delta \subset \Gamma, \Gamma - \Delta$ always contains three special symbols — $\square, \rhd$, and $\lhd$. Furthermore,

$$R \subseteq (Q \cup \Gamma) \times (Q \cup \Gamma),$$

where each $(x, y) \in R$ has one of the following forms:

(I) $\{x, y\} \subseteq \{\rhd\}Q$, or
(II) $\{x, y\} \subseteq (\Gamma - \{\rhd, \lhd\})Q \cup Q(\Gamma - \{\rhd, \lhd\})$, or
(III) $x \in Q\{\lhd\}$ and $y \in Q\{\square\lhd, \lhd\}$.

In M, the components Q, Δ, Γ, s, and F are referred to as the *set of states*, the *input alphabet*, the *tape alphabet*, the *start state*, and the *set of final states*, respectively. $\rhd$ and $\lhd$ are referred to as the *left* and *right bounders*, respectively, and $\square$ is the blank symbol. R is called the set of rules of M, with each $(x, y) \in R$ written as $x \to y \in R$ in what follows. For brevity, we often denote $x \to y$ by the unique label m as $m : x \to y$, and we briefly use m instead of $x \to y$ under this denotation.

A *configuration* of M is any string of the form uqv, where $u \in \{\rhd\}\Gamma^*, q \in Q$, and $v \in \Gamma^*\{\lhd\}$; X_M denotes the set of all configurations of M. Over X_M, we define the binary relation $\Rightarrow$ as $\beta \Rightarrow \chi$ in M, for all $\beta, \chi \in X_M$ such that $\beta = uxv, \chi = uyv, x \to y \in R$. As usual, $\Rightarrow^*$ denotes the transitive-reflexive closure of $\Rightarrow$. If $\beta \Rightarrow \chi$ in M, where $\beta, \chi \in X_M$, we say that M makes a *move* from β to χ. M makes a *computation* from β to χ if $\beta \Rightarrow^* \chi$ in M, where $\beta, \chi \in X_M$.

M accepts $w \in \Delta^*$ if $\triangleright sw\triangleleft \Rightarrow^* \triangleright ufv\triangleleft$ in M, where $u, v \in \Gamma^*, f \in F$. The *language accepted by M* or, briefly, the *language of M* is denoted by $L(M)$ and defined as the set of all strings that M accepts; formally,

$$L(M) = \{w \mid w \in \Delta^*, \triangleright sw\triangleleft \Rightarrow^* \triangleright ufv\triangleleft, u, v \in \Gamma^*, f \in F\}.$$

We denote the set of all Turing machines by **TM**. $\mathbf{L_{TM}}$ denotes the family of Turing languages, also known as the family of recursively enumerable languages, defined as $\mathbf{L_{TM}} = \{L(M) \mid M \in \mathbf{TM}\}$. ■

To rephrase this definition less formally, in X_M, every configuration is a string of the form $\triangleright uqv\triangleleft, u, v \in \Gamma^*, q \in Q$, and it captures the current situation M occurs in. That is, $\triangleright uqv\triangleleft$ means that q is the current state of M, uv is the current content of the workspace bounded by $\triangleright$ and $\triangleleft$, and the read–write head occurs in between u and v. If $\beta \Rightarrow \chi$ in M, where $\beta, \chi \in X_M$, we say that M makes a move from β to χ. Moves involving the bounders $\triangleright$ and $\triangleleft$ deserve our special attention. Concerning $\triangleright$, all the rules containing the left bounder are of the form $\triangleright p \rightarrow \triangleright q$, where $p, q \in Q$. Therefore, from $\triangleright pv\triangleleft, M$ cannot make any left move, which would place a state in front of $\triangleright$. Concerning $\triangleleft$, all the rules containing the right bounder are of the form $p\triangleleft \rightarrow q\triangleleft$ or $p\triangleleft \rightarrow q\Box\triangleleft$, where $p, q \in Q$ (see (III)). Thus, from $\triangleright up\triangleleft$, by using $p\triangleleft \rightarrow q\triangleleft, M$ just changes p to q in front of $\triangleleft$. By using $p\triangleleft \rightarrow q\Box\triangleleft$, M changes p to q and, in addition, extends the workspace by inserting $\Box$ in front of $\triangleleft$ as $\triangleright up\triangleleft \Rightarrow \triangleright uq\Box\triangleleft$. Consequently, any computation executed by M always takes place within the workspace delimited by $\triangleright$ from the left and by $\triangleleft$ from the right.

Convention 5.1. For every $M \in \mathbf{TM}$, we automatically assume that Q, Δ, Γ, R, s, and F denote its total alphabet, the input alphabet, the workspace alphabet, the set of rules, the start state, and the set of final states, respectively. If there exists any danger of confusion, we mark Q, Δ, Γ, R, s, and F with M as $Q_M, \Delta_M, \Gamma_M, R_M, s_M$, and F_M, respectively, in order to explicitly relate these components to M. ■

Example 5.1. Consider $L = \{x \mid x \in \{a, b, c\}^*, occur(x, a) = occur(x, b) = occur(x, c)\}$. Less formally, x is in L iff x has an equal number of as, bs, and cs; for instance, $babcca \in L$, but $babcc \notin L$. In this example, we construct a TM M such that $L(M) = L$.

Basic idea. M records symbols it has read by using its states from $power(\{a, b, c\})$ (see Section 1.2). That is, M moves on the tape in any direction. Whenever it reads an input symbol that is not already recorded in the current state, M can add this symbol to its current state while simultaneously changing it to $\blacksquare$ on the tape. M can any time change the state that records all three symbols to the state that records no symbol at all. By using a special state, $\dashv$, M can scan the entire tape so it starts from $\lhd$ and moves left toward $\rhd$ in order to find out whether the tape is completely blank, and if this is the case, M accepts.

Construction. Define $M = (Q, \Delta, \Gamma, R, \blacktriangleright, F)$, where $\Delta = \{a, b, c\}, \Gamma = \Delta \cup \{\rhd, \lhd, \Box\}, Q = \{\blacktriangleright, \dashv, \blacksquare\} \cup W$, with $W = \{\langle O\rangle \mid O \subseteq \{a, b, c\}\}$ and $F = \{\blacksquare\}$. Construct the rules of R by performing (1)–(5), given in the following: (As stated in Section 1.2, $\{\}$ denotes the empty set just like $\emptyset$ does. In this example, we use $\{\}$ for this purpose.)

(1) add $\rhd \blacktriangleright \rightarrow \rhd\langle\{\}\rangle$ to R;
(2) for every $\langle O\rangle \in W$ and every $d \in \Delta \cup \{\Box\}$, add $\langle O\rangle d \rightarrow d\langle O\rangle$ and $d\langle O\rangle \rightarrow \langle O\rangle d$ to R;
(3) for every $\langle O\rangle \in W$ such that $O \subset \{a, b, c\}$ and every $d \in \Delta - O$, add $\langle O\rangle d \rightarrow \langle O \cup \{d\}\rangle\Box$ to R;
(4) add $\langle\{a, b, c\}\rangle d \rightarrow \langle\{\}\rangle d$ to R, where $d \in \Delta \cup \{\Box, \lhd\}$;
(5) add $\langle\{\}\rangle\lhd \rightarrow \dashv\lhd, \Box\dashv \rightarrow \dashv\Box$, and $\rhd\dashv \rightarrow \rhd\blacksquare$ to R.

Computation. Consider both the informal and formal descriptions of M. Observe that by (1), M starts every computation. By (2), M moves on its tape. By (3), M adds the input symbol to its current state from $power(\Delta)$ and, simultaneously, changes the input symbol to $\Box$ on the tape. By (4), M empties $\{a, b, c\}$ so it changes this state to the state equal to the empty set. By (5), M makes a final scan of the tape, starting from $\lhd$ and moving left toward $\rhd$ in order to make sure that the tape is completely blank, and if it is, M accepts.

For instance, in this way, M accepts *babcca* as follows:

$$
\begin{aligned}
\rhd \blacktriangleright babcca\lhd \quad &\Rightarrow \quad \rhd \langle\{\}\rangle babcca\lhd\\
&\Rightarrow^* \quad \rhd babc\langle\{\}\rangle ca\lhd\\
&\Rightarrow \quad \rhd babc\langle\{c\}\rangle\Box a\lhd\\
&\Rightarrow^* \quad \rhd ba\langle\{c\}bc\Box a\lhd\\
&\Rightarrow \quad \rhd ba\langle\{b,c\}\rangle\Box c\Box a\lhd\\
&\Rightarrow^* \quad \rhd ba\Box c\Box\langle\{b,c\}\rangle a\lhd\\
&\Rightarrow \quad \rhd ba\Box c\Box\langle\{a,b,c\}\rangle\Box\lhd\\
&\Rightarrow^* \quad \rhd b\langle\{a,b,c\}\rangle a\Box c\Box\Box\lhd\\
&\Rightarrow \quad \rhd b\langle\{\}\rangle a\Box c\Box\Box\lhd\\
&\Rightarrow^* \quad \rhd \Box\Box\Box\Box\Box\Box\langle\{\}\rangle\lhd\\
&\Rightarrow^* \quad \rhd \Box\Box\Box\Box\Box\Box\text{↲}\lhd\\
&\Rightarrow \quad \rhd \Box\Box\Box\Box\Box\text{↲}\Box\lhd\\
&\Rightarrow^* \quad \rhd \text{↲}\Box\Box\Box\Box\Box\Box\lhd\\
&\Rightarrow \quad \rhd \blacksquare\Box\Box\Box\Box\Box\Box\lhd .
\end{aligned}
$$

Note, however, that M accepts the same string in many other ways, including

$$
\begin{aligned}
\rhd \blacktriangleright babcca\lhd \quad &\Rightarrow \quad \rhd \langle\{\}\rangle babcca\lhd\\
&\Rightarrow \quad \rhd \langle\{b\}\rangle\Box abcca\lhd\\
&\Rightarrow \quad \rhd \Box\langle\{b\}\rangle abcca\lhd\\
&\Rightarrow \quad \rhd \Box\langle\{a,b\}\rangle\Box bcca\lhd\\
&\Rightarrow^* \quad \rhd \Box\Box b\langle\{a,b\}\rangle cca\lhd\\
&\Rightarrow \quad \rhd \Box\Box b\langle\{a,b,c\}\rangle\Box ca\lhd\\
&\Rightarrow \quad \rhd \Box\Box b\langle\{\}\rangle\Box ca\lhd\\
&\Rightarrow \quad \rhd \Box\Box\langle\{\}\rangle b\Box ca\lhd
\end{aligned}
$$

$$\Rightarrow \quad \triangleright \square\square\langle\{b\}\rangle\square\square ca\triangleleft$$

$$\Rightarrow^* \quad \triangleright \square\square\square\square\square\square\hookleftarrow\triangleleft$$

$$\Rightarrow^* \quad \triangleright \hookleftarrow\square\square\square\square\square\square\triangleleft$$

$$\Rightarrow \quad \triangleright \blacksquare\square\square\square\square\square\square \triangleleft .$$

Working on the same string in several different ways, M represents a non-deterministic TM. We explain how to turn any TM into an equivalent TM that works deterministically later in this chapter (see Theorem 5.1). ■

As illustrated in Example 5.1, the strictly *formal description* of a TM spells out the states, symbols, and rules of the TM under discussion. It is the most detailed and, thereby, the most rigorous description. At the same time, this level of description tends to be tremendously lengthy and tedious. Thus, paradoxically, this fully detailed description frequently obscures what the TM is actually designed for. For instance, without any intuitive comments included in Example 5.1, we would find it somewhat difficult to figure out the way the TM accepts its language. Therefore, in the sequel, we prefer an *informal description* of TMs. That is, we describe them as procedures, omitting various details concerning their components. Crucially, the Church–Turing thesis makes both ways of description perfectly legitimate because it assures us that every procedure is identifiable with a TM defined in a rigorously mathematical way. As a matter of fact, whenever describing TMs in an informal way, we always make sure that the translation from this informal description into the corresponding formal description represents a straightforward task; unfortunately, this task is usually unbearably time-consuming too. To illustrate an informal description of TMs, we give the following example that informally describes a TM as a Pascal-like procedure which explains the changes of the tape but omits the specification of states or rules.

Convention 5.2. When informally describing a TM as a pseudocode procedure, we express that the machine accepts or rejects its input by **ACCEPT** or **REJECT**, respectively. ■

Example 5.2. Consider $L = \{a^i \mid i \text{ is a prime number}\}$. This example constructs the TM M satisfying $L(M) = L$. Therefore, from a more general viewpoint, TMs are able to recognize the primes as opposed to PDAs. M is defined as a pseudocode in the following way.

Input: Let a^i be the input string on the tape, where $i \in \mathbb{N}$;
Method:

```
begin
  If i ≤ 1, then REJECT
  change a^i to AAa^(i-2) on the tape
  while A^k a^h occurs on the tape with k ≤ h and i = k + h do
    on the tape, change A^k a^h to the unique string y satisfying i = |y|
    and y ∈ A^k{a^k A^k}* z with z ∈ prefix(a^k A^(k-1))
    if |z| = 0 or |z| = k then
      REJECT
    else
      change y to A^(k+1) a^(h-1) on the tape
  ACCEPT
end
```

Observe that i is not a prime iff an iteration of the **while** loop obtains $y = A^k a^k A^k \cdots a^k A^k$. Indeed, at this point, i is divisible by k, so M rejects a^i. On the other hand, if during every iteration, $y = A^k a^k A^k \cdots a^k A^k z$ such that $z \in prefix(a^k A^{k-1}) - \{\varepsilon, a^k\}$, then after exiting from this loop, M accepts the input string because i is a prime.

In the **while** loop, consider the entrance test whether $A^k a^h$ occurs on the tape, with $k \leq h$ and $i = k + h$. By using several states, tape symbols, and rules, we can easily reformulate this test to its strictly formal description in a straightforward way. However, a warning is in order: This reformulation also represents a painfully tedious task. As is obvious, a strictly mathematical definition of the other parts of M is lengthy as well. ■

Even more frequently and informally, we just use English prose to describe procedures representing TMs under consideration. As a matter of fact, this highly informal description of TMs is used in most proofs of theorems given in the sequel.

5.2 Restrictions

In this section, we restrict TMs so that, compared to their general versions (see Definition 5.1), the resulting restricted TMs are easier to deal with, yet they are equally powerful. In essence, we classify all the restrictions into: (i) restrictions placed upon the way TMs perform their computations; and (ii) restrictions placed upon the size of TMs.

Computational restrictions

Perhaps most importantly, we want TMs to work deterministically; that is, from any configuration, they can make no more than one move. We simply define deterministic TMs in the following way.

Definition 5.2. A TM M is *deterministic* if for every $\beta \in X_M$, R_M contains no more than one rule, according to which M can make a move from β. ■

In the proof of the following theorem, we demonstrate how to turn any TM into an equivalent deterministic TM (TMs are equivalent if they define the same language). As an exercise, give this proof in greater detail.

Theorem 5.1. *From every TM I, we can construct an equivalent deterministic TM, O.*

Proof. Let I be any TM. From I, we obtain an equivalent deterministic TM, O, so that on every input string $x \in \Delta_M^*, O$ works as follows. First, O saves x somewhere on the tape, so this string is available whenever needed. Then, O systematically produces the sequences of the rules from R_I on its tape, for instance, in lexicographical order. Always, after producing a sequence of the rules from R_I in this way, O simulates the moves that I performs on x according to this sequence. If the sequence causes I to accept, O accepts as well; otherwise, it proceeds to the simulation according to the next sequence of rules. If there exists a sequence of moves according to which I accepts x, O eventually produces this sequence and accepts x too. □

Next, without affecting the power of TMs, we place further reasonable restrictions on the way deterministic TMs work.

Definition 5.3. Let M be a TM. If from $\chi \in X_M, M$ can make no move, then χ is a *halting configuration* of M. ■

Theorem 5.2. *From every deterministic TM I, we can construct an equivalent deterministic TM, $O = (Q_O, \Delta_O, \Gamma_O, R_O, s_O, F_O)$, such that Q_O contains two new states, $\lozenge$ and $\blacksquare$, which do not occur on the left-hand side of any rule in $R_O, F_O = \{\blacksquare\}$, and:*

(I) *every halting configuration $\chi \in X_O$ has the form $\chi = \rhd qu \lhd$, with $q \in \{\lozenge, \blacksquare\}$ and $u \in \Gamma_O^*$, and every non-halting configuration $\upsilon \in X_O$ satisfies $\{\lozenge, \blacksquare\} \cap symbols(\upsilon) = \emptyset$;*

(II) *on every input string $x \in \Delta_O^*, O$ performs one of the following three kinds of computations:*

  (i) $\rhd \blacktriangleright x \lhd \Rightarrow^* \rhd \blacksquare u \lhd$, *where* $u \in \Gamma_O^*$;

  (ii) $\rhd \blacktriangleright x \lhd \Rightarrow^* \rhd \lozenge v \lhd$, *where* $v \in \Gamma_O^*$;

  (iii) *never enters any halting configuration.*

Proof. Let I be a deterministic TM. From I, construct O satisfying the properties of Theorem 5.2 as follows. In both I and $O, \blacktriangleright$ is the start state. Introduce $\lozenge$ and $\blacksquare$ as two new states into Q_O. Define $\blacksquare$ as the only final state in O; formally, set $F_O = \{\blacksquare\}$. On every input string x, O works as follows:

(1) runs I on x;

(2) if I halts in $\rhd yqv \lhd$, where $y, v \in \Gamma_I^*$ and $q \in Q_I, O$ continues from $\rhd yqv \lhd$ and computes $\rhd yqv \lhd \Rightarrow^* \rhd qyv \lhd$;

(3) if $q \in F_I, O$ computes $\rhd qyv \lhd \Rightarrow \rhd \blacksquare yv \lhd$ and halts, and if $q \in Q_I - F_I, O$ computes $\rhd qyv \lhd \Rightarrow \rhd \lozenge yv \lhd$ and halts.

As is obvious, O satisfies the properties stated in Theorem 5.2. □

We use the following convention in almost all proofs throughout the discussion concerning TMs in this book, so pay special attention to it.

Convention 5.3. In what follows, we automatically assume that every TM has the properties satisfied by O stated in Theorems 5.1 and 5.2. We denote the set of all these TMs by **TM**. We set

$\mathbf{L_{TM}} = \{L(M) \mid M \in \mathbf{TM}\}$ and refer to $\mathbf{L_{TM}}$ as the *family of Turing languages.*

Consider the three ways of computation described in part (II) of Theorem 5.2: (i), (ii), and (iii). Let $M \in \mathbf{TM}$ and $x \in \Delta_M^*$. We say that M *accepts* x iff on x, M makes a computation of the form (i). M *rejects* x iff on x, M makes a computation of the form (ii). M *halts* on x iff it accepts or rejects x; otherwise, M *loops* on x; in other words, M loops on x iff it performs a computation of the form (iii). States $\blacksquare$ and $\Diamond$ are referred to as the *accepting* and *rejecting states*, respectively; accordingly, configurations of the form $\triangleright\blacksquare u\triangleleft$ and $\triangleright\Diamond u\triangleleft$, where $u \in \Gamma_M^*$, are referred to as *accepting* and *rejecting configurations*, respectively.

We assume that Δ denotes the input alphabet of all TMs in what follows. Under this assumption, for brevity, we usually simply state that $M \in \mathbf{TM}$ works on an input string x instead of stating that M works on an input string x, where $x \in \Delta^*$. $\blacksquare$

According to Convention 5.3, we restrict our attention strictly to **TM** and $\mathbf{L_{TM}}$ in the sequel (a single exception is made in Section 7.3). Observe that this restriction is without any loss of generality because $\mathbf{L_{TM}}$ is also characterized by the general versions of TMs (see Definition 5.1) as follows from Theorems 5.1 and 5.2.

Next, we prove that every $L \in \mathbf{L_{TM}}$ is accepted by $O \in \mathbf{TM}$ that never rejects any input x, that is, either O accepts x or O loops on x. It is worth noting that we cannot reformulate this result so that O never loops on any input. In other words, $\mathbf{L_{TM}}$ contains languages accepted only by TMs that loop on some inputs; we prove this important result and explain its crucial consequences in computer science as a whole in Chapter 7.

Theorem 5.3. *From any $I \in$ **TM**, we can construct $O \in$ **TM** such that $L(I) = L(O)$ and O never rejects any input.*

Proof. Consider any $I \in \mathbf{TM}$. In I, replace every rule with $\Diamond$ on its right-hand side with a set of rules that cause the machine to keep looping in the same configuration. Let O be the TM resulting from this simple modification. Clearly, $L(I) = L(O)$ and O never rejects any input. A fully rigorous proof of this theorem is left to the reader. $\square$

Size restrictions

By Theorem 5.4 and Corollary 5.1, given in the following, we can always place a limit on the number of tape symbols in TMs.

Theorem 5.4. *From any* $I \in$ ***TM*** *with* $|\Delta_I| \geq 2$, *we can construct* $O \in$ ***TM*** *with* $\Gamma_O = \Delta_I \cup \{\square\}$.

Proof. Let $I = (Q_I, \Delta_I, \Gamma_I, R_I, s_I, F_I) \in$ **TM** such that $a, b \in \Delta_I$. Let $2^{k-1} \leq |\Gamma_I| \leq 2^k$, for some $k \in \mathbb{N}$. We encode every symbol in $\Gamma_I - \{\square\}$ as a unique string of length k over $\{a, b\}$ by a mapping f from $\Gamma_I - \{\square\}$ to $\{a, b\}^k$. Based on f, define the homomorphism g from $(Q_I \cup \Delta_I \cup \Gamma_I)^*$ to $(Q_I \cup \{\triangleright, \triangleleft, \square, a, b\})^*$ so that for all $Z \in \Gamma_I - \{\square\}, g(Z) = f(Z)$, and for all $Z \in (Q_I \cup \{\triangleright, \triangleleft, \square\}), g(Z) = Z$ (see Section 1.3 for the definition of homomorphism). Next, we construct a TM, O, that simulates I over the configurations encoded by f in the following way:

(1) *Initialization*: Let $w = c_1c_2 \cdots c_n$ be an input string, where $c_1, \ldots, c_n$ are input symbols from Δ_I. O starts its computation on w by changing w to $g(w)$; in greater detail, it changes $\triangleright \blacktriangleright c_1c_2 \cdots c_n\triangleleft$ to $g(\triangleright \blacktriangleright c_1c_2 \cdots c_n\triangleleft)$, which equals $\triangleright \blacktriangleright f(c_1)f(c_2) \cdots f(c_n)\triangleleft$.

(2) *Simulation of a move*: If $\triangleright d_1d_2 \cdots d_{i-1}qd_i \cdots d_m\triangleleft \in X_I$ is the current configuration of I, where $q \in Q_I$ and $d_i \in \Gamma_I, 1 \leq i \leq m$, then the corresponding configuration in X_O encoded by g is $g(\triangleright d_1d_2 \cdots d_{i-1}qd_i \cdots d_m\triangleleft) = \triangleright f(d_1d_2 \cdots d_{i-1})qf(d_i \cdots d_m)\triangleleft$. Let $\chi, \kappa \in X_I$, and let I compute $\chi \Rightarrow \kappa$ by using $r \in R_I$. Then, O simulates $\chi \Rightarrow \kappa$ so that it computes $g(\chi) \Rightarrow g(\kappa)$, during which it changes $g(lhs(r))$ to $g(rhs(r))$ by performing several moves.

(3) *Simulation of a computation*: O continues the simulation of moves in I one by one. If I makes a move by which it accepts, O also accepts; otherwise, O continues the simulation. □

Note that we can apply the encoding technique employed in the proof of Theorem 5.4 even if a or b are not in Δ_I, which gives rise to the following corollary.

Corollary 5.1. *Let $I \in$ **TM**. Then, there exists $O \in$ **TM** with $\Gamma_O = \{a, b, \square\} \cup \Delta_I$.* ∎

The following theorem, whose proof is left as an exercise, says we can also place a limit on the number of states in TMs without affecting their powers.

Theorem 5.5. *Let $I \in$ **TM**. Then, there exists $O \in$ **TM** with $|Q_M| \leq 3$.* ∎

The bottom line of all the restricted versions of TMs discussed in this section is that they are as powerful as the general versions of TMs, according to Definition 5.1. Of course, there also exist restrictions placed on TMs that decrease their powers. As a matter of fact, whenever we simultaneously place a limit on both the number of non-input tape symbols and the number of states, we decrease the power of TMs (a proof of this result is omitted because it is beyond the scope of this introductory text). Furthermore, later in this book, we introduce some more restricted versions of TMs, such as Turing deciders and linear bounded automata in Chapter 7, which are less powerful than TMs.

5.3 Universality

A formally described Turing machine in **TM** resembles the machine code of a program executed by a computer, which thus acts as a universal device that executes all possible programs of this kind. Considering the subject of this chapter, we obviously want to know whether there also exists a Turing machine acting as such a universal device, which simulates all machines in **TM**. The answer is yes, and in this section, we construct a *universal Turing machine* $U \in$ **TM** that does the job; that is, U simulates every $M \in$ **TM** working on any input w. However, since the input of any Turing machine, including U, is always a string, we first show how to encode every $M \in$ **TM** as a string, symbolically denoted by $\langle M \rangle$, from which U interprets M before it simulates its computation. To be quite precise, as its input, U has the code of M followed by the code of w, denoted by $\langle M, w \rangle$, from which U decodes M and w to simulate M working on w, so U accepts $\langle M, w \rangle$ iff M accepts w. As a result, before the construction of U, we explain how to obtain $\langle M \rangle$ and $\langle M, w \rangle$ for every $M \in$ **TM** and every input w.

Turing machine codes

Any reasonable encoding for Turing machines over a fixed alphabet $\vartheta \subseteq \Delta$ is acceptable, provided that for every $M \in$ **TM**, U can mechanically and uniquely interpret $\langle M \rangle$ as M. Mathematically, this encoding should represent a total function *code* from **TM** to ϑ^* such that $code(M) = \langle M \rangle$ for all $M \in$ **TM**. In addition, we select an arbitrary but fixed $Z \in$ **TM** and define the decoding of Turing machines, *decode*, so that for every $x \in range(code), decode(x) = inverse(code(M))$, and for every $y \in \vartheta^* - range(code), decode(y) = Z$, so $range(decode) =$ **TM**. As a result, *decode* is a total surjection because it maps every string in ϑ^*, including the strings that *code* maps to no machine in **TM** and to a machine in **TM**. Note, on the other hand, that several binary strings in ϑ^* may be decoded to the same machine in **TM**; mathematically, *decode* may not be an injection. From a more practical viewpoint, we just require that the mechanical interpretation of both *code* and *decode* is relatively easily performable. Apart from encoding and decoding all machines in **TM**, we also use *code* and *decode* to encode and decode the pairs consisting of Turing machines and input strings. Next, we illustrate *code* and *decode* in binary.

A Binary code for Turing machines. Consider any $M \in$ **TM**. Consider Q as the set of states of M. Rename these states to $q_1, q_2, q_3, q_4, \ldots, q_m$, so $q_1 = \blacktriangleright, q_2 = \blacksquare, q_3 = \Diamond$, where $m = |Q|$. Rename the symbols of $\{\triangleright, \triangleleft\} \cup \Gamma$ to $a_1, a_2, \ldots, a_n$, so $a_1 = \triangleright, a_2 = \triangleleft, a_3 = \blacksquare$, where $n = |\Gamma| + 2$. Introduce the homomorphism h from $Q \cup \Gamma$ to $\{0, 1\}^*$ as $h(q_i) = 10^i, 1 \leq i \leq m$, and $h(a_j) = 110^j, 1 \leq j \leq n$ (homomorphism is defined in Section 1.3). Extend h so that it is defined from $(\Gamma \cup Q)^*$ to $\{0, 1\}^*$ in the standard way, i.e., $h(\varepsilon) = \varepsilon$, and $h(X_1 \cdots X_k) = h(X_1)h(X_2) \cdots h(X_k)$, where $k \geq 1$ and $X_l \in \Gamma \cup Q, 1 \leq l \leq k$. Based on h, we now define the mapping *code* from R to $\{0, 1\}^*$ so that for each rule $r : x \to y \in R, code(r) = h(xy)$. Then, write the rules of R one after the other in an order as $r_1, r_2, \ldots, r_o$,with $o = |R|$; for instance, order them lexicographically. Set $code(R) = code(r_1)111code(r_2)111 \cdots code(r_o)111$. Finally, from $code(R)$, we obtain the desired $code(M)$ by setting $code(M) = 0^m10^n1code(R)1$.

Taking a closer look at $code(M) = 0^m10^n1code(R)1, 0^m1$ and 0^n1 state that $m = |Q|$ and $n = |\Gamma|$, respectively, and $code(R)$ encodes the rules of R. Seen as a mapping from **TM** to $\{0, 1\}^*$, *code* obviously represents a total function **TM** to ϑ^*. On the other hand, there are binary strings that represent no legal code of any machine in **TM**; mathematically, $inverse(code)$ is a partial mapping, not a total mapping. For example, ε, any string in $\{0\}^* \cup \{1\}^*$, or any string that starts with 1 are illegal codes, so their inverses are undefined. Select an arbitrary but fixed $Z \in$ **TM**; for instance, take Z as a Turing machine without any rules. Extend $inverse(code)$ to the total mapping *decode* from $\{0, 1\}^*$ to **TM** so that *decode* maps all binary strings that represent no code of any Turing machine in **TM** to Z. More precisely, for every $x \in \{0, 1\}^*$, if x is a legal code of a Turing machine K in **TM**, *decode* maps x to K, but if it is not, *decode* maps x to Z; equivalently and briefly, if x encodes $K \in$ **TM** and, therefore, $x \in range(code), decode(x) = K$, and if $x \in \{0, 1\}^* - range(code), decode(x) = Z$ (note that *decode* represents a surjection).

To encode every $w \in \Delta^*$, we simply set $code(w) = h(w)$, where h is the homomorphism defined above. Select an arbitrary but fixed $y \in \Delta^*$; for instance, take $y = \varepsilon$. Define the total surjection *decode* from $\{0, 1\}^*$ to Δ^*, so for every $x \in \{0, 1\}^*$, if $x \in range(code), decode(x) = inverse(code(w))$; otherwise, $decode(x) = y$.

For every $(M, w) \in \mathbf{TM} \times \Delta^*$, define $code(M, w) = code(M)$ $code(w)$. Viewed as a mapping from $\mathbf{TM} \times \Delta^*$ to $\{0, 1\}^*$, *code* obviously represents a total function from $\mathbf{TM} \times \Delta^*$ to ϑ^*. Define the total surjection *decode* from $\{0, 1\}^*$ to $\mathbf{TM} \times \Delta^*$, so $decode(xy) = decode(x)decode(y)$, where $decode(x) \in$ **TM** and $decode(y) \in \Delta^*$.

Example 5.3. Consider this trivial Turing machine $M \in$ **TM**, where $M = (Q, \Delta, \Gamma, R, s, S, F), Q = \{\blacktriangleright, \blacksquare, \blacklozenge, A, B, C, D\}, \Gamma = \Delta \cup \{\rhd, \lhd, \square\}, \Delta = \{b\}$, and R contains the following rules:

$$\blacktriangleright \lhd \to \blacksquare\lhd, \blacktriangleright b \to bA, Ab \to bB, Bb \to bA, A\lhd \to C \lhd,$$

$$B\lhd \to D\lhd, bD \to D\square, bC \to C\square, \rhd C \to \rhd\blacklozenge, \rhd D \to \rhd\blacksquare.$$

Leaving a simple proof that $L(M) = \{b^i \mid i \geq 0, i \textit{ is even}\}$ as an exercise, we next obtain the binary code of M by applying the encoding method described above. Introduce the homomorphism h from $Q \cup \{\triangleright, \triangleleft\} \cup \Gamma$ to $\{0,1\}^*$ as $h(q_i) = 10^i, 1 \leq i \leq 7$, where $q_1, q_2, q_3, q_4, q_5, q_6$, and q_7 coincide with $\blacktriangleright, \blacksquare, \blacklozenge, A, B, C$, and D, respectively, and $h(a_i) = 110^j, 1 \leq j \leq 4$, where a_1, a_2, a_3, and a_4 coincide with $\triangleright, \triangleleft, \square$, and b, respectively. Extend h so it is defined from $(Q \cup \{\triangleright, \triangleleft\} \cup \Gamma)^*$ to $\{0,1\}^*$ in the standard way. Based on h, define the mapping *code* from R to $\{0,1\}^*$, so for each rule $x \rightarrow y \in R, code(x \rightarrow y) = h(xy)$. For example, $code(\blacktriangleright b \rightarrow bA) = 1011000011000010000$. Take, for instance, the above order of the rules from R, and set

$$
\begin{aligned}
code(R) = {} & code(\blacktriangleright \triangleleft \rightarrow \blacksquare\triangleleft)111code(\blacktriangleright b \rightarrow bA)111 \\
& code(Ab \rightarrow bB)111code(Bb \rightarrow bA)111 \\
& code(A\triangleleft \rightarrow C\triangleleft)111code(B\triangleleft \rightarrow D\triangleleft)111 \\
& code(bD \rightarrow D\square)111code(bC \rightarrow C\square)111 \\
& code(\triangleright C \rightarrow \triangleright\blacklozenge)111code(\triangleright D \rightarrow \triangleright\blacksquare)111 \\
= {} & 10110010011001111011000011000010000111 \\
& 1000011000011000010000011110000011000011000010000111 \\
& 100001100100000011001111000001100100000001100111 \\
& 1100001000000010000000110001111100001000000100000011000111 \\
& 1101000000110100011111010000000110100111.
\end{aligned}
$$

To encode M as a whole, set

$$
\begin{aligned}
code(M) = {} & 0^7 10^4 1 code(R)1 \\
= {} & 0000000100001 \\
& 10110010011001111011000011000010000111 \\
& 1000011000011000010000011110000011000011000010000111 \\
& 100001100100000011001111000001100100000001100111 \\
& 1100001000000010000000110001111100001000000100000011000111 \\
& 1101000000110100011111010000000110100111 \\
& 1.
\end{aligned}
$$

Take $w = bb$, whose $code(bb) = 110000110000$. As a result, the binary string denoted by $code(M, bb)$ is

0000000100001101100100110011110110000110000100001111000011000011000
0100000011110000001100001100001000011110000110010000000110011110000
0110010000000110011111000010000000100000001100011111000010000001000
00011000111110100000011010001111101000000001101001111110000110000.

■

Convention 5.4. In what follows, we suppose that there exist a fixed encoding and a fixed decoding of all Turing machines in **TM**. We just require that both be uniquely and mechanically interpretable; otherwise, they may differ from *code* and *decode* (in fact, they may not even be in binary). As already stated in the beginning of this section, we denote the code of $M \in \mathbf{TM}$ by $\langle M \rangle$. Similarly, we suppose that there exist analogical encoding and decoding of the members of Δ^*, $\mathbf{TM} \times \Delta^*$, $\mathbf{TM} \times \mathbf{TM}$, and $\mathbf{TM} \times {}_0\mathbb{N}$. Again, for brevity, we denote the codes of $w \in \Delta^*$, $(M, w) \in \mathbf{TM} \times \Delta^*$, $(M, N) \in \mathbf{TM} \times \mathbf{TM}$, and $(M, i) \in \mathbf{TM} \times {}_0\mathbb{N}$ by $\langle w \rangle$, $\langle M, w \rangle$, $\langle M, N \rangle$, and $\langle M, i \rangle$, respectively (as an exercise, encode and decode the members of ${}_0\mathbb{N}$ similarly to encoding the machine in **TM**). Even more generally, for any automaton, X, discussed earlier in this book, $\langle X \rangle$ represents its code analogical to $\langle M \rangle$. ■

Out of all the terminology introduced in Convention 5.4, we just need $\langle M \rangle$ and $\langle M, w \rangle$ in the rest of this chapter. In Chapter 7, however, we also make use of the other abbreviations, as always explicitly pointed out therein.

Construction of universal Turing machines

We are now ready to construct U, i.e., a universal Turing machine (see the beginning of this section). As a matter of fact, we construct two versions of U. The first version, denoted by ${}_{\text{TM-Acceptance}}U$, simulates every $M \in \mathbf{TM}$ on $w \in \Delta^*$, so ${}_{\text{TM-Acceptance}}U$ accepts $\langle M, w \rangle$ iff

M accepts w. In other words, $L({}_{\text{TM-Acceptance}}U) = {}_{\text{TM-Acceptance}}L$, with

$${}_{\text{TM-Acceptance}}L = \{\langle M, w\rangle \mid M \in \mathbf{TM}, w \in \Delta^*, M \text{ accepts } w\}.$$

The other version, denoted by ${}_{\text{TM-Halting}}U$, simulates every $M \in \mathbf{TM}$ on $w \in \Delta^*$ in such a way that ${}_{\text{TM-Halting}}U$ accepts $\langle M, w\rangle$ iff M halts on w (see Convention 5.4). To rephrase this in terms of formal languages, $L({}_{\text{TM-Halting}}U) = {}_{\text{TM-Halting}}L$ with

$${}_{\text{TM-Halting}}L = \{\langle M, w\rangle \mid M \in \mathbf{TM}, w \in \Delta^*, M \text{ halts on } w\}.$$

Convention 5.5. Precisely, in the following proof, we should state that ${}_{\text{TM-Acceptance}}U$ works on $\langle M, w\rangle$, so it first interprets $\langle M, w\rangle$ as M and w; then, it simulates the moves of M on w. However, instead of a long and obvious statement like this, we just state that ${}_{\text{TM-Acceptance}}U$ runs M on w. In a similar manner, we shorten the other proofs of results concerning Turing machines in the sequel whenever no confusion exists. ■

Theorem 5.6. *There exists* ${}_{\text{TM-Acceptance}}U \in \boldsymbol{TM}$ *such that* $L({}_{\text{TM-Acceptance}}U) = {}_{\text{TM-Acceptance}}L$.

Proof. On every input $\langle M, w\rangle$, ${}_{\text{TM-Acceptance}}U$ runs M on w. ${}_{\text{TM-Acceptance}}U$ accepts $\langle M, w\rangle$ if it finds out that M accepts w; otherwise, ${}_{\text{TM-Acceptance}}U$ keeps simulating the moves of M in this way. □

Observe that ${}_{\text{TM-Acceptance}}U$ represents a procedure, not an algorithm, because if M loops on w, so does ${}_{\text{TM-Acceptance}}U$ on $\langle M, w\rangle$. As a matter of fact, in Chapter 7, we demonstrate that no Turing machine can halt on every input and, simultaneously, act as a universal Turing machine (see Theorem 7.13). To reformulate this in terms of formal languages, no Turing machine accepts ${}_{\text{TM-Acceptance}}L$ in such a way that it halts on all strings. Indeed, for all $X \in \mathbf{TM}$ satisfying ${}_{\text{TM-Acceptance}}L = L(X), \Delta^* - {}_{\text{TM-Acceptance}}L$ necessarily contains a string on which X loops.

By analogy with the proof of Theorem 5.6, we next obtain ${}_{\text{TM-Halting}}U$ that accepts ${}_{\text{TM-Halting}}L$, as defined above.

Theorem 5.7. *There exists* ${}_{\text{TM-Halting}}U \in \boldsymbol{TM}$ *such that* $L({}_{\text{TM-Halting}}U) = {}_{\text{TM-Halting}}L$.

Proof. On every $\langle M, w\rangle$, ${}_{\text{TM}-\text{Halting}}U$ works, so it runs M on w. ${}_{\text{TM}-\text{Halting}}U$ accepts $\langle M, w\rangle$ iff M halts w, which means that M either accepts w or rejects w (see Convention 5.3). Thus, ${}_{\text{TM}-\text{Halting}}U$ loops on $\langle M, w\rangle$ iff M loops on w. Observe that $L({}_{\text{TM}-\text{Halting}}U) = {}_{\text{TM}-\text{Halting}}L$. $\square$

Chapter 6

Automata and Their Grammatical Equivalents

As already explained in Chapter 2, apart from automata, formal languages are also often defined by grammars, which are based upon finitely many rules by which they generate their languages. This two-section chapter presents the fundamental types of grammars that represent counterparts to the automata covered in the previous two chapters. Most importantly, it demonstrates that their power coincides with that of the corresponding automata.

Section 6.1 deals with grammars that make their language generation contextually independent. That is, they make each step of this generation process regardless of the context in which the applied rule is used, hence their name — context-free grammars (CFG). These grammars are of particular interest both in formal language theory and its applications because they define the same language family as pushdown automata do, so they represent their grammatical counterparts. That is also why Section 6.1 pays special attention to these grammars as well as their modified versions, such as state grammars and scattered context grammars.

Section 6.2 defines general grammars and demonstrates their equivalence with Turing machines. Then, it introduces special cases of general grammars, called context-sensitive grammars, and special cases of Turing machines, called linear bounded automata, and it shows that both are equally powerful.

In its conclusion, the current chapter summarizes the relations between major language families previously discussed in Part 2.

6.1 Context-Free Grammars and Pushdown Automata

Definition 6.1. A *context-free grammar* (CFG for short) is a quadruple

$$G = (V, T, P, S),$$

where V is an alphabet, $T \subseteq V, P \subseteq (V - T) \times V^*$ is finite, and $S \in V - T$. Set $N = V - T$. In G, V, N, and T are referred to as the *total alphabet*, *alphabet of nonterminal symbols*, and the *alphabet of terminal symbols*, respectively. S is the *start symbol*. P is a finite *set of rules*; every rule $(u, v) \in P$ is written as $u \rightarrow v$ in what follows. For brevity, we often denote $u \rightarrow v$ by a unique label p as $p : u \rightarrow v$, and we briefly use p instead of $u \rightarrow v$ under this denotation.

Define the relation $\Rightarrow$ over V^* so that for all $u \rightarrow v \in P$ and $x, y \in V^*, xuy \Rightarrow xvy$; in words, G *directly derives* xvy from xuy. As usual, $\Rightarrow^*$ denotes the transitive-reflexive closure of $\Rightarrow$. If $S \Rightarrow^* w$, where $w \in V^*, G$ *derives* w, and w is a *sentential form*. $F(G)$ denotes the set of all sentential forms derived by G. The *language generated by* G, symbolically denoted by $L(G)$, is defined as $L(G) = F(G) \cap T^*$; in other words, $L(G) = \{w \in T^* \mid S \Rightarrow^* w\}$. The members of $L(G)$ are called *sentences*. If $S \Rightarrow^* w$ and w is a sentence, $S \Rightarrow^* w$ is a *successful derivation* in G. ■

By $u \Rightarrow v$ $[r]$, where $u, v \in V^*$ and $r \in P$, we say that G directly rewrites u to v by r or, as we customarily say in terms of CFGs, G makes a *derivation step* from u to v by r. Furthermore, to express that G makes $u \Rightarrow^* w$ according to a sequence of rules, $r_1 r_2 \cdots r_n$, we write $u \Rightarrow^* v$ $[r_1 r_2 \cdots r_n]$, which is read as a *derivation from* u *to* v *by using* $r_1 r_2 \cdots r_n$. On the other hand, whenever the information regarding the applied rules is immaterial, we omit these rules. In other words, we often simplify u $\Rightarrow v$ $[r]$ and $u \Rightarrow^* v$ $[r_1 r_2 \cdots r_n]$ to $u \Rightarrow v$ and $u \Rightarrow^* v$, respectively.

Convention 6.1. For any CFG G, we automatically assume that V, N, T, S, and P denote the total alphabet, the alphabet of nonterminal symbols, the alphabet of terminal symbols, the start symbol, and the set of rules, respectively. If there exists a danger of confusion, we mark V, N, T, S, P with G as V_G, N_G, T_G, S_G, P_G, respectively, in order to clearly relate these components to G (in particular, we make these marks when several CFGs are simultaneously discussed). For brevity, we often abbreviate nonterminal symbols and terminal symbols to *nonterminals* and *terminals*, respectively. If we want to express that a nonterminal A forms the left-hand side of a rule, we refer to this rule as an *A-rule*. If we want to specify that A is rewritten during a derivation step, $xAy \Rightarrow xuy$, we underline this A as $x\underline{A}y \Rightarrow xuy$.

For brevity, G is often defined by simply listing its rules together with specifying its nonterminals and terminals, usually denoted by uppercases and lowercases, respectively.

We denote the set of all CFGs by **CFG**. We set $\mathbf{L_{CFG}} = \{L(G) \mid G \in \mathbf{CFG}\}$, and we refer to $\mathbf{L_{CFG}}$ as the *family of context-free languages*, so any language in $\mathbf{L_{CFG}}$ is called a *context-free language* (CFL for short). ■

The following example demonstrates how to determine the language generated by a CFG in a rigorous way. The CFG considered in the example is relatively simple; in fact, it represents a linear grammar, in which every rule has no more than one occurrence of a nonterminal on its right-hand side.

As a result, every sentential form contains no more than one occurrence of a nonterminal, and this property makes the determination of the generated language relatively simple. Before giving the example, we define the notion of a *linear grammar*, LG for short, as a CFG, $H = (V_H, T_H, P_H, S_H)$, in which the right-hand side of every rule contains no more than one occurrence of a nonterminal.

Definition 6.2. Let $G = (V, T, P, S)$ be a CFG such that each $A \rightarrow x \in P$ satisfies $x \in T^*(N \cup \{\varepsilon\})T^*$. Then, G is said to be a *linear* grammar (LG for short). ■

Example 6.1. Consider $L = \{a^k b^k \mid k \geq 1\}$. In principle, we generate the strings from L using a CFG, G, so G derives sentential forms

$$S, aSb, aaSbb, aaaSbbb, \ldots,$$

in which G rewrites S with ab during the very last derivation step. Formally, $G = (V, T, P, S)$, where $V = \{a, b, S\}$ and

$P = \{S \to aSb, S \to ab\}$. In G, $T = \{a, b\}$ is the alphabet of terminals and $N = \{S\}$ is the alphabet of nonterminals, where S is the start symbol of G. Under Convention 6.1, we can specify G simply as

$$1 : S \to aSb$$
$$2 : S \to ab$$

Consider $aaSbb$. By using rule 1, G rewrites S with aSb in this string, so $aaSbb \Rightarrow aaaSbbb$ [1]. By using rule 2, G rewrites S with ab, so $aaSbb \Rightarrow aaabbb$ [2]. By using the sequence of rules 112, G makes

$$\begin{array}{lcll} aaSbb & \Rightarrow & aaaSbbb & [1] \\ & \Rightarrow & aaaaSbbbb & [1] \\ & \Rightarrow & aaaaabbbbb & [2] \end{array}$$

Briefly, we write $aaSbb \Rightarrow^* aaaaabbbbb$ [112] or, even more simply, $aaSbb \Rightarrow^* aaaaabbbbb$.

To verify that G generates $\{a^k b^k \mid k \geq 1\}$, recall that by using rule 1, G replaces S with aSb, and by rule 2, G replaces S with ab. Consequently, every successful derivation has the form

$$S \Rightarrow aSb \Rightarrow aaSbb \Rightarrow \cdots \Rightarrow a^k Sb^k \Rightarrow a^{k+1}b^{k+1},$$

which G makes according to a sequence of rules of the form $1^k 2$, for some $k \geq 0$. In symbols, $S \Rightarrow^* a^{k+1}b^{k+1}[1^k 2]$. From these observations, we see $L(G) = \{a^k b^k \mid k \geq 1\}$; a detailed verification of this identity is left as an exercise. ■

The two-rule linear CFG considered in the previous example generates every sentence by a unique derivation, so the analysis of its derivation process is rather easy. As a result, the determination of its generated language represents a simple task as well. In a general case, however, $G \in$ **CFG** may contain several occurrences of nonterminals on the right-hand sides of their rules and generate their sentences through a variety of different derivations. Under these circumstances, the determination of $L(G)$ may be more complicated, as shown in the following example. This example also illustrates a typical two-phase approach to achieving complicated results: (1) first, a more general result is established, and then (2) as its straightforward consequence, the desired result is derived.

Example 6.2. Consider L from Example 6.1. Let K be the set of all permutations of strings in L. Equivalently, K contains all non-empty strings consisting of an equal number of as and bs. Formally,

$$K = \{w \mid w \in \{a,b\}^+ \text{ and } occur(w,a) = occur(w,b)\}.$$

In this example, we prove that K is generated by the CFG G defined as

$$1 : S \to aB, \quad 2 : S \to bA, \quad 3 : A \to a, \quad 4 : A \to aS,$$

$$5 : A \to bAA, \quad 6 : B \to b, \quad 7 : B \to bS, \quad 8 : B \to aBB.$$

We first prove the following claim, which says something more than we actually need to establish $K = L(G)$. From this claim, we subsequently obtain $K = L(G)$ as a straightforward consequence of this claim.

Claim. For all $w \in \{a,b\}^*$, the following three equivalences hold:

(I) $S \Rightarrow^* w$ iff $occur(a,w) = occur(b,w)$;
(II) $A \Rightarrow^* w$ iff $occur(a,w) = occur(b,w) + 1$;
(III) $B \Rightarrow^* w$ iff $occur(b,w) = occur(a,w) + 1$.

■

Proof. This claim is proved by induction on $|w| \geq 1$.

Basis. Let $|w| = 1$.

(1) From S, G generates no sentence of length one. On the other hand, no sentence of length one satisfies $occur(a,w) = occur(b,w)$. Thus, in this case, the basis holds vacuously.
(2) Examine G to see that if $A \Rightarrow^* w$ with $|w| = 1$, then $w = a$. For $w = a, A \Rightarrow^* w$ [3]. Therefore, (II) holds in this case.
(3) Prove (III) by analogy with the proof of (II).

Consequently, the basis holds.

Induction hypothesis. Assume that there exists a positive integer $n \geq 1$ such that the claim holds for every $w \in \{a,b\}^*$ satisfying $1 \leq |w| \leq n$.

Induction step. Let $w \in \{a, b\}^*$ with $|w| = n + 1$.

Consider (I) in the claim. To prove its only-if part, consider any derivation of the form $S \Rightarrow^* w\ [\rho]$, where ρ is a sequence of rules. This derivation starts from S. As only rules 1 and 2 have S on the left-hand side, express $S \Rightarrow^* w\ [\rho]$ as $S \Rightarrow^* w\ [r\pi]$, where $\rho = r\pi$ and $r \in \{1, 2\}$.

(i) If $r = 1, S \Rightarrow^* w\ [1\pi]$, where $1 : S \to aB$. At this point, $w = av$, and $B \Rightarrow^* v\ [\pi]$, where $|v| = n$. By the induction hypothesis, (III) holds for v, so $occur(b, v) = occur(a, v) + 1$. Therefore, $occur(a, w) = occur(b, w)$.
(ii) If $r = 2, S \Rightarrow^* w\ [2\pi]$, where $2 : S \to bA$. Thus, $w = bv$, and $A \Rightarrow^* v\ [\pi]$, where $|v| = n$. By the induction hypothesis, (II) holds for v, so $occur(a, v) = occur(b, v) + 1$. As $w = bv, occur(a, w) = occur(b, w)$.

To prove the if part of (I), suppose that $occur(a, w) = occur(b, w)$. Clearly, $w = av$ or $w = bv$, for some $v \in \{a, b\}^*$ with $|v| = n$.

(i) Let $w = av$. Then, $|v| = n$ and $occur(a, v) + 1 = occur(b, v)$. As $|v| = n$, by the induction hypothesis, we have $B \Rightarrow^* v$ iff $occur(b, v) = occur(a, v) + 1$ from (III). By using $1 : S \to aB$, we obtain $S \Rightarrow aB\ [1]$. Putting $S \Rightarrow aB$ and $B \Rightarrow^* v$ together, we have $S \Rightarrow aB \Rightarrow^* av$, so $S \Rightarrow^* w$ because $w = av$.
(ii) Let $w = bv$. Then, $|v| = n$ and $occur(a, v) = occur(b, v) + 1$. By the induction hypothesis, we have $A \Rightarrow^* v$ iff $occur(a, v) = occur(b, v) + 1$ (see (II)). By $2 : S \to bA, G$ makes $S \Rightarrow bA$. Thus, $S \Rightarrow bA$ and $A \Rightarrow^* v$, so $S \Rightarrow^* w$.

Consider (II) in the claim. To prove its only-if part, consider any derivation of the form $A \Rightarrow^* w\ [\rho]$, where ρ is a sequence of rules in G. Express $A \Rightarrow^* w\ [\rho]$ as $A \Rightarrow^* w\ [r\pi]$, where $\rho = r\pi$ and $r \in \{3, 4, 5\}$ because rules $3 : A \to a, 4 : A \to aS$, and $5 : A \to bAA$ are all the A-rules in G.

(i) If $r = 3, A \Rightarrow^* w\ [r\pi]$ is a one-step derivation $A \Rightarrow a\ [3]$, so $w = a$, which satisfies $occur(a, w) = occur(b, w) + 1$.

(ii) If $r = 4, A \Rightarrow^* w$ $[4\pi]$, where $4 : A \to aS$. Thus, $w = av$, and $S \Rightarrow^* v$ $[\pi]$, where $|v| = n$. By the induction hypothesis, from I, $occur(a, v) = occur(b, v)$, so $occur(a, w) = occur(b, w) + 1$.

(iii) If $r = 5, A \Rightarrow^* w$ $[5\pi]$, where $5 : A \to bAA$. Thus, $w = buv, A \Rightarrow^* u, A \Rightarrow^* v$, where $|u| \leq n, |v| \leq n$. By the induction hypothesis, from (II), $occur(a, u) = occur(b, u) + 1$ and $occur(a, v) = occur(b, v) + 1$, so $occur(a, uv) = occur(b, uv) + 2$. Note that $occur(b, uv) = occur(b, w) - 1$ implies $occur(a, uv) - 2 = occur(b, w) - 1$. Furthermore, from $occur(a, uv) - 2 = occur(b, w) - 1$, it follows that $occur(a, uv) = occur(a, w)$, so $occur(a, w) = occur(b, w) + 1$.

To prove the if part of (II), suppose that $occur(a, w) = occur(b, w) + 1$. Obviously, $w = av$ or $w = bv$, for some $v \in \{a, b\}^*$ with $|v| = n$.

(i) Let $w = av$. At this point, $|v| = n$ and $occur(a, v) = occur(b, v)$. As $|v| = n$, by the induction hypothesis, we have $S \Rightarrow^* v$. By using $4 : A \to aS, A \Rightarrow aS$ [4]. Putting $A \Rightarrow aS$ and $S \Rightarrow^* v$ together, we obtain $A \Rightarrow aS \Rightarrow^* av$, so $A \Rightarrow^* w$ because $w = av$.

(ii) Let $w = bv$. At this point, $|v| = n$ and $occur(a, v) = occur(b, v) + 2$. Express v as $v = uz$ so that $occur(a, u) = occur(b, u) + 1$ and $occur(a, z) = occur(b, z) + 1$. As an exercise, we leave a proof that $occur(a, v) = occur(b, v) + 2$ implies that v can always be expressed in this way. Since $|v| = n, |u| \leq n \geq |z|$. Thus, by the induction hypothesis (see (II)), we have $A \Rightarrow^* u$ and $A \Rightarrow^* z$. By using $5 : A \to bAA, A \Rightarrow bAA$ [5]. Putting $A \Rightarrow bAA, A \Rightarrow^* u$, and $A \Rightarrow^* z$ together, we obtain $A \Rightarrow bAA \Rightarrow^* buz$, so $A \Rightarrow^* w$ because $w = bv = buz$.

Prove (III) by analogy with the proof of the inductive step of (II), given above. □

Having established this claim, we easily obtain the desired equation $L(G) = \{w \mid w \in \{a, b\}^+ \text{ and } occur(w, a) = occur(w, b)\}$ as a consequence of Equivalence I. Indeed, this equivalence says that for all $w \in \{a, b\}^*, S \Rightarrow^* w$ iff $occur(a, w) = occur(b, w)$. Consequently, $w \in L(G)$ iff $occur(a, w) = occur(b, w)$. As G has no ε-rules, $\varepsilon \notin L(G)$, so $L(G) = \{w \mid w \in \{a, b\}^+ \text{ and } occur(w, a) = occur(w, b)\}$.

Restricted context-free grammars

In this section, we introduce several reasonable restrictions placed on CFGs to simplify their investigation in theory and use in practice. We struggle to introduce all the restricted versions so they are as powerful as their general versions, defined in the previous section.

First, this section reduces the derivation multiplicity of CFGs by restricting its attention only to canonical derivations and derivation trees. This section also explores the phenomenon of ambiguity in CFGs and their languages, and it points out the existence of CFLs that are always generated in an ambiguous way. Then, the current section explains how to remove all redundant symbols from CFGs, after which it describes how to turn any CFG to an equivalent CFG in which no rule has ε as its right-hand side. Furthermore, this section transforms any CFG to an equivalent CFG in which a single nonterminal does not form the right-hand side of any rule. Continuing with the same topic in a more restrictive way, the section describes how to convert any grammar in **CFG** into an equivalent Chomsky normal form grammar, in which every rule has on its right-hand side a terminal or two nonterminals. It also discusses the grammatical phenomenon of left recursion, which causes the grammars to go into an infinite loop, and explains how to remove this phenomenon. Finally, this section describes how to transform any grammar in **CFG** into an equivalent Greibach normal form grammar, in which every rule has on its right-hand side a terminal followed by zero or more nonterminals.

In its conclusion, this section demonstrates the equivalence between CFGs and PDAs.

Canonical derivations and derivation trees

As illustrated by Example 6.2, in a general case, $G \in$ **CFG** may generate the same sentence by many different derivations, and this derivation multiplicity obviously complicates the discussion of G and $L(G)$ in both theory and practice. To reduce this derivation multiplicity, we first introduce two special types of *canonical* *derivations*, namely leftmost derivations and rightmost derivations, and demonstrate that $x \in L(G)$ iff G generates x by either of these two canonical derivations. In addition, in terms of graph theory, we simplify the discussion concerning grammatical derivations by *derivation trees*, which represent derivations by graphically displaying rules but suppressing the order of their applications.

During this section, the reader should carefully keep in mind that we frequently and automatically make use of Convention 6.1.

Leftmost derivations

A derivation is *leftmost* if during its every single derivation step, the leftmost occurrence of a nonterminal is rewritten in the sentential form.

Definition 6.3. Let $G = (V, T, P, S)$ be a CFG.

(I) Let $r : A \to z \in P, t \in T^*, o \in V^*$. Then, G makes the *leftmost derivation step* from tAo to tzo according to r, symbolically written as $tAo\ {}_{lm}{\Rightarrow}\ tzo\ [r]$.

(II) Leftmost derivations in G are defined recursively as follows:

(i) for all $u \in V^*$, G makes the *leftmost derivation from* u *to* u *according to* ε, symbolically written as $u\ {}_{lm}{\Rightarrow}^*\ u\ [\varepsilon]$;

(ii) if $u, w, v \in V^*, \rho = \sigma r, \sigma \in P^*, r \in P, u\ {}_{lm}{\Rightarrow}^*\ w\ [\sigma]$ and $w\ {}_{lm}{\Rightarrow}\ v\ [r]$ in G, then G makes the leftmost derivation from u to v according to ρ, symbolically written as $u\ {}_{lm}{\Rightarrow}^*\ v\ [\rho]$ in G. ■

To point out the crucial parts of Definition 6.3, note that in (I), A is the leftmost occurrence of a nonterminal in the rewritten string tAo. According to (II), $u\ {}_{lm}{\Rightarrow}^*\ u\ [\varepsilon]$ in G, for every $u \in V^*$. If $\rho = r_1 r_2 \cdots r_n$, where $r_j \in P$, for some $n \in \mathbb{N}$, and there are $w_0, w_1, \ldots, w_n \in V^*$ such that $w_{j-1}\ {}_{lm}{\Rightarrow}\ w_j[r_j]$ in G (see (I)), for all $1 \leq j \leq n$, then G makes the leftmost derivation from w_0 to w_n according to ρ, $w_0\ {}_{lm}{\Rightarrow}^*\ w_n[\rho]$. If ρ represents an immaterial piece of information, we omit it and simplify $w_0\ {}_{lm}{\Rightarrow}^*\ w_n[\rho]$ to $w_0\ {}_{lm}{\Rightarrow}^*\ w_n$. It is worth noting that apart from $w_0\ {}_{lm}{\Rightarrow}^*\ w_n[\rho]$, there may exist $\sigma \in P^*$ such that $\rho \neq \sigma$ and $w_0\ {}_{lm}{\Rightarrow}^*\ w_n[\sigma]$ in G as well. In fact, in a general case, G can make $w_0 \ {}_{lm}\Rightarrow^* w_n$ according to several different sequences of rules from P^*.

Next, we demonstrate that every CFG can generate each sentence by a leftmost derivation.

Theorem 6.1. *Let* $G \in \mathbf{CFG}$*. Then,* $w \in L(G)$ *iff* $S \ {}_{lm}\Rightarrow^* w$ *in* G*.*

Proof. The *if* part of the proof says that $S \ {}_{lm}\Rightarrow^* w$ implies $w \in L(G)$, for every $w \in T^*$. As $S \ {}_{lm}\Rightarrow^* w$ is a special case of a derivation from S to w, this implication surely holds. Therefore, we only need to prove the *only-if* part that says that $w \in L(G)$ implies $S \ {}_{lm}\Rightarrow^* w$ in G. This implication straightforwardly follows from the next claim.

Claim. For every $w \in L(G), S \Rightarrow^n w$ implies $S \ {}_{lm}\Rightarrow^n w$, for all $n \geq 0$.

Proof. (by induction on $n \geq 0$).

Basis. For $n = 0$, this implication is trivial.
Induction hypothesis. Assume that there exists an integer $n \geq 0$ such that the claim holds for all derivations of length n or less.

Induction step. Let $S \Rightarrow^{n+1} w\ [\rho]$, where $w \in L(G), \rho \in P^+$, and $|\rho| = n + 1$. If $S \Rightarrow^{n+1} w\ [\rho]$ is leftmost, the induction step is completed. Assume that this derivation is not leftmost. Express $S \Rightarrow^{n+1} w\ [\rho]$ as

$$\begin{aligned} S \ {}_{lm}\Rightarrow^* &\ uAvBx && [\sigma] \\ \Rightarrow &\ uAvyx && [r : B \to y] \\ \Rightarrow^* &\ w && [\theta], \end{aligned}$$

where $\sigma, \theta \in P^*, \rho = \sigma r \theta, r : B \to y \in P, u \in \mathit{prefixes}(w), A \in N$, and $v, x, y \in V^*$. In other words, $S \ {}_{lm}\Rightarrow^* uAvBx$ is the longest leftmost derivation that begins with $S \Rightarrow^{n+1} w$. As $w \in L(G)$ and $L(G) \subseteq T^*$ (see Definition 6.1), $w \in T^*$. Thus, $A \notin \mathit{symbols}(w)$ because $A \in N$. Hence, A is surely rewritten during $uAvyx \Rightarrow^* w$. Express

$S \Rightarrow^{n+1} w$ as

$$\begin{aligned} S\ {}_{lm}\Rightarrow^* &\ uAvBx && [\sigma] \\ \Rightarrow &\ uAvyx && [r : B \to y] \\ \Rightarrow^* &\ uAz && [\pi] \\ {}_{lm}\Rightarrow &\ utz && [p : A \to t] \\ \Rightarrow^* &\ w && [o], \end{aligned}$$

where $\pi, o \in P^*, \theta = \pi po, p : A \to t \in P, vyx \Rightarrow^* z$, and $z \in V^*$. Rearrange this derivation so that the derivation step according to p is made right after the initial part $S\ {}_{lm} \Rightarrow^* uAvBx\ [\sigma]$; more formally,

$$\begin{aligned} S\ {}_{lm}\Rightarrow^* &\ uAvBx && [\sigma] \\ {}_{lm}\Rightarrow &\ utvBx && [p : A \to t] \\ \Rightarrow &\ utvyx && [r : B \to y] \\ \Rightarrow^* &\ utz && [\pi] \\ \Rightarrow^* &\ w && [o]. \end{aligned}$$

The resulting derivation $S \Rightarrow^* w\ [\sigma pr\pi o]$ begins with at least $|\sigma p|$ leftmost steps, so its leftmost beginning is definitely longer than the leftmost beginning of the original derivation $S \Rightarrow^{n+1} w\ [\rho]$. If $S \Rightarrow^* w\ [\sigma pr\pi o]$ is leftmost, the induction step is completed. If not, apply the derivation rearrangement described above to $S \Rightarrow^* w\ [\sigma pr\pi o]$. After no more than $n - 2$ repetitions of this rearrangement, we necessarily obtain $S\ {}_{lm} \Rightarrow^* w$, which completes the induction step, so the proof of the claim is completed.

By this claim, we see that $w \in L(G)$ implies $S\ {}_{lm}\Rightarrow^* w$ in G, so the theorem holds. □

We might naturally ask whether for every $G = (V, T, P, S) \in$ **CFG**, $S \Rightarrow^* w$ iff $S\ {}_{lm}\Rightarrow^* w$, for all $w \in V^*$. To rephrase this question less formally, we might ask whether Theorem 6.1 can be generalized in terms of all sentential forms, not just sentences. Surprisingly, the answer is no. Indeed, to give a trivial counterexample, consider a two-rule CFG defined as $S \to AA$ and $A \to a$. Observe that this grammar makes $S \Rightarrow AA \Rightarrow Aa$; however, there is no leftmost derivation of Aa in G.

Rightmost derivations

As their name indicates, in rightmost derivations, the rightmost occurrence of a nonterminal is rewritten during its every derivation step.

Definition 6.4. Let $G = (V, T, P, S)$ be a CFG.

(I) Let $r : A \to z \in P, t \in T^*, o \in V^*$. Then, G makes the *rightmost derivation step* from oAt to ozt according to r, symbolically written as $oAt \;_{rm}\!\Rightarrow ozt \; [r]$.

(II) Rightmost derivations in G are defined recursively as follows:

   (i) for all $u \in V^*$, then G makes the *rightmost derivation from* u *to* u *according to* ε, symbolically written as $u \;_{rm}\!\Rightarrow^* u \; [\varepsilon]$;

   (ii) if $u, w, v \in V^*, \rho = \sigma r, \sigma \in P^*, r \in P, u \;_{rm}\!\Rightarrow^* w \; [\sigma]$ and $w \;_{rm}\!\Rightarrow v \; [r]$ in G, then G makes the *rightmost derivation from* u *to* v *according to* ρ, symbolically written as $u \;_{rm}\!\Rightarrow^* v \; [\rho]$ in G. ■

Let $u, v \in V^*, \rho \in P^*$, and $u \;_{rm}\!\Rightarrow^* v \; [\rho]$ in G. If ρ represents an immaterial piece of information, we usually simplify $u \;_{rm}\!\Rightarrow^* v \; [\rho]$ to $u \;_{rm}\!\Rightarrow^* v$. Of course, in a general case, G can make $u \;_{rm}\!\Rightarrow^* v$ according to several different sequences of rules from P^*. As an exercise, by analogy with Theorem 6.1, prove the following theorem.

Theorem 6.2. *Let $G \in$ **CFG**. Then, $w \in L(G)$ iff $S \;_{rm}\!\Rightarrow^* w$.* ■

Derivation trees

Apart from the canonical derivations, we often simplify the discussion concerning grammatical derivations graphically by *derivation trees*, composed of rule trees. As their names indicate, rule trees describe grammatical rules, while derivation trees represent derivations by specifying the rules, expressed as rule trees, according to which they are combined with the nonterminals to which the rules are applied. On the other hand, derivation trees suppress the order of the application of the rules, so we make use of these trees when this order is immaterial.

A derivation tree in a CFG, $G = (V, T, P, S)$, is any tree such that its root is from V, and for each of its elementary subtrees (see

Section 1.4), e, there is a rule, $l \in P$, such that e represents l (see (I) in Definition 6.5). Apart from the notion of a derivation tree, the following definition also specifies its correspondence to the derivation it represents. Let us note that this definition makes use of the terminology concerning trees introduced in Section 1.4.

Definition 6.5. Let $G = (V, T, P, S)$ be a CFG.

(I) For $l : A \to x \in P, A\langle x\rangle$ is the *rule tree that represents* l.

(II) The *derivation trees representing derivations in* G are defined recursively as follows:

  (i) One-node tree X is the derivation tree corresponding to $X \Rightarrow^0 X$ in G, where $X \in V$;

  (ii) Let d be the derivation tree representing $A \Rightarrow^* uBv\ [\rho]$ with $frontier(d) = uBv$, and let $l : B \to z \in P$. The derivation tree that represents

$$\begin{array}{llll} A & \Rightarrow^* & uBv & [\rho] \\ & \Rightarrow & uzv & [l] \end{array}$$

  is obtained by replacing the $(|u| + 1)$st leaf in d, B, with the rule tree corresponding to $l, B\langle z\rangle$.

(III) A *derivation tree* in G is any tree t for which there is a derivation represented by t (see (II)). ■

Convention 6.2. Let $G = (V, T, P, S)$ be a CFG. For any $l : A \to x \in P, {}_G♣(l)$ denotes the rule tree corresponding to l. For any $A \Rightarrow^* x\ [\rho]$ in G, where $A \in N, x \in V^*$, and $\rho \in P^*, {}_G♣(A \Rightarrow^* x\ [\rho])$ denotes the derivation tree corresponding to $A \Rightarrow^* x\ [\rho]$. Just like we often write $A \Rightarrow^* x$ instead of $A \Rightarrow^* x\ [\rho]$ (see Convention 6.1), we sometimes simplify ${}_G♣(A \Rightarrow^* x\ [\rho])$ to ${}_G♣(A \Rightarrow^* x)$ in what follows if there is no danger of confusion. Finally, ${}_G♣_{all}$ denotes the set of all derivation trees for G. If G is automatically understood, we often drop the subscript G from ${}_G♣_{all}$, and simply write $♣_{all}$. ■

Theorem 6.3. *Let $G \in$ **CFG**$, G = (V, T, P, S), A \in N$, and $x \in V^*$. Then, $A \Rightarrow^* x$ in G iff $t \in {}_G♣_{all}$ with $root(t) = A$ and $frontier(t) = x$.*

Proof. Consider any CFG, $G = (V, T, P, S)$. The *only-if* part of the equivalence says that for every derivation $A \Rightarrow^* x$, where $A \in N$ and $x \in V^*$, there exists $t \in {}_G\clubsuit_{all}$ such that $root(t) = A$ and $frontier(t) = x$. From Definition 6.5, we know how to construct ${}_G\clubsuit(A \Rightarrow^* x)$, which satisfies these properties.

The *if* part says that for every $t \in {}_G\clubsuit_{all}$ with $root(t) = A$ and $frontier(t) = x$, where $A \in N$ and $x \in V^*$, there exists $A \Rightarrow^* x$ in G. We prove the if part by induction on $depth(t) \geq 0$.

Basis. Consider any $t \in {}_G\clubsuit_{all}$ such that $depth(t) = 0$. As $depth(t) = 0, t$ is a tree consisting of one node, so $root(t) = frontier(t) = A$, where $A \in N$. Observe that $A \Rightarrow^0 A$ in G; therefore, the basis holds.

Induction hypothesis. Suppose that the if part holds for all trees of depth n or less, where $n \in {}_0\mathbb{N}$.

Induction step. Consider any $t \in {}_G\clubsuit_{all}$ with $depth(t) = n + 1$, $root(t) = A, frontier(t) = x, A \in N, x \in V^*$. Consider the topmost rule tree, ${}_G\clubsuit(p)$, occurring in t. That is, ${}_G\clubsuit(p)$ is the rule tree whose root coincides with $root(t)$. Let $p : A \to u \in P$. Distinguish these two cases as (a) $u = \varepsilon$ and (b) $u \neq \varepsilon$:

(a) If $u = \varepsilon, t$ has actually the form $A\langle\rangle$, which means $u = \varepsilon$ and $depth(t) = 1$, and at this point, $A \Rightarrow \varepsilon\ [p]$, so the induction step is completed.
(b) Assume $u \neq \varepsilon$. Let $u = X_1X_2 \cdots X_m$, where $m \geq 1$. Thus, t is of the form $A\langle t_1t_2 \cdots t_m\rangle$, where each t_i is in ${}_G\clubsuit_{all}$ and satisfies $root(t_i) = X_i, 1 \leq i \leq m$, with $depth(t_i) \leq n$. Let $frontier(t_i) = y_i$, where $y_i \in V^*$, so $x = y_1y_2 \cdots y_m$. As $depth(t_i) \leq n$, by the induction hypothesis, $X_i \Rightarrow^* y_i$ in $G, 1 \leq i \leq m$. Since $A \to u \in P$ with $u = X_1X_2 \cdots X_m$, we have $A \Rightarrow X_1X_2 \cdots X_m$. Putting together $A \Rightarrow X_1X_2 \cdots X_m$ and $X_i \Rightarrow^* y_i$, for all $1 \leq i \leq m$, we obtain

$$\begin{aligned} A &\Rightarrow X_1X_2 \cdots X_m \\ &\Rightarrow^* y_1X_2 \cdots X_m \\ &\Rightarrow^* y_1y_2 \cdots X_m \\ &\ \ \vdots \\ &\Rightarrow^* y_1y_2 \cdots y_m. \end{aligned}$$

Thus, $A \Rightarrow^* x$ in G, the induction step is completed, and the if part of the equivalence holds true. □

Corollary 6.1. *Let $G \in$ **CFG**. Then, $w \in L(G)$ iff ${}_G\clubsuit_{all}$ contains t such that $root(t) = S$ and $frontier(t) = w$.*

Proof. This corollary follows from Theorem 6.3 for $S \Rightarrow^* w$, with $w \in T_G^*$. □

Theorem 6.3 and Corollary 6.1 imply the following important corollary, which says that without any loss of generality, we can always restrict our attention to the canonical derivations or derivation trees when discussing the language generated by CFGs.

Corollary 6.2. *For every $G \in$ **CFG**, (I)–(III), given in the following, coincide with $L(G) = \{w \in T_G^* \mid S \Rightarrow^* w\}$:*

(I) $\{w \in T_G^* \mid S\ {}_{lm}\!\Rightarrow^* w\}$;
(II) $\{w \in T_G^* \mid S\ {}_{rm}\!\Rightarrow^* w\}$;
(III) $\{w \in T_G^* \mid w = frontier(t),\ where\ t \in {}_G\clubsuit_{all}\ with\ root(t) = S\}$. ■

Ambiguity

Unfortunately, even if we reduce our attention only to canonical derivations or derivation trees, we may still face a derivation multiplicity of some sentences. Indeed, some CFGs make several different canonical derivations of the same sentences; even worse, some languages in $\mathbf{L}_{\mathbf{CFG}}$ are generated only by CFGs of this kind.

Definition 6.6. Let $G \in$ **CFG**:

(I) $G \in$ **CFG** is *ambiguous* if $L(G)$ contains a sentence w such that $S\ {}_{lm}\!\Rightarrow^* w\ [\rho]$ and $S\ {}_{lm}\!\Rightarrow^* w\ [\sigma]$ for some $\rho, \sigma \in P^*$, with $\rho \neq \sigma$; otherwise, G is *unambiguous*.
(II) $L \in \mathbf{L}_{\mathbf{CFG}}$ is *inherently ambiguous* if every $G \in$ **CFG** such that $L(G) = L$ is ambiguous. ■

Less formally, according to (I) in Definition 6.6, G is ambiguous if it generates a sentence by two different leftmost derivations. To rephrase (I) in terms of rightmost derivations, G is ambiguous if there exist $S\ {}_{rm}\!\Rightarrow^* w\ [\rho]$ and $S\ {}_{rm}\!\Rightarrow^* w\ [\sigma]$, with $\rho \neq \sigma$, for some

$w \in L(G)$. In terms of ${}_G♣_{all}$ (see Convention 6.2), G is ambiguous if ${}_G♣_{all}$ contains t and u such that $t \neq u$ while $frontier(t) = frontier(u)$.

We close this section by illustrating its key notions: canonical derivations, rule trees, derivation trees, and ambiguity.

Example 6.3. Consider

$$Z = \{a^i b^j c^k | i, j, k \in \mathbb{N}, i = j \text{ or } j = k\}.$$

Observe that $Z = L(G)$, where $G \in \mathbf{CFG}$ is defined by the following ten rules:

$$0 : S \to AB, \quad 1 : A \to aAb, \quad 2 : A \to ab, \quad 3 : B \to cB, \quad 4 : B \to c,$$

$$5 : S \to CD, \quad 6 : C \to aC, \quad 7 : C \to a, \quad 8 : D \to bDc, \quad 9 : D \to bc.$$

Indeed, G uses rules 0–4 to generate $\{a^i b^j c^k \mid i, j, k \in \mathbb{N}, i = j\}$. By using rules 5–9, it generates $\{a^i b^j c^k \mid i, j, k \in \mathbb{N}, j = k\}$. As the union of these two languages coincides with $Z, L(G) = Z$.
Note that G can generate every sentence by a variety of different derivations. For instance, consider $aabbcc \in L(G)$. Observe that G generates this sentence by the 12 different derivations, I–XII, listed in Table 6.1 (according to Convention 6.1, we specify the rewritten symbols by underlining them).

Table 6.2 describes the rule trees ${}_G♣(0)$ – ${}_G♣$ (9), corresponding to the 10 rules in G. In addition, ${}_G♣$ (0) is pictorially shown in Figure 6.1.

Consider, for instance, the first derivation in Table 6.1. Table 6.3 presents this derivation together with its corresponding derivation tree constructed in a step-by-step way. In addition, the resulting derivation tree, $S\langle A\langle aA\langle ab\rangle b\rangle B\langle cB\langle c\rangle\rangle\rangle$, is pictorially shown in Figure 6.2.

In Table 6.1, derivations I and VII represent two different leftmost derivations that generate $aabbcc$. Thus, G is an ambiguous CFG.

As a matter of fact, Z is inherently ambiguous because every CFG that generates Z is ambiguous. Leaving a fully rigorous proof of this inherent ambiguousness as an exercise, we next only sketch its two-step gist. Consider any $H \in \mathbf{CFG}$ satisfying $L(H) = Z$. Take any $l \in \mathbb{N}$ satisfying $l > |rhs(r)|$ for all rules $r \in P_H$.

Table 6.1. Twelve derivations of *aabbcc*.

I		II		III		IV	
$\underline{S}$		$\underline{S}$		$\underline{S}$		$\underline{S}$	
$\Rightarrow \underline{A}B$	[0]	$\Rightarrow \underline{A}B$	[0]	$\Rightarrow A\underline{B}$	[0]	$\Rightarrow A\underline{B}$	[0]
$\Rightarrow a\underline{A}bB$	[1]	$\Rightarrow aAb\underline{B}$	[1]	$\Rightarrow aAb\underline{B}$	[1]	$\Rightarrow \underline{A}cB$	[3]
$\Rightarrow aabb\underline{B}$	[2]	$\Rightarrow aAbc\underline{B}$	[3]	$\Rightarrow aAbc\underline{B}$	[3]	$\Rightarrow aa\underline{A}bcB$	[1]
$\Rightarrow aabbc\underline{B}$	[3]	$\Rightarrow aabbc\underline{B}$	[2]	$\Rightarrow a\underline{A}bcc$	[4]	$\Rightarrow aabbc\underline{B}$	[2]
$\Rightarrow aabbcc$	[4]	$\Rightarrow aabbcc$	[4]	$\Rightarrow aabbcc$	[2]	$\Rightarrow aabbcc$	[4]

V		VI		VII		VIII	
$\underline{S}$		$\underline{S}$		$\underline{S}$		$\underline{S}$	
$\Rightarrow A\underline{B}$	[0]	$\Rightarrow A\underline{B}$	[0]	$\Rightarrow \underline{C}D$	[5]	$\Rightarrow \underline{C}D$	[5]
$\Rightarrow Ac\underline{B}$	[3]	$\Rightarrow \underline{A}cB$	[3]	$\Rightarrow a\underline{C}D$	[6]	$\Rightarrow aC\underline{D}$	[6]
$\Rightarrow \underline{A}cc$	[4]	$\Rightarrow aAbc\underline{B}$	[1]	$\Rightarrow aa\underline{D}$	[7]	$\Rightarrow a\underline{C}bDc$	[8]
$\Rightarrow a\underline{A}bcc$	[1]	$\Rightarrow a\underline{A}bcc$	[4]	$\Rightarrow aab\underline{D}c$	[8]	$\Rightarrow aab\underline{D}c$	[7]
$\Rightarrow aabbcc$	[2]	$\Rightarrow aabbcc$	[2]	$\Rightarrow aabbcc$	[9]	$\Rightarrow aabbcc$	[9]

IX		X		XI		XII	
$\underline{S}$		$\underline{S}$		$\underline{S}$		$\underline{S}$	
$\Rightarrow \underline{C}D$	[5]	$\Rightarrow C\underline{D}$	[5]	$\Rightarrow C\underline{D}$	[5]	$\Rightarrow C\underline{D}$	[5]
$\Rightarrow aC\underline{D}$	[6]	$\Rightarrow \underline{C}bDc$	[8]	$\Rightarrow Cb\underline{D}c$	[8]	$\Rightarrow \underline{C}bDc$	[8]
$\Rightarrow aCb\underline{D}c$	[8]	$\Rightarrow a\underline{C}bDc$	[6]	$\Rightarrow \underline{C}bbcc$	[9]	$\Rightarrow aCb\underline{D}c$	[6]
$\Rightarrow a\underline{C}bbcc$	[9]	$\Rightarrow aab\underline{D}c$	[7]	$\Rightarrow a\underline{C}bbcc$	[6]	$\Rightarrow a\underline{C}bbcc$	[9]
$\Rightarrow aabbcc$	[7]	$\Rightarrow aabbcc$	[9]	$\Rightarrow aabbcc$	[7]	$\Rightarrow aabbcc$	[7]

(1) Show that H necessarily contains two disjoint subsets of rules, X and Y, such that by rules from X, H generates $\{a^ib^jc^k \mid i,j,k \in \mathbb{N}, i = j\}$, and by rules from Y, it generates $\{a^ib^jc^k \mid i,j,k \in \mathbb{N}, j = k\}$; otherwise, H would generate $a^ib^jc^k$, with i $\neq j \neq k$.

(2) Consider $a^lb^lc^l \in$ Z. By using rules from both X and Y, H necessarily makes two different leftmost derivations of $a^lb^lc^l$; consequently, Z is inherently ambiguous. ■

The conclusion of the previous example implies a pragmatically negative result, saying that some CFLs are generated only by ambiguous CFGs.

Table 6.2. ${}_G\clubsuit(0) - {}_G\clubsuit(9)$.

Rule	Rule Tree
$0: S \to ABS$	$\langle AB \rangle$
$1: A \to aAbA$	$\langle aAb \rangle$
$2: A \to abA$	$\langle ab \rangle$
$3: B \to cBB$	$\langle cB \rangle$
$4: B \to cB$	$\langle c \rangle$
$5: S \to CDS$	$\langle CD \rangle$
$6: C \to aCC$	$\langle aC \rangle$
$7: C \to aC$	$\langle a \rangle$
$8: D \to bDcD$	$\langle bDc \rangle$
$9: D \to bcD$	$\langle bc \rangle$

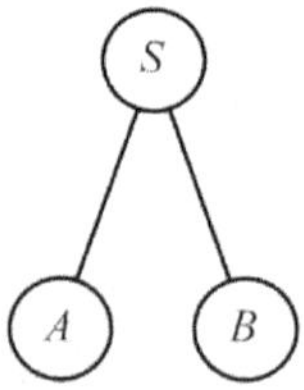

Figure 6.1. Rule tree ${}_G\clubsuit(0)$ of the form $S\langle AB \rangle$.

Corollary 6.3. *Unambiguous CFGs generate a proper subfamily of* $\mathbf{L}_{\mathbf{CFG}}$. ■

Removal of useless symbols

CFGs may contain some symbols that are of no use regarding the generated languages. As a result, useless symbols like these only unnecessarily increase the size of CFGs and, thereby, obscure their specification. Therefore, we next explain how to remove these superfluous symbols from CFGs.

As completely useless, we obviously consider all symbols from which no terminal string is derivable, so we eliminate them first.

Definition 6.7. Let $G \in$ **CFG**. A symbol $X \in V_G$ is *terminating* if $X \Rightarrow^* w$ in G for some $w \in T_G^*$; otherwise, X is *non-terminating*. ■

Basic idea. Let $G = (V_G, T_G, P_G, S_G)$ be a CFG. We construct the set W containing all terminating symbols in G in the following way.

Table 6.3. Derivation I and its corresponding derivation tree.

Derivation		Derivation tree
S		S
$\Rightarrow AB$	[0]	$S\langle AB\rangle$
$\Rightarrow aAbB$	[1]	$S\langle A\langle aAb\rangle B\rangle$
$\Rightarrow aabbB$	[2]	$S\langle A\langle aA\langle ab\rangle b\rangle B\rangle$
$\Rightarrow aabbcB$	[3]	$S\langle A\langle aA\langle ab\rangle b\rangle B\langle cB\rangle\rangle$
$\Rightarrow aabbcc$	[4]	$S\langle A\langle aA\langle ab\rangle b\rangle B\langle cB\langle c\rangle\rangle\rangle$

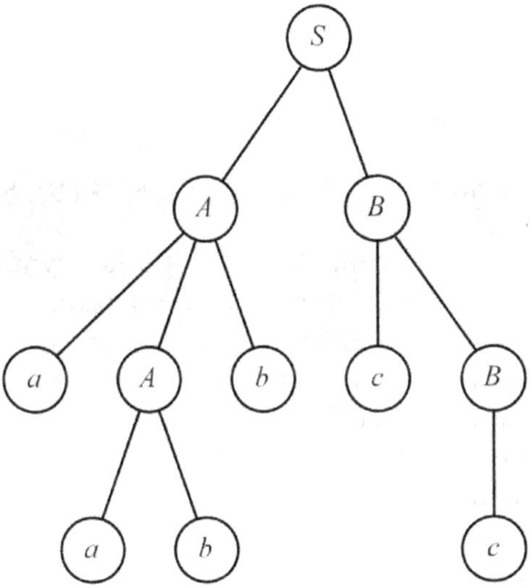

Figure 6.2. Derivation tree $S\langle A\langle aA\langle ab\rangle b\rangle B\langle cB\langle c\rangle\rangle\rangle$.

First, set W to T_G because every terminal $a \in T_G$ satisfies $a \Rightarrow^0 a$, so a is surely terminating by Definition 6.7. If $A \rightarrow x \in P_G$ satisfies $x \in V_G^*$, then $A \Rightarrow x \Rightarrow^* w$ in G, for some $w \in T_G^*$; therefore, add A to W because $A \Rightarrow x \Rightarrow^* w$, and consequently, A is terminating. In this way, keep extending W until no further terminating symbol can be added to W. The resulting set W contains only terminating symbols in G.

The following algorithm, whose formal verification is left as an exercise, constructs W based upon this idea.

Algorithm 6.1. Terminating symbols.

Input: A CFG $G = (V, T, P, S)$.
Output: The subalphabet, $W \subseteq V$, that contains all terminating symbols in G.
Method:

```
begin
  set W to T
  repeat
    if A → x ∈ P and x ∈ V* then
      add A to W
  until no change
end
```

Example 6.4. Consider CFG G:

$$S \to SoS, S \to SoA, S \to A, A \to AoA, S \to (S), S \to i, B \to i,$$

where $o, i,$ (, and) are terminals and the other symbols are nonterminals. Intuitively, o and i stand for an operator and an identifier, respectively. Note, that G generates the set of expressions that can be built up using the terminal symbols; for instance, $S \Rightarrow^* io(ioi)$.

With G as its input, Algorithm 6.1 first sets $W = \{o, i, (,)\}$. Then, it enters the **repeat** loop.

As $B \to i$, with $i \in W$, it adds B to W. For the same reason, $S \to i$ leads to the inclusion of S to W, so $W = \{o, i, (,), B, S\}$. At this point, the **repeat** loop cannot further increase W, so it exits. As a result, A is non-terminating because $A \notin W$. ■

Apart from non-terminating symbols, a symbol is considered useless in a CFG if, starting from the start symbol, the grammar makes no derivation of a string that contains the symbol. To put it more formally and briefly, for $G \in$ **CFG**, $X \in V_G$ is inaccessible and, therefore, useless in G if $X \notin symbols(F(G))$; recall that $F(G)$ denotes the set of all sentential forms of G (see Definition 6.1), and $symbols(F(G))$ denotes the set of all symbols occurring in $F(G)$.

Definition 6.8. Let $G = (V, T, P, S)$ be a CFG, and let $X \in V$. X is *accessible* if $X \in symbols(F(G))$; otherwise, X is *inaccessible*. ■

Basic idea. Let $G = (V, T, P, S)$ be a CFG. To construct the alphabet, $W \subseteq V$, that contains all accessible symbols, initialize W with S. Indeed, the start symbol S is always accessible because $S \Rightarrow^0 S$. If $A \to x \in P$ with $A \in W$, we include $symbols(x)$ into W because we can always change A to x by $A \to x$ in any sentential form, so $symbols(x) \subseteq symbols(F(G))$. Keep extending W in this way until no further symbols can be added to W to obtain the set of all accessible symbols. The following algorithm constructs W in this way.

Algorithm 6.2. Accessible symbols.

Input: A CFG $G = (V, T, P, S)$.
Output: The subalphabet, $W \subseteq V$, that contains all accessible symbols in G.
Method:

```
begin
  set W to {S}
  repeat
    if lhs(r) ∈ W for some r ∈ P then
      add symbols(rhs(r)) to W
  until no change
end
```

Example 6.5. Consider the same CFG as in Example 6.4, i.e.,

$$S \to SoS, S \to SoA, S \to A, A \to AoA, S \to (S), S \to i, B \to i.$$

With this CFG as its input, Algorithm 6.2 first sets $W = \{S\}$. As $S \to SoS$, the **repeat** loop adds o to W. Furthermore, since $S \to A$, this loop also adds A there. Continuing in this way, this loop exits with W containing all symbols but B, so B is the only inaccessible symbol in this CFG. ■

Definition 6.9. Let $G = (V, T, P, S)$ be a CFG. A symbol $X \in V$ is *useful* in G if $X \in symbols(F(G))$ and $X \Rightarrow^* w$, with $w \in T^*$; otherwise, X is *useless*. ■

In other words, X is *useful* if it is both accessible and terminating. Making use of the previous two algorithms, we next explain how to turn any CFG to an equivalent CFG that contains only useful symbols.

Algorithm 6.3. Useful symbols.

Input: A CFG $I = (V_I, T_I, P_I, S_I)$.
Output: A CFG $O = (V_O, T_O, P_O, S_O)$ such that $L(I) = L(O)$, and all symbols in V_O are useful.
Method:

begin
(1) by using Algorithm 6.1, find all terminating symbols in V_I then, eliminate all non-terminating symbols and the rules that contain them from I;
(2) consider the CFG obtained in (1); by using Algorithm 6.2, determine all its accessible symbols; then, remove all inaccessible symbols and the rules that contain them from the CFG; the resulting CFG is O.
end

Theorem 6.4. *Algorithm 6.3 is correct.*

Proof. By contradiction, prove that every nonterminal in O is useful. Assume that $X \in V_O$ and X is a useless symbol. Consequently: (i) for every $y \in V^*$ such that $S \Rightarrow^* y, X \notin symbols(y)$, or (ii) for every $x \in V^*$ such that $X \Rightarrow^* x, x \notin T_O^*$. Case (i) is ruled out because Algorithm 6.2 would eliminate A. Case (ii) is ruled out as well. Indeed, if for every $x \in V^*$ such that $X \Rightarrow^* x, x \notin T_O^*$, then X would be eliminated by Algorithm 6.1. Thus, $X \notin V_O$, which contradicts $X \in V_O$. As a result, every symbol in V_O is useful.

As an exercise, complete the proof. That is, prove that $L(I) = L(O)$. □

Observe that the order of the two transformations in Algorithm 6.3 is crucially important. Indeed, if we reverse them, this algorithm does not work properly. To give a trivial counterexample, consider the CFG I defined as $S \rightarrow a, S \rightarrow A, A \rightarrow AB$, and $B \rightarrow a$. Note that $L(I) = \{a\}$. If we apply the transformations in Algorithm 6.3 properly, we obtain an equivalent one-rule grammar O defined as $S \rightarrow a$. That is, in this way, Algorithm 6.3 rightly detects and eliminates A and B as useless symbols. However, if we improperly apply Algorithm 6.2 before Algorithm 6.1, we obtain a two-rule grammar $S \rightarrow a$ and $B \rightarrow a$, in which B is useless.

Example 6.6. Once again, return to the CFG G:

$$S \to SoS,\ S \to SoA,\ S \to A,\ A \to AoA,\ S \to (S),\ S \to i,\ B \to i,$$

discussed in the previous two examples. Eliminate the non-terminating symbols from G by Algorithm 6.3. From Example 6.4, we already know that A is the only non-terminating symbol, so the elimination of all rules containing A produces the grammar defined as

$$S \to SoS, \quad S \to (S), \quad S \to i, \quad B \to i.$$

Apply Algorithm 6.2 to G to find out that B is the only inaccessible symbol in it. By removing $B \to i$, we obtain

$$S \to SoS, \quad S \to (S), \quad S \to i$$

as the resulting equivalent CFG in which all symbols are useful. ■

Removal of erasing rules

It is often convenient to eliminate all erasing rules — the rules that have ε on their right-hand sides — in a CFG. Indeed, without these rules, the CFG can never make any sentential form shorter during a derivation step, and this property obviously simplifies its exploration as well as application. In this section, we explain how to make this elimination.

Definition 6.10. Let $G = (V, T, P, S)$ be a CFG. A rule of the form $A \to \varepsilon \in P$ is called an *erasing rule* or, briefly, an *ε-rule*. ■

Before eliminating all ε-rules in any $G \in$ **CFG**, we explain how to determine all the nonterminals from which G can derive ε.

Definition 6.11. Let $G \in$ **CFG**. A nonterminal $A \in N$ is *ε-nonterminal* in G if $A \Rightarrow^* \varepsilon$ in G. ■

Basic idea. Let $G = (V, T, P, S) \in$ **CFG**. To determine the set $E \subseteq N$ containing all ε-nonterminals in G, we initialize E with the left-hand sides of all these ε-rules. Indeed, if for $A \in N$ and $A \to \varepsilon \in P$, then $A \Rightarrow \varepsilon$ in G. Then, we extend E by every $B \in N$ for which there is $B \to x \in P$ with $x \in E^*$, which obviously implies $B \Rightarrow^* \varepsilon$. Repeat this extension until no further ε-nonterminals can be added to E.

Algorithm 6.4. Determination of ε-nonterminals.

Input: A CFG $I = (V, T, P, S)$.
Output: The subalphabet, $E \subseteq N_I$, containing all ε-nonterminals in I.
Method:

```
begin
  initialize E with {A | A → ε ∈ P}
  repeat
    if B → x ∈ P with x ∈ E* then
      add B to E
  until no change
end
```

As an exercise, prove that Algorithm 6.4 is correct.

Example 6.7. Consider Algorithm 6.4. As the input CFG I, take

$$S \to AB, \quad A \to aAb, \quad B \to cBd, \quad A \to \varepsilon, \quad B \to \varepsilon.$$

As is obvious, $L(I) = \{a^n b^n \mid n \geq 0\}\{c^m d^m \mid m \geq 0\}$. Algorithm 6.4 initializes E with A and B because both nonterminals occur as the left-hand sides of the two ε-rules in I, i.e., $A \to \varepsilon$ and $B \to \varepsilon$. Then, it enters the **repeat** loop. As $S \to AB$ with $AB \in E^*$, it includes S into E. After this inclusion, the **repeat** loop cannot further increase E; therefore, it exits. ■

We are now ready to eliminate all ε-rules from any CFG. More exactly, since no CFG without ε rules can generate ε, we explain how to turn any $I \in$ **CFG** to $O \in$ **CFG** so that O generates $L(I) - \{\varepsilon\}$ without possessing any ε-rules.

Basic idea. Let $I = (V_I, T_I, P_I, S_I)$ be a CFG. Determine all its ε-nonterminals by Algorithm 6.4. Take any $B \to y \in P_I$ with $y = x_0 A_1 x_1 \cdots A_n x_n$, where $n \in \mathbb{N}$, so that A_1 through A_n are $\varepsilon-$nonterminals. Add all possible rules of the form $B \to x_0 X_1 x_1 \cdots X_n x_n$ to P_O, where $X_i \in \{\varepsilon, A_i\}, 1 \leq i \leq n$, and $X_1 X_2 \cdots X_n \neq \varepsilon$, because each A_i can be erased by $A_i \Rightarrow^* \varepsilon$ in I. Keep extending P_O in this way until no further ε-rules can be added to it.

Algorithm 6.5. Elimination of ε-rules.

Input: A CFG $I = (V_I, T_I, P_I, S_I)$.
Output: A CFG $O = (V_O, T_O, P_O, S_O)$, such that $L(O) = L(I) - \{\varepsilon\}$ and P_O contains no ε-rules.
Method:

begin
  set $P_O = \{A \to y \mid A \to y \in P_I, y \neq \varepsilon\}$
  use Algorithm 6.4 to determine $E_I \subseteq N_I$ containing all
    ε-nonterminals in I
  repeat
    if $B \to x_0A_1x_1 \cdots A_nx_n$ in P_I, where $A_i \in E_I, x_j \in (V_I - E_I)^*$,
    for all $1 \leq i \leq n, 0 \leq j \leq n$, where $n \in \mathbb{N}$ **then**
      extend P_O by
      $\{B \to x_0X_1x_1 \cdots X_nx_n \mid X_i \in \{\varepsilon, A_i\}, 1 \leq i \leq n, |X_1X_2 \cdots X_n| \geq 1\}$
  until no change
end

Theorem 6.5. *Algorithm 6.5 is correct. Therefore, for every $L \in$ **L_{CFG}**, there exists a CFG, $G = (V, T, P, S)$, such that $L(G) = L - \{\varepsilon\}$ and P contains no ε-rules.* ■

Example 6.8. Reconsider the CFG G defined as

$$S \to AB, \quad A \to aAb, \quad B \to cBd, \quad A \to \varepsilon, \quad \text{and} \quad B \to \varepsilon$$

(see Example 6.7). Take G as I in Algorithm 6.5. Initially, this algorithm sets $P_O = \{S \to AB, A \to aAb, B \to cBd\}$, then it enters the **repeat** loop. Consider $S \to AB$. Both A and B are ε-nonterminals, so Algorithm 6.5 adds $S \to AB, S \to B$, and $S \to A$ to P_O. Analogically, from $A \to aAb$ and $B \to cBd$, this algorithm constructs $A \to aAb, B \to cBd, A \to ab$, and $B \to cd$, respectively. In this way, as the resulting CFG O without ε-rules, Algorithm 6.5 produces

$$S \to AB, S \to A, S \to B, A \to aAb, B \to cBd, A \to ab, B \to cd.$$

Observe that O generates $\{a^nb^n \mid n \geq 0\}\{c^nd^n \mid n \geq 0\} - \{\varepsilon\}$, so $L(O) = L(I) - \{\varepsilon\}$. ■

Before closing this section, we make two final remarks. First, we make a remark concerning the generation of ε. Then, we sketch an alternative method of eliminating erasing rules.

Generation of ε. As already pointed out, if a CFG contains no ε-rules, it cannot generate ε. That is, let $L = L(I)$, where $I = (V_I, T_I, P_I, S_I)$ be a CFG; then, Algorithm 6.5 converts I to a CFG, $O = (V_O, T_O, P_O, S_O)$, so $L(O) = L(I) - \{\varepsilon\}$. To generate L, including ε, we can easily change O to $G = (V_G, T_G, P_G, S_G)$, so $V_G = V_O \cup \{S_G\}$ and $P_G = P_O \cup \{S_G \to S_O, S_G \to \varepsilon\}$, where the start symbol S_G is a newly introduced symbol, which is not in V_O. As is obvious, $L(G) = L$, which gives rise to the following theorem, whose straightforward proof is left as an exercise.

Theorem 6.6. *For every* $L \in \boldsymbol{L_{CFG}}$ *such that* $\varepsilon \in L$, *there is a CFG,* $G = (V_G, T_G, P_G, S_G)$, *such that* G *simultaneously satisfies properties (I)–(III), given in the following.*

(I) $L(G) = L$;
(II) $S_G \to \varepsilon$ *is the only* ε*-rule in* P_G, *where* S_G *is the start symbol of* G*;*
(III) S_G *does not occur on the right-hand side of any rule in* P_G. ■

An alternative removal of ε*-rules.* The previously described method that removes ε-rules consists, in fact, of two algorithms. Indeed, Algorithm 6.5, which performs this removal, makes use of Algorithm 6.4, which determines the set of all ε-nonterminals in the input CFG. Next, we describe an alternative removal of ε-rules, which does not require any predetermination of ε nonterminals.

Basic idea. Let $I \in \mathbf{CFG}$ and $A \in N_I$. If A derives ε in I, then a derivation like this can be expressed in the following step-by-step way:

$$A \Rightarrow x_1 \Rightarrow x_2 \Rightarrow \cdots \Rightarrow x_n \Rightarrow \varepsilon,$$

where $x_i \in N_G^*$, for all $1 \leq i \leq n$, for some $n \in {}_0\mathbb{N}$ ($n = 0$ means $A \Rightarrow \varepsilon$). If a sentential form contains several occurrences of A, each of them can be erased in this way, although there may exist many alternative ways of erasing A. Based upon these observations, during an application of a rule, the next algorithm introduces a compound nonterminal of the form $\langle X, W\rangle$, in which X is a symbol X that is not erased during the derivation and W is a set of nonterminals that is erased. Within the compound nonterminal, the algorithm simulates the erasure of nonterminals in W in the way sketched above. Observe that, as W is a set, W contains no more than one occurrence of any nonterminal because there is no need to record several occurrences of the same nonterminal; indeed, as already pointed out, all these occurrences can be erased in the same way.

Algorithm 6.6. Alternative elimination of ε-rules.

Input: A CFG, $I = (V_I, T_I, P_I, S_I)$.
Output: A CFG $O = (V_O, T_O, P_O, S_O)$, such that $L(O) = L(I) - \{\varepsilon\}$ and P_O contains no ε-rules.
Method:

begin
  set $V_O = \{\langle X, U\rangle \mid X \in V_I, U \subseteq N_I\} \cup T_I, T_O = T_I$, and $S_O = \langle S_I, \emptyset\rangle$
  set P_O to $\emptyset$
  for all $a \in T_O$, add $\langle a, \emptyset\rangle \to a$ to P_O **repeat**
    if $B \to x_0 X_1 x_1 X_2 x_2 \cdots X_n x_n$ in P_I, where $X_i \in V_I, x_j \in N_I^*$, for all $1 \le i \le n, 0 \le j \le n$, where $n \in \mathbb{N}$ **then**
      add $\langle B, \emptyset\rangle \to \langle X_1, symbols(x_0 x_1 \cdots x_n)\rangle X_2 \cdots X_n$ to P_O
    if $\langle X, U\rangle \in N_O$, where $X \in V_I, U \subseteq N_I, U \neq \emptyset$, and $C \to z \in P_I$ with $C \in U$ and $z \in N_I^*$ **then**
      add $\langle X, U\rangle \to \langle X, (U - \{C\}) \cup symbols(z)\rangle$ to P_O
  until no change
end

As an exercise, prove that Algorithm 6.6 is correct.

Example 6.9. Consider the CFG G defined as

$$S \to aSb \text{ and } S \to \varepsilon.$$

Algorithm 6.6 converts G to the next equivalent grammar without ε-rules:

$$\langle S,\emptyset\rangle \to \langle a,\emptyset\rangle\langle S,\emptyset\rangle\langle b,\emptyset\rangle,\ \langle S,\emptyset\rangle \to \langle a,\{S\}\rangle\langle b,\emptyset\rangle,$$
$$\langle a,\{S\}\rangle \to \langle a,\emptyset\rangle,\ \langle a,\emptyset\rangle \to a,\ \langle b,\emptyset\rangle \to b.$$

A detailed description of this conversion is left as an exercise. ■

As a rule, Algorithm 6.6 produces O, with many rules having a single nonterminal on their right-hand sides, which often makes the definition of O clumsy, as the previous example illustrates. Therefore, in the following section, we explain how to eliminate these rules without affecting the generated language.

Removal of single rules

By using rules with a single nonterminal on the right-hand side, CFGs only rename their nonterminals; otherwise, they fulfill no role at all. Therefore, it is sometimes desirable to remove them. This section explains how to make this removal.

Definition 6.12. Let $G = (V, T, P, S)$ be a CFG. A rule of the form $A \to B \in P$, where A and B are in N, is called a *single rule.* ■

Basic idea. To transform a CFG I into an equivalent CFG O without single rules, observe that according to a sequence of single rules, every derivation is of the form $A \Rightarrow^* B$ in I, where A and B are nonterminals. Furthermore, note that for any derivation of the form $A \Rightarrow^* B$, there exists a derivation from A to B during which no two identical rules are applied. Consider the set of all derivations that have the form $A \Rightarrow^* B \Rightarrow x$, where $x \notin N_I$, and $A \Rightarrow^* B$ is made according to a sequence of single rules so that this sequence contains no two identical rules. Note that this set is finite because every derivation of the above form consists of no more than $|P_I|$ steps, and P_I is finite. For any derivation $A \Rightarrow^* B \Rightarrow x$ in the set, which satisfies the above requirements, introduce $A \to x$ to obtain the resulting equivalent output CFG O without single rules.

Algorithm 6.7. Elimination of single rules.

Input: A CFG, $I = (V_I, T_I, P_I, S_I)$.
Output: A CFG $O = (V_O, T_O, P_O, S_O)$ such that $L(O) = L(I)$ and P_O contains no single rules.
Method:

begin
  set $V_O = V_I, T_O = T_I, S_O = S_I$ and $P_O = \emptyset$
  repeat
    if $A \Rightarrow^n B \Rightarrow x$ in I, where $A, B \in N_I, x \in V_I^* - N_I, 1 \leq n \leq |P_I|$, and $A \Rightarrow^n B$ is made by n single rules **then**
      add $A \to x$ to P_O
  until no change
end

Theorem 6.7. *Algorithm 6.7 is correct. Therefore, for every $L \in$ $\mathbf{L_{CFG}}$, there exists a CFG, $G = (V, T, P, S)$, such that $L(G) = L$ and P contains no single rules.* ■

Example 6.10. Reconsider the CFG G obtained in Example 6.8. Recall its set of rules:

$$S \to AB, S \to A, S \to B, A \to aAb, B \to cBd, A \to ab, B \to cd$$

As is obvious, $S \to A$ and $S \to B$ are single rules. Consider G as I in Algorithm 6.7 to transform it into an equivalent CFG O without single rules. From $S \Rightarrow A \Rightarrow aAb$, the algorithm constructs $S \to aAb$. Similarly, from $S \Rightarrow A \Rightarrow ab$, it makes $S \to ab$. As the resulting CFG O, it produces

$$S \to AB, S \to aAb, S \to ab, S \to cBd, S \to cd, A \to aAb, B \to cBd,$$
$$A \to ab, B \to cd.$$ ■

We close this section by summarizing all the useful grammatical transformations given earlier in this chapter to demonstrate how to obtain a properly defined CFG from any CFG. Of course, first, we state what we mean by this proper definition.

Definition 6.13. A CFG, $G = (V, T, P, S)$, is proper if

(I) V contains no useless symbols;
(II) P contains neither ε-rules nor single rules. ■

Theorem 6.8. *For every CFG, I, there exists a proper CFG, O, such that $L(O) = L(I) - \{\varepsilon\}$.*

Proof. Let $I = (V_I, T_I, P_I, S_I)$ be any CFG. Remove all useless symbols from V_I (see Algorithm 6.3). Apply Algorithm 6.5 to the CFG without useless symbols to an equivalent CFG containing no ε rules; then, by Algorithm 6.7, convert the CFG without ε-rules to a CFG without any single rules. Take the resulting CFG as O. Observe that O is proper and $L(O) = L(I) - \{\varepsilon\}$. □

Chomsky normal form

Next, we explain how to transform any CFG into an equivalent proper grammar in Chomsky normal form, in which each of its rules has on its right-hand side either a terminal or two nonterminals. We often make use of the Chomsky normal form to simplify proofs, as demonstrated later in this chapter (see, for instance, Algorithm 6.9 and the proof of Theorem 6.10).

Definition 6.14. A CFG, $G = (V, T, P, S)$, is in Chomsky normal form if it is proper, and in addition, every rule $A \to x \in P$ satisfies $x \in T \cup NN$. ■

Basic idea. Let $I = (V_I, T_I, P_I, S_I)$ be a CFG. Without any loss of generality, suppose that I is proper (see Theorem 6.8). Start the transformation of I into an equivalent CFG, $O = (V_O, T_O, P_O, S_O)$, in Chomsky normal form by introducing nonterminal subalphabet $W = \{a' \mid a \in T_I\}$, together with the bijection β from V_I to $W \cup N_I$ that maps every $a \in T_I$ to the nonterminal a' and every $A \in N_I$ to itself. Set $V_O = W \cup V_I$. For every $a \in T_I$, include $a' \to a$ into P_O, and for every $A \to a \in P_I$, move $A \to a$ from P_I to P_O. Furthermore, for each $A \to XY \in P_I$, where X and Y are in V_I, add $A \to \beta(X_1)\beta(X_2)$ to P_O and, simultaneously, eliminate $A \to X_1X_2$ in P_I. Finally, for every $A \to X_1X_2X_3 \cdots X_{n-1}X_n \in P_I$ with $n \geq 3$, include new nonterminals $\langle X_2 \cdots X_n\rangle, \langle X_3 \cdots X_n\rangle, \ldots, \langle X_{n-2}X_{n-1}X_n\rangle, \langle X_{n-1}X_n\rangle$ in V_O, and add the rules $A \to \beta(X_1)\langle X_2 \cdots X_n\rangle$, $\langle X_2 \cdots X_n\rangle \to \beta(X_2)$ $\langle X_3 \cdots X_n\rangle, \ldots, \langle X_{n-2}X_{n-1}X_n\rangle \to \beta(X_{n-2})\langle X_{n-1}X_n\rangle$, $\langle X_{n-1}X_n\rangle \to$ $\beta(X_{n-1})\beta(X_n)$ to P_O; note that the added rules satisfy the Chomsky normal form. In this way, $A \Rightarrow X_1X_2X_3 \cdots X_{n-1}X_n[A \to X_1X_2X_3 \cdots X_{n-1}X_n]$ in I is simulated in O as

$$\begin{aligned} A &\Rightarrow \beta(X_1)\langle X_2 \cdots X_n\rangle \\ &\Rightarrow \beta(X_1)\beta(X_2)\langle X_3 \cdots X_n\rangle \\ &\ \ \vdots \\ &\Rightarrow \beta(X_1)\beta(X_2) \cdots \beta(X_{n-2})\langle X_{n-1}X_n\rangle \\ &\Rightarrow \beta(X_1)\beta(X_2) \cdots \beta(X_{n-2})\beta(X_{n-1})\beta(X_n) \end{aligned}$$

and $\beta(X_j) \Rightarrow X_j[\beta(X_j) \to X_j]$ in O, for every $X_j \in T_I$, where $1 \leq j \leq n$.

O, constructed in this way, may contain some useless nonterminals; if it does, remove these useless symbols and all rules that contain them by using Algorithm 6.3.

Algorithm 6.8. Chomsky normal form.

Input: A proper CFG $I = (V_I, T_I, P_I, S_I)$.
Output: A CFG $O = (V_O, T_O, P_O, S_O)$, in Chomsky normal form such that $L(I) = L(O)$.
Method:

begin
  introduce $W = \{a' \mid a \in T_I\}$ and the bijection β from V_I to $W \cup N_I$ defined as $\beta(a) = a'$ for all $a \in T_I$, and $\beta(A) = A$ for all $A \in N_I$
  set $V_O = W \cup V_I, T_O = T_I$ and $P_O = \emptyset$
  for all $a \in T_I$ **do**
    add $a' \to a$ to P_O
  for all $A \to a \in P_I, A \in N_I, a \in T_I$ **do**
    move $A \to a$ from P_I to P_O
  for all $A \to X_1X_2 \in P_I, A \in N_I, X_i \in V_I, i = 1, 2$ **do**
    add $A \to \beta(X_1)\beta(X_2)$ to P_O
    remove $A \to X_1X_2$ from P_I
  repeat
    if for some $n \geq 3$,
    $A \to X_1X_2X_3 \cdots X_{n-1}X_n \in P_I, A \in N_I, X_i \in V_I$, i = 1, ..., n **then**
      introduce new nonterminals $\langle X_2 \cdots X_n\rangle, \langle X_3 \cdots X_n\rangle, \ldots,$ $\langle X_{n-2}X_{n-1}X_n\rangle, \langle X_{n-1}X_n\rangle$ into V_O
      add $A \to \beta(X_1)\langle X_2 \cdots X_n\rangle, \langle X_2 \cdots X_n\rangle \to \beta(X_2)\langle X_3 \cdots X_n\rangle, \ldots,$
      $\langle X_{n-2}X_{n-1}X_n\rangle \to \beta(X_{n-2})\langle X_{n-1}X_n\rangle,$
      $\langle X_{n-1}X_n\rangle \to \beta(X_{n-1})\beta(X_n)$ to P_O
      remove $A \to X_1X_2X_3 \cdots X_{n-1}X_n$ from P_I
  until no change
  remove all useless symbols and rules that contain them from O by Algorithm 6.3
end

Following the basic idea above, prove the following theorem as an exercise.

Theorem 6.9. *Algorithm 6.8 is correct. Therefore, for every $L \in$ **L_{CFG}**, there exists a CFG, $G = (V, T, P, S)$, such that $L(G) = L$ and G satisfies the Chomsky normal form.* ■

Example 6.11. Return to the CFG G obtained in Example 6.6, i.e.,

$$S \to SoS, \quad S \to (S), \quad S \to i,$$

which obviously represents a proper CFG. Consider G as I in Algorithm 6.8, which converts G into an equivalent CFG, $O = (V_O, T_O, P_O, S_O)$, in the Chomsky normal form as follows. Initially, Algorithm 6.8 introduces four new nonterminals: o',$('$,$)'$, and i', and the bijection β from $\{S, o, (,), i\}$ to $\{S, o', (',)', i'\}$ that maps $S, o, (,)$, and i to $S, o', (',)'$, and i', respectively. Then, it includes $o' \to o, (' \to (,)' \to)$, and $i' \to i$ in P_O. After this, it places $S \to i$ in P_O. From $S \to SoS$, this algorithm subsequently constructs $S \to S\langle oS\rangle, \langle oS\rangle \to o'S$. Analogously, from $S \to (S)$, it constructs $S \to ('\langle S)\rangle, \langle S)\rangle \to S)'$. Algorithm 6.8 exits from the repeat loop, with O defined as

$$o' \to o, (' \to (,)' \to), i' \to i, S \to i, S \to S\langle oS\rangle, \langle oS\rangle \to o'S,$$
$$S \to ('\langle S)\rangle, \langle S)\rangle \to S)'.$$

This CFG contains an inaccessible symbol i'. Algorithm 6.3 detects this symbol as useless and removes it together with the rule $i' \to i$, which contains it. Rename $S, \langle oS\rangle, o', (', \langle S)\rangle$, and $)'$ to A_1, A_2, A_3, A_4, A_5, and A_6, respectively, and order the rules according to their left-hand sides as follows:

$$A_1 \to A_1A_2, \quad A_1 \to A_4A_5, \quad A_1 \to i, \quad A_2 \to A_3A_1, \quad A_3 \to o,$$
$$A_4 \to (, A_5 \to A_1A_6, A_6 \to),$$

where A_1 is the start symbol. This CFG represents the resulting version of the CFG in the Chomsky normal form. ■

Elimination of left recursion

A CFG is left-recursive if it can make a derivation of the form $A \Rightarrow^+ Ax$ for a nonterminal A and a string x. Left recursion causes the CFG to enter an infinite loop during leftmost derivations, which often represents an undesirable grammatical phenomenon. In fact, applying many methods necessitates the prior removal of left recursion from CFGs. Indeed, from a theoretical viewpoint, this removal is needed to turn CFGs to their Greibach normal form in the following section. From a more practical point of view, most top-down parsing methods, discussed in Section 11.4, work only with non-left-recursive CFGs. Therefore, this section explains how to make this removal.

Definition 6.15. Let $G = (V, T, P, S)$ be a CFG and $A \in N$.

(I) A rule of the form $A \to Ay \in P$, where $y \in V^*$, is a *directly left-recursive rule*, and A is a *directly left-recursive nonterminal.* G is *directly left-recursive* if N contains a directly left-recursive nonterminal.
(II) A derivation of the form $A \Rightarrow^+ Ax$, where $x \in V^*$, is a *left-recursive derivation*, and A is a *left-recursive nonterminal* in G. G is *left-recursive* if N contains a left-recursive nonterminal. ■

Next, we give an insight into a transformation that converts any left-recursive CFG to an equivalent CFG that is not left-recursive.

Basic idea. As is obvious, directly left-recursive nonterminals are special cases of left-recursive nonterminals. We first sketch how to remove them from any CFG without affecting the generated language.

Elimination of direct left recursion. Let $G = (V, T, P, S)$ be a directly left-recursive CFG. Without any loss of generality, suppose that G is proper (see Definition 6.13 and Theorem 6.8). Observe that for every nonterminal $A \in N$, there surely exists an A-rule that is not directly left-recursive; otherwise, A would be a useless symbol, which contradicts that G is proper.

For every directly left-recursive symbol A, introduce a new nonterminal B into N, and for any pair of rules

$$A \to Aw \quad \text{and} \quad A \to u,$$

where $u \in V^*$ and $A \to u$ is not a left-recursive rule, introduce

$$A \to u, \quad A \to uB, \quad B \to wB, \quad \text{and} \quad B \to w$$

in P. Repeat this extension of G until nothing can be added to N or P in this way. Then, eliminate all the directly left-recursive A-rules in P_G, so the resulting CFG is not a directly left-recursive. Consider, for instance,

$$A \quad \Rightarrow \quad Aw \quad \Rightarrow \quad Aww \quad \Rightarrow \quad uww$$

made by two applications of $A \to Aw$, followed by one application of $A \to u$ in the original version of G. The modified version of G simulates the above derivation as

$$A \quad \Rightarrow \quad uB \quad \Rightarrow \quad uwB \quad \Rightarrow \quad uww$$

by applying $A \to u, B \to wB$, and $B \to w$. To illustrate this elimination by an example, take

$$S \to SoS, \quad S \to (S), \quad S \to i$$

as G (see Example 6.6). From this CFG, the transformation sketched above produces

$$S \to (S), \quad S \to i, \quad S \to (S)B, \quad S \to iB, \quad B \to oSB, \quad B \to oS.$$

For instance, $S \Rightarrow SoS \Rightarrow Soi \Rightarrow ioi$ in the original CFG is simulated by the transformed non-directly-left-recursive CFG as $S \Rightarrow iB \Rightarrow ioS \Rightarrow ioi$.

Elimination of general left recursion. We are now ready to eliminate general left recursion, which represents a more hidden grammatical trap. In essence, we make this elimination by the elimination of direct left recursion combined with a modification of rules, so their right-hand sides start with nonterminals ordered in a way that rules out left recursion.

More precisely, without any loss of generality (see Theorem 6.9), we consider a CFG, $I = (V_I, T_I, P_I, S_I)$, in Chomsky normal form with $N_I = \{A_1, \ldots, A_n\}$, where $|N_I| = n$ (of course, we can always rename the nonterminals in I in this way). We next sketch how to turn I to an equivalent CFG, $O = (V_O, T_O, P_O, S_O)$, with $N_O = \{A_1, \ldots, A_n\} \cup \{B_1, \ldots, B_m\}$, where B_1 through B_m are new nonterminals ($m = 0$ actually means that $N_O = N_I$), and P_O containing rules of the following three forms:

(I) $A_i \to au$ with $a \in T_O$ and $u \in N_O^*$;
(II) $A_i \to A_jv$, for some $A_i, A_j \in \{A_k \mid 1 \leq k \leq n\}$ such that $i < j$, and $v \in N_O^+$;

(III) $B_i \to Cw$ with $C \in \{A_k \mid 1 \leq k \leq n\} \cup \{B_l \mid 1 \leq l \leq i-1\}$ and $w \in N_O^*$.

(I)–(III) imply that O cannot make a derivation of the form $A_k \Rightarrow^+ A_k x$ or $B_l \Rightarrow^+ B_l x$, where $1 \leq k \leq n$ and $1 \leq l \leq i-1$; in other words, O is a non-left-recursive CFG (of course, as opposed to I, O may not be in Chomsky normal form).

The construction of O is based on repeatedly changing I by (1) and (2), given in the following, until I cannot be further modified. Then, O is defined as the final version of I, resulting from this repeated modification. To start with, set $k = 0$ and $i = 1$:

(1) In P_I, for every rule of the form $A_i \to A_j y$, where $j < i, y \in N_I^*$, extend P_I by all A_i-rules of the form $A_i \to zy$, where $A_j \to z \in P_I, z \in N_I^+$ (A_i-rule is a rule with A_i on its left-hand side). After this extension, remove $A_i \to A_j y$ from P_I.
(2) Rephrase and perform the elimination of direct recursion described above in terms of I instead of G. That is, let $A_i \in N_I$ be a directly left-recursive nonterminal in I. Rephrase the above-described elimination of direct left recursion in terms of A_i as follows. Increase k by one and introduce a new nonterminal B_k into N_I. First, for every pair of rules,

$$A_i \to A_i w \quad \text{and} \quad A_i \to u,$$

where $A_i \to u$ is not a left-recursive rule, add

$$A_i \to u, \quad A_i \to uB_k, \quad B_k \to wB_k, \quad \text{and} \quad B_k \to w$$

to P_I. Repeat this extension of P_I until no more rules can be inserted into P_I in this way. After this extension is completed, remove all directly left-recursive A_i-rules from P_I.

When P_I cannot be modified by further repetition of (1) or (2), increase i by 1. If $i \leq n$, repeat (1) and (2) again in the same way; otherwise, set V_O and P_O to V_I and P_I, respectively, in order to obtain $O = (V_O, T_O, P_O, S_O)$ as the resulting non-left-recursive CFG equivalent with the original version of I (of course, as opposed to I, O may not satisfy the Chomsky normal form).

Based on this basic idea, we next give Algorithm 6.9 that turns I to O, satisfying the above properties.

Algorithm 6.9. Elimination of left recursion.

Input: A left-recursive CFG $I = (V_I, T_I, P_I, S_I)$ in Chomsky normal form with $N_I = \{A_1, \ldots, A_n\}$, for some $n \geq 1$.

Output: A non-left-recursive CFG $O = (V_O, T_O, P_O, S_O)$ such that $L(I) = L(O), N_O = \{A_k \mid 1 \leq k \leq n\} \cup \{B_l \mid 1 \leq l \leq m\}$, for some $m \geq 0$, and each rule in P_O has one of these three forms

(I) $A_i \to au$ with $a \in T_O$, and $u \in N_O^*$;

(II) $A_i \to A_j v$, for some $A_i, A_j \in \{A_k \mid 1 \leq k \leq n\}$ such that $i < j$, and $v \in N_O^+$;

(III) $B_i \to Cw$ with $C \in \{A_k \mid 1 \leq k \leq n\} \cup \{B_l \mid 1 \leq l \leq i-1\}$ and $w \in N_O^*$.

Method:

begin
  set $k = 0$
  for i = 1, …, n **do**
    repeat
      if $A_i \to A_j y \in P_I$, where $j < i$ and $y \in N_I^*$ **then**
        repeat
          if $A_j \to z \in P_I$, where $z \in N_I^+$ **then**
            add $A_i \to zy$ to P_I
        until no change
        remove $A_i \to A_j y$ from P_I
      if $A_i \in N_I$ is a directly left-recursive nonterminal in I **then**
        $k = k + 1$
        introduce a new nonterminal B_k into N_I
        repeat
          if $A_i \to A_i w$ and $A_i \to u$ are in P_I, where $w, u \in N_I^+$
          and $A_i \to u$ is not a directly left-recursive rule **then**
            add $A_i \to uB_k, B_k \to wB_k$, and $B_k \to w$ to P_I
        until no change
        remove all directly left-recursive A_i-rules from P_I
        (note that $A_i \to u$ remains in P_I)
    until no change
  end of the **for** loop
  define $O = (V_O, T_O, P_O, S_O)$ with $V_O = V_I, T_O = T_I, P_O = P_I$,
  and $S_O = S_I$
end

Theorem 6.10. *Algorithm 6.9 is correct. Therefore, for every CFG I in Chomsky normal form, there exists an equivalent non-left-recursive CFG O.*

Proof. By a straightforward examination of Algorithm 6.9, we see that each rule in P_O has one of the forms (I)–(III). Next, by contradiction, we prove that O is non-left-recursive. Suppose that O is left-recursive. We distinguish two cases: (a) $\{A_k \mid 1 \leq k \leq n\}$ contains a left-recursive nonterminal and (b) $\{B_l \mid 1 \leq l \leq m\}$ contains a left-recursive nonterminal.

(a) Let A_i be left-recursive, for some $A_i \in \{A_1, \ldots, A_n\}$. That is, there exists $x \in N_O^+$ such that $A_i \Rightarrow^+ A_i x$ in O. Recall that the right-hand side of every A_i-rule starts with a terminal or a nonterminal from $\{A_k \mid i < k \leq n\}$ (see (I) and (II) in Algorithm 6.9), which rules out $A_i \Rightarrow^+ A_i x$ in O — a contradiction.
(b) Let B_i be left-recursive, for some $B_i \in \{B_1, \ldots, B_m\}$. That is, there exists $x \in N_O^+$ such that $B_i \Rightarrow^+ B_i x$ in O. Recall that the right-hand side of every B_i-rule starts with a nonterminal from $\{A_k \mid 1 \leq k \leq n\} \cup \{B_l \mid 1 \leq l < i\}$ (see (III) in Algorithm 6.9). Furthermore, no right-hand side of any A_i-rule starts with a nonterminal from $\{B_l \mid 1 \leq l \leq m\}$, for all $1 \leq i \leq n$ (see (II)). Thus, O cannot make $B_i \Rightarrow^+ B_i x$ — a contradiction.

Thus, O is a non-left-recursive CFG. As an exercise, complete this proof by demonstrating that $L(I) = L(O)$. □

Example 6.12. Reconsider the CFG G in the Chomsky normal form

$$A_1 \to A_1A_2, \quad A_1 \to A_4A_5, \quad A_1 \to i, \quad A_2 \to A_3A_1, \quad A_3 \to o,$$
$$A_4 \to (, A_5 \to A_1A_6, \quad A_6 \to)$$

from Example 6.11. Observe that G is left-recursive because A_1 is a directly left-recursive nonterminal. Consider G as the input grammar I in Algorithm 6.9. Observe that I satisfies all the properties required by Algorithm 6.9. Apply this algorithm to convert I into an equivalent non-left-recursive grammar O, satisfying the output properties described in Algorithm 6.9. In this application, $n = 6$ because I has six nonterminals: A_1–A_6. For $i = 1$, the for loop replaces $A_1 \to A_1A_2$

with

$$A_1 \to A_4A_5, A_1 \to A_4A_5B_1, A_1 \to i, A_1 \to iB_1, B_1 \to A_2B_1, B_1 \to A_2.$$

After this replacement, for $i = 2, \ldots, 4$, the for loop does not change any rule. For $i = 5$, this loop replaces $A_5 \to A_1A_6$ with

$$A_5 \to A_4A_5A_6, A_5 \to A_4A_5B_1A_6, A_5 \to iA_6, A_5 \to iB_1A_6.$$

After this replacement, it replaces $A_5 \to A_4A_5A_6$ with $A_5 \to (A_5A_6$. Finally, it replaces $A_5 \to A_4A_5B_1A_6$ with $A_5 \to (A_5B_1A_6$. For $k = 6$, the for loop does not change any rule. Consequently, the resulting output grammar O produced by Algorithm 6.9 is defined as

(I) $A_1 \to i, A_1 \to iB_1, A_3 \to o, A_4 \to (, A_5 \to (A_5A_6, A_5 \to (A_5B_1A_6, A_5 \to iA_6, A_5 \to iB_1A_6, A_6 \to)$;
(II) $A_1 \to A_4A_5, A_1 \to A_4A_5B_1, A_2 \to A_3A_1$;
(III) $B_1 \to A_2B_1, B_1 \to A_2$. ■

Theorem 6.11. *For every $L \in \mathbf{L_{CFG}}$, there exists a non-left-recursive $G \in$ **CFG** satisfying $L(G) = L$.*

Proof. If $\varepsilon \notin L$, this theorem follows from Theorems 6.9 and 6.10. Suppose that $\varepsilon \in L$. Let $G \in$ **CFG** be non-left-recursive, and let $L(G) = L - \{\varepsilon\}$. Define $H = (V_H, T_H, P_H, S_H)$ in **CFG**, so $V_H = V_G \cup \{S_H\}$ and $P_H = P_G \cup \{S_H \to S_G, S_H \to \varepsilon\}$, where the start symbol S_H is a newly introduced symbol, $S_H \notin V_G$. Observe that $H \in$ **CFG** is non-left-recursive, and $L(H) = L$. □

Right recursion. By analogy with left recursion, we define its right counterpart.

Definition 6.16. $G \in$ **CFG** is right-recursive if $A \Rightarrow^+ xA$ in G, for some $A \in N_G$ and $x \in V_G^*$. ■

Theorem 6.12. *For every $L \in \mathbf{L_{CFG}}$, there exists a non-right-recursive $G \in$ **CFG** satisfying $L(G) = L$.* ■

Greibach normal form

A CFG in Greibach normal form has the right-hand side of every rule start with a terminal followed by zero or more nonterminals.

In theory, this form often simplifies proofs of results concerning CFGs. In practice, some parsing methods also make use of this form.

Definition 6.17. A CFG, $G = (V, T, P, S)$, is in Greibach normal form if every rule $A \to x \in P$ satisfies $x \in TN^*$. ■

Basic idea. Without any loss of generality, suppose that a CFG, $I = (V_I, T_I, P_I, S_I)$, satisfies the properties of the output CFG produced by Algorithm 6.9 (see Theorem 6.10). That is, $N_I = \{A_k \mid 1 \leq k \leq n\} \cup \{B_l \mid 1 \leq l \leq m\}$, for some $n, m \geq 0$, and each rule in P_I has one of the following three forms:

(I) $A_i \to au$ with $a \in T_I$, and $u \in N_I^*$;
(II) $A_i \to A_j v$, for some $A_i, A_j \in \{A_k \mid 1 \leq k \leq n\}$ such that $i < j$ and $v \in N_I^+$;
(III) $B_i \to Cw$ with $C \in \{Ak \mid 1 \leq k \leq n\} \cup \{B_l \mid 1 \leq l \leq i-1\}$ and $w \in N_I^*$.

Consider all rules of the form $A_i \to au$ with $a \in T_I$, and $u \in N_I^*$ (see (I)). As obvious, they are in Greibach normal form.

Consider any rule of the form $A_n \to u$ (see (I) and (II)). As n is the greatest number in $\{1, \ldots, n\}$, $A_n \to u$ is not of the form (II), so it is of the form (I). That is, u starts with a terminal; therefore, $A_n \to u$ is in Greibach normal form. For every pair of rules $A_{n-1} \to A_n v$ and $A_n \to u$, introduce a rule $A_{n-1} \to uv$ into P_O, which is in Greibach normal form because so is $A_n \to u$. As a result, all the newly introduced A_{n-1}-rules are in Greibach normal form. Then, for every pair of rules $A_{n-2} \to A_j v$ and $A_j \to u$ in (II) with $n-2 < j$, i.e., $j = n-1$ or $j = n$, make an analogical introduction of new rules. Proceeding down toward $n = 1$ in this way, we eventually obtain all A_k-rules in Greibach normal form.

Consider any rule $B_1 \to u$ from (III). For B_1, u starts with A_j for some $j = 1, \ldots, n$, and all the A_j-rules in P_O are in Greibach normal form. Therefore, for every pair of rules $B_1 \to A_j y$ in P_I and $A_j \to v$ in P_O, add $B_1 \to vy$ to P_O. As a result, all the newly introduced B_1-rules in P_O are in Greibach normal form. Then, for every pair of rules $B_2 \to Cw$ in P_I, with $C \in \{A_k \mid 1 \leq k \leq n\} \cup \{B_1\}$, and a C-rule $C \to z$ in P_O, which is already in Greibach normal form, add $B_2 \to zw$ to P_O. Proceeding from 1 to m in this way, we eventually obtain all the B_l-rules in Greibach normal form, $1 \le l \le m$. The resulting CFG is in Greibach normal form and generates $L(I)$.

Algorithm 6.10. Greibach normal form.

Input: A non-left-recursive CFG $I = (V_I, T_I, P_I, S_I)$ such that $N_I = \{A_k \mid 1 \le k \le n\} \cup \{B_l \mid 1 \le l \le m\}$, for some $n, m \ge 0$, and each rule in P_I has one of these three forms

(I) $A_i \to au$ with $a \in T_I, u \in N_I^*$;
(II) $A_i \to A_j v$, for some $A_i, A_j \in \{A_k \mid 1 \le k \le n\}$ such that $i < j$, and $v \in N_I^+$;
(III) $B_i \to Cw$ with $C \in \{A_k \mid 1 \le k \le n\} \cup \{B_l \mid 1 \le l \le i-1\}$ and $w \in N_I^*$.

Output: A CFG $O = (V_O, T_O, P_O, S_O)$ such that $L(O) = L(I)$ and O is in Greibach normal form.

Method:

begin
  set $V_O = V_I$ and $T_O = T_I$
  set $P_O = \{A_i \to au \in P_I \mid a \in T_I \text{ and } u \in N_I^*\}$
  for $i = n, n-1, \ldots, 1$ **do**
    for each $A_i \to A_j y \in P_I$, where $A_i, A_j \in \{A_k \mid 1 \le k \le n\}$ such that $i < j$, and $y \in N_I^+$ **do**
      $P_O = P_O \cup \{A_i \to zy \mid A_j \to z \in P_O, z \in T_I N_I^*\}$
  for $i = 1, 2, \ldots, m$ **do**
    for each $B_i \to Cw \in P_I$ with $C \in \{A_k \mid 1 \le k \le n\} \cup \{B_l \mid 1 \le l \le i-1\}$ and $w \in N_I^*$ **do**
      $P_O = P_O \cup \{B_i \to zw \mid C \to z \in P_O, z \in T_I N_I^*\}$
end

Based on the basic idea preceding Algorithm 6.10, prove the following theorem.

Theorem 6.13. *Algorithm 6.10 is correct. Therefore, for every $L \in$ $\boldsymbol{L_{CFG}}$, there exists $G \in$ **CFG** in Greibach normal form satisfying $L(G) = L - \{\varepsilon\}$.* ■

Example 6.13. Consider the non-left-recursive CFG G:

(I) $A_1 \to i, A_1 \to iB_1, A_5 \to (A_5A_6, A_5 \to (A_5B_1A_6, A_5 \to iA_6, A_5 \to iB_1A_6, A_3 \to o, A_4 \to (, A_6 \to)$,
(II) $A_1 \to A_4A_5, A_1 \to A_4A_5B_1, A_2 \to A_3A_1$,
(III) $B_1 \to A_2B_1, B_1 \to A_2$,

obtained in Example 6.12. Let G be I in Algorithm 6.10. Note that it satisfies all the input requirements stated in this algorithm.

Consider (I). All these rules are in Greibach normal form. Thus, initialize P_O with them.

Consider (II). Note that the first for loop works with these rules. For $i = 2$, as $A_2 \to A_3A_1$ is in (II) and P_O contains $A_3 \to o$, this loop adds $A_2 \to oA_1$ to P_O. For $i = 1$, since $A_1 \to A_4A_5$ and $A_1 \to A_4A_5B_1$ are in (II) and $A_4 \to ($ is in P_O, this loop also adds $A_1 \to (A_5$ and $A_1 \to (A_5B_1$ to P_O.

Consider (III). The second for loop works with the two B_1-rules listed in (III). As $B_1 \to A_2B_1$ and $B_1 \to A_2$ are in (III) and $A_2 \to oA_1$ is in P_O, the second for loop includes $B_1 \to oA_1B_1$ and $B_1 \to oA_1$ into P_O.

Consequently, as O, Algorithm 6.10 produces this CFG in Greibach normal form:

$$A_1 \to i, \quad A_1 \to iB_1, \quad A_5 \to (A_5A_6, A_5 \to (A_5B_1A_6, A_5 \to iA_6,$$

$$A_5 \to iB_1A_6, \quad A_3 \to o,$$

$$A_4 \to (, A_6 \to), \quad A_2 \to oA_1, \quad A_1 \to (A_5, A_1 \to (A_5B_1,$$

$$B_1 \to oA_1B_1, \quad B_1 \to oA_1.$$

As an exercise, verify that $L(I) = L(O)$. ■

If a Greibach normal form CFG has n or fewer nonterminals on the right-hand side of every rule, where $n \in {}_0\mathbb{N}$, then we say that it is in n-standard Greibach normal form.

Definition 6.18. Let $G = (V, T, P, S)$ be a CFG in Greibach normal form. G is in *n-standard Greibach normal form* if every rule $A \to ax \in P$ satisfies $a \in T, x \in N^*$, and $|x| \leq n$. ■

In the rest of this section, we take a closer look at CFGs in n-standard Greibach normal form with $n \leq 2$. As is obvious, any zero-standard Greibach normal form CFG generates $\emptyset$ or a language that, in fact, represents an alphabet, i.e., a finite language in which every string is of length one.

Theorem 6.14. *A language L is generated by a CFG in zero-standard Greibach normal form iff L represents a finite set of symbols.* ■

In one-standard Greibach normal form CFGs, every rule is of the form $A \rightarrow aB$ or $A \rightarrow a$. These CFGs are usually referred to as regular CFGs because of the following result concerning their generative power.

Theorem 6.15. *A language L is generated by a CFG in one-standard Greibach normal form iff $L - \{\varepsilon\}$ is regular.*

Proof. Leaving a fully detailed version of this proof as an exercise, we only give its sketch. First, convert any CFG in one-standard Greibach normal form into an equivalent FA (see Definition 3.1). Then, turn any DFA M (see Definition 3.3) to a one-standard Greibach normal form CFG that generates $L(M) - \{\varepsilon\}$. By Theorem 3.8, Theorem 6.15 holds true. □

Consider two-standard Greibach normal form CFGs. For instance, return to the CFG discussed in Example 6.2, i.e.,

$$S \rightarrow aB, S \rightarrow bA, A \rightarrow a, A \rightarrow aS, A \rightarrow bAA, B \rightarrow b, B \rightarrow bS, B \rightarrow aBB.$$

This CFG represents a CFG that satisfies the two-standard Greibach normal form. Interestingly enough, for every CFL L, there is a two-standard Greibach normal form CFG that generates $L - \{\varepsilon\}$ as the following theorem says.

Theorem 6.16. *For every CFL L, there exists a CFG G in the two-standard Greibach normal form satisfying $L(G) = L - \{\varepsilon\}$. Consequently, let $n \geq 2$, and let K be a CFL; then, $K - \{\varepsilon\}$ is generated by a CFG in the n-standard Greibach normal form.*

Proof. By Theorem 6.13, for every CFL L, there exists a CFG H in Greibach normal form that generates $L - \{\varepsilon\}$. Consider the basic idea underlying the transformation to the Chomsky normal form (see Algorithm 6.8). As an exercise, modify this idea to convert H into an equivalent CFG G that satisfies the two-standard Greibach normal form. □

Equivalence with pushdown automata

Next, we demonstrate the equivalence between CFGs and PDAs.

From context-free grammars to pushdown automata

We begin by explaining how to turn any $I \in \mathbf{CFG}$ to $O \in \mathbf{PDA}$ so that $L(I) = L(O)$.

Basic idea. O uses its pushdown to simulate every leftmost derivation in I. In greater detail, this simulation is performed in the following way. If a terminal a occurs as the pushdown top symbol, O removes a from the pushdown and, simultaneously, reads a as the input symbol; in this way, it verifies their coincidence with each other. If a nonterminal A occurs as the pushdown top symbol, O simulates a leftmost derivation step made by a rule $A \to X_1X_2 \cdots X_n \in P_I$, where each $X_i \in V_I, 1 \leq i \leq n$, for some $n \in {}_0\mathbb{N}$($n = 0$ means $X_1X_2 \cdots X_n = \varepsilon$), so that it replaces the pushdown top A with $X_n \cdots X_1$. In somewhat greater detail, consider $S_I \; {}_{lm}\Rightarrow^* \; w$ in I, where $w \in T_I^*$, expressed as

$$\begin{aligned} S \; {}_{lm}\Rightarrow^* & \; vAy && [\rho] \\ {}_{lm}\Rightarrow & \; vX_1X_2 \ldots X_ny && [A \to X_1X_2 \ldots X_n] \\ {}_{lm}\Rightarrow^* & \; vu, \end{aligned}$$

where $w = vu$, so $v, u \in T_I^*$. Suppose that I has just simulated the first portion of this derivation, $S \; {}_{lm}\Rightarrow^* vAy$; at this point, the pushdown contains Ay in reverse while having u as the remaining input to read. In symbols, O occurs in the configuration $reversal(y)Asu$, from which it simulates $vAy \; {}_{lm}\Rightarrow vX_1X_2 \cdots X_ny \; [A \to X_1X_2 \cdots X_n]$ as

$$reversal(y)Asu \Rightarrow reversal(y)X_n \cdots X_1su$$

by using $As \to X_n \cdots X_1s$ from R_O. Note that

$$reversal(y)X_n \cdots X_1 = reversal(X_1 \cdots X_ny).$$

In this step-by-step way, O simulates $S \; {}_{lm}\Rightarrow^* w$ in I. Next, we give Algorithm 6.11, which describes the construction of O from I in a rigorous way.

Algorithm 6.11. CFG-to-PDA conversion.

Input: A CFG $I = (V_I, T_I, P_I, S_I)$.
Output: A PDA $O = (Q_O, \Delta_O, \Gamma_O, R_O, s_O, S_O, F_O)$, such that $L(O) = L(I)$.
Method:

begin
set $Q_O = F_O = \{s_O\}$
set $\Gamma_O = V_I, \Delta_O = T_I$ and $S_O = S_I$
set $R_O = \{As \rightarrow reversal(x)s \mid A \rightarrow x \in P_I\} \cup \{asa \rightarrow s \mid a \in T_I\}$
end

As an exercise, give a fully detailed proof of the following theorem.

Theorem 6.17. *Algorithm 6.11 is correct. Therefore,* $\mathbf{L_{CFG}} \subseteq \mathbf{L_{PDA}}$. ■

Note that Algorithm 6.11 produces its output PDA so that by using each of its rules, the PDA changes precisely one symbol on the pushdown top. In other words, we have proved the following corollary, which fulfills an important role later in this section (see Theorem 6.19).

Corollary 6.4. *For every* $L \in \mathbf{L_{CFG}}$, *there exists a PDA,* $M = (Q, \Delta, \Gamma, R, s, S, F)$, *such that* $L(M) = L$ *and every rule* $r \in R$ *is of the form* $r : Aqa \rightarrow yp$, *where* $q, p \in Q, a \in \Delta \cup \{\varepsilon\}, y \in \Gamma^*$ *and, most importantly,* $A \in \Gamma$. ■

From pushdown automata to context-free grammars

First, we show that every $L \in \mathbf{L_{PDA}}$ is accepted by a PDA that satisfies the properties stated in Corollary 6.4.

Theorem 6.18. $L \in \mathbf{L_{PDA}}$ *iff there exists a PDA,* $M = (Q, \Delta, \Gamma, R, s, S, F)$, *such that* $L(M) = L$ *and every rule* $r \in R$ *is of the form* $r : Aqa \rightarrow yp$, *where* $q, p \in Q, a \in \Delta \cup \{\varepsilon\}, y \in \Gamma^*, A \in \Gamma$.

Proof. The *if* part of this theorem is clearly true. Indeed, if $L(M) = L$, where M is a PDA that satisfies the required properties, then it obviously holds that $L(M) \in \mathbf{L_{PDA}}$.

To prove the *only-if* part of the equivalence in Theorem 6.18, consider any $L \in \mathbf{L_{PDA}}$. Let $N = (Q_N, \Delta_N, \Gamma_N, R_N, s_N, S_N, F_N)$ be a PDA such that $L(N) = L$. From N, we construct a PDA, $M = (Q_M, \Delta_M, \Gamma_M, R_M, s_M, S_M, F_M)$, satisfying the required properties so that M simulates every move in N by several moves during which it records the top pushdown symbols of N in its states. More precisely, suppose that N makes a move according to a rule of the form $xqa \to yp \in R_N$, where $q, p \in Q_N, a \in \Delta_N \cup \{\varepsilon\}, x, y \in \Gamma_N^*$. Assume that $|x| \geq 1$, so $x = Xz$, for some $X \in \Gamma_N$ and $z \in \Gamma_N^*$. M simulates this move in the following two-phase way:

(1) Starting from q, M makes $|z|$ moves, during which, in a symbol-by-symbol way, it stores the string u consisting of the top $|z|$ pushdown symbols into its state of the form $\langle uq \rangle$.
(2) From $\langle uq \rangle$, by applying a rule of the form $X\langle zq \rangle a \to yp \in R_M$, M verifies that $z = u$, reads a, pushes y onto its pushdown, and moves to state p. In other words, M completes the simulation of this move in N, and it is ready to simulate another move of N.

As an exercise, explain how to make this simulation under the assumption that $|x| = 0$. Then, rephrase this entire proof in a fully rigorous way. □

Next, we explain how to turn any PDA that satisfies the properties stated in Theorem 6.18 to an equivalent CFG.

Basic idea. Suppose that $I = (Q_I, \Delta_I, \Gamma_I, R_I, s_I, S_I, F_I)$ is a PDA in which every rule has the form $Aqa \to yp \in R_I$, where $q, p \in Q_I, a \in \Delta_I \cup \{\varepsilon\}, y \in \Gamma_I^*$, and $A \in \Gamma_I$. To transform I into an equivalent $O \in \mathbf{CFG}$, the following algorithm constructs O so that a leftmost derivation of $w \in T_O^*$ in O simulates a sequence of moves made by I on w. O performs this simulation by using nonterminals of the form $\langle qAp \rangle$, where $q, p \in Q_I$ and $A \in \Gamma_I$. More precisely, $\langle qAp \rangle\ {}_{lm}\Rightarrow^* w$ in O iff $Aqw \Rightarrow^* p$ in I. In addition, we introduce a new symbol, S_O, as the start symbol of O and define all S_O-rules as $\{S_O \to \langle s_I S_I f \rangle \mid f \in F_I\}$. Thus, $S_O \Rightarrow \langle s_I S_I f \rangle\ {}_{lm}\Rightarrow^* w$ in O iff $S_I s_I w\ {}_{lm}\Rightarrow^* f$ in I, with $f \in F_I$. As a result, $L(O) = L(I)$.

Algorithm 6.12. PDA-to-CFG Conversion.

Input: A PDA $I = (Q_I, \Delta_I, \Gamma_I, R_I, s_I, S_I, F_I)$, in which every rule is of the form $Aqa \rightarrow yp \in R_I$, where $q, p \in Q_I, a \in \Delta_I \cup \{\varepsilon\}, y \in \Gamma_I^*$, and $A \in \Gamma_I$.
Output: A CFG $O = (V_O, T_O, P_O, S_O)$, such that $L(O) = L(I)$.
Method:

begin
set $V_O = \{\langle pAq\rangle \mid p, q \in Q_I, A \in \Gamma_I\} \cup \Delta_I \cup \{S_O\}$
set $T_O = \Delta_I$
set $P_O = \{S_O \rightarrow \langle s_I S_I f\rangle \mid f \in F_I\}$
add $\langle q_0 A q_{n+1}\rangle \rightarrow a\langle q_1 X_1 q_2\rangle\langle q_2 X_2 q_3\rangle \cdots \langle q_n X_n q_{n+1}\rangle$ to P_O for all $Aq_0 a \rightarrow X_n \cdots X_1 q_1 \in R_I$, where $q_0, q_{n+1} \in Q_I, a \in \Delta_I, A \in \Gamma_I$, and $q_j \in Q_I, X_j \in \Gamma_I, \langle q_j X_j q_{j+1}\rangle \in (V_O - T_O), 1 \leq j \leq n$, for some $n \in {}_0\mathbb{N}$ ($n = 0$ means $X_n \cdots X_1 = \langle q_1 X_1 q_2\rangle \cdots \langle q_n X_n q_{n+1}\rangle = \varepsilon$)
end

Lemma 6.1. *Algorithm 6.12 is correct.*

Proof. We open this proof by establishing the following claim concerning I and O from Algorithm 6.12.

Claim. For all $w \in \Delta_I^*, A \in \Gamma_I$, and $p, q \in Q_I$,

$$\langle qAp\rangle \ {}_{lm}\Rightarrow^* w \text{ in } O \text{ iff } Aqw \Rightarrow^* p \text{ in } I.$$

Proof of the Claim. First, we establish the only-if part of this equivalence. That is, by induction on $i \geq 0$, we prove that $\langle qAp\rangle \ {}_{lm}\Rightarrow^i w$ in O implies $Aqw \Rightarrow^* p$ in I.

Basis. For $i = 0$, $\langle qAp\rangle \ {}_{lm}\Rightarrow^0 w$ never occurs in O, for any $w \in \Delta_I^*$, so the basis holds true.

Induction hypothesis. Assume that the implication holds for all derivations consisting of no more than j steps, for some $j \in {}_0\mathbb{N}$.

Induction step. Consider any derivation of the form $\langle qAp\rangle \ {}_{lm}\Rightarrow^{j+1} w$ $[p\pi]$ in O, where $|\pi| = j$. Let this derivation start by the application of a rule of the form

$$\langle q_0 A q_{n+1}\rangle \rightarrow a\langle q_1 X_1 q_2\rangle\langle q_2 X_2 q_3\rangle \cdots \langle q_n X_n q_{n+1}\rangle$$

from R_O, where $a \in \Delta_I, A \in \Gamma_I, X_k \in \Gamma_I, 1 \leq k \leq n, q = q_0$, $p = q_{n+1}$, and $q_l \in Q_I, 0 \leq j \leq n+1$, for some $n \in {}_0\mathbb{N}$. Thus, we can express $\langle qAp\rangle\ {}_{lm}{\Rightarrow}^{i+1}\ w$ as

$$\begin{aligned}\langle q_0 A q_{n+1}\rangle\ {}_{lm}{\Rightarrow}\ & a\langle q_1 X_1 q_2\rangle\langle q_2 X_2 q_3\rangle\cdots\langle q_n X_n q_{n+1}\rangle\\ {}_{lm}{\Rightarrow}^{i}\ & a w_1 w_2 \cdots w_n,\end{aligned}$$

where for all $1 \leq j \leq n, \langle q_j X_j q_{j+1}\rangle \Rightarrow^* w_j$, and each of these n derivations consists of i or fewer steps. Thus, by the induction hypothesis, $X_i q_i w_i \Rightarrow^* q_{i+1}$ in $I, 1 \leq i \leq n$. Algorithm 6.12 constructs $\langle q_0 A q_{n+1}\rangle \rightarrow a\langle q_1 X_1 q_2\rangle\langle q_2 X_2 q_3\rangle\cdots\langle q_n X_n q_{n+1}\rangle \in R_O$ from $Aq_0 a \rightarrow X_n \cdots X_2 X_1 q_1 \in R_I$, so

$$\begin{aligned} Aq_0 w \quad & \Rightarrow^* \quad X_n \cdots X_2 X_1 q_1 w_1 w_2 \cdots w_n\\ & \Rightarrow^* \quad X_n \cdots X_2 q_2 w_2 \cdots w_n\\ & \Rightarrow^* \quad X_n \cdots q_3 \cdots w_n\\ & \ \vdots\\ & \Rightarrow^* \quad X_n q_n w_n\\ & \Rightarrow^* \quad q_{n+1}\end{aligned}$$

in I. Because $q = q_0$ and $p = q_{n+1}$, $Aqw \Rightarrow^* p$ in I, and the inductive step is completed.

Next, we establish the if part of the equivalence stated in the claim, so we show that $Aqw \Rightarrow^i p$ in I implies $\langle qAp\rangle\ {}_{lm}{\Rightarrow}\ w$ in O by induction on $i \geq 0$.

Basis. The basis holds vacuously because $Aqw \Rightarrow^0 p$ in I is false, for all $w \in \Delta_I^*$; indeed, I cannot remove A from the pushdown top by making zero moves.

Induction hypothesis. Assume that the implication holds for all sequences consisting of no more than j moves, for some $j \in {}_0\mathbb{N}$.

Induction step. Let $Aqw \Rightarrow^{j+1} p\ [r\rho]$ in I, where $|\rho| = j$. Let r be a rule of the form $Aq_0 a \rightarrow X_n \cdots X_2 X_1 q_1 \in R_I$, where

$a \in \Delta_I, A \in \Gamma_I, X_j \in \Gamma_I, 1 \leq j \leq n$, for some $n \in {}_0\mathbb{N}$. Express $Aqw \Rightarrow^{j+1} p$ as

$$\begin{aligned} Aq_0a \Rightarrow & \; X_n \cdots X_2X_1q_1w_1w_2 \cdots w_n \\ \Rightarrow^* & \; X_n \cdots X_2q_2w_2 \cdots w_n \\ \Rightarrow^* & \; X_n \cdots q_3 \cdots w_n \\ & \vdots \\ \Rightarrow^* & \; X_nq_nw_n \\ \Rightarrow^* & \; q_{n+1} \end{aligned}$$

in I, where $q = q_0, p = q_{n+1}$, and $w = aw_1w_2 \cdots w_n$. Clearly, for all $1 \leq k \leq n, X_n \cdots X_{k+1}X_kq_kw_kw_{k+1} \cdots w_n \Rightarrow^* X_n \cdots X_{k+1}q_{k+1}w_{k+1} \cdots w_n$ consists of no more than j steps, so by the induction hypothesis, $\langle q_iX_iq_{i+1}\rangle \Rightarrow^* w_i$, for $1 \leq i \leq n$. From $Aq_0a \to X_n \cdots X_2X_1q_1 \in R_I$, Algorithm 6.12 constructs $\langle q_0Aq_{n+1}\rangle \to a\langle q_1X_1q_2\rangle\langle q_2X_2q_3\rangle \cdots \langle q_nX_nq_{n+1}\rangle \in R_O$. Thus, O makes

$$\begin{aligned} \langle q_0Aq_{n+1}\rangle \;{}_{lm}\Rightarrow & \; a\langle q_1X_1q_2\rangle\langle q_2X_2q_3\rangle \cdots \langle q_nX_nq_{n+1}\rangle \\ {}_{lm}\Rightarrow^* & \; aw_1w_2 \cdots w_n, \end{aligned}$$

with $w = aw_1w_2 \cdots w_n$, and the induction step is completed. Thus, the claim holds.

Considering the claim for $S_I = A$ and $s_I = q$, we have for all $w \in \Delta_I^*$ and $p \in Q_I$, $\langle s_IS_Ip\rangle \;{}_{lm}\Rightarrow^* w$ in O iff $S_Is_Iw \Rightarrow^* p$ in I. As follows from the algorithm, O starts every derivation by applying a rule of the form $S_O \to \langle s_IS_If\rangle$, with $f \in F_I$. Consequently, $S_O \;{}_{lm}\Rightarrow \langle s_IS_If\rangle \;{}_{lm}\Rightarrow^* w$ in O iff $S_Is_Iw \Rightarrow^* f$ in I, so $L(I) = L(O)$.

Thus, the algorithm is correct. As a result, every language in $\mathbf{L_{PDA}}$ also belongs to $\mathbf{L_{CFG}}$, so $\mathbf{L_{PDA}} \subseteq \mathbf{L_{CFG}}$. □

Example 6.14. Reconsider $M \in \mathbf{PDA}$ from Example 4.1. Recall that M is defined as

$$Ssa \to as, \quad asa \to aas, \quad asb \to f, \quad afb \to f,$$

where f is a final state. Recall that $L(M) = \{a^nb^n \mid n \geq 1\}$. With $I = M$, Algorithm 6.12 constructs $O \in \mathbf{CFG}$ satisfying $L(O) = L(I)$

in the following way. Denote the new start symbol of O by Z. Initially, set

$$N_O = \{\langle pAq\rangle \mid p, q \in \{s, f\}, A \in \{S, a\}\} \cup \{Z\}, \text{ and}$$
$$R_O = \{Z \rightarrow \langle sSf\rangle\}.$$

From $Ssa \rightarrow as$, Algorithm 6.12 produces these two rules

$$\langle sSs\rangle \rightarrow a\langle sas\rangle \quad \text{and} \quad \langle sSf\rangle \rightarrow a\langle saf\rangle.$$

From $asa \rightarrow aas$, it produces

$$\langle sas\rangle \rightarrow a\langle sas\rangle\langle sas\rangle \qquad \langle sas\rangle \rightarrow a\langle saf\rangle\langle fas\rangle,$$
$$\langle saf\rangle \rightarrow a\langle sas\rangle\langle saf\rangle \qquad \langle saf\rangle \rightarrow a\langle saf\rangle\langle faf\rangle.$$

From $asb \rightarrow f$, the algorithm makes

$$\langle sas\rangle \rightarrow b, \langle saf\rangle \rightarrow b.$$

Construct the rest of the output grammar as an exercise. In addition, by using Algorithm 6.3, remove from the grammar all useless symbols to obtain the resulting grammar O defined as

$$Z \rightarrow \langle sSf\rangle, \langle sSf\rangle \rightarrow a\langle saf\rangle, \langle saf\rangle \rightarrow a\langle saf\rangle\langle fbf\rangle, \langle saf\rangle \rightarrow b,$$
$$\langle fbf\rangle \rightarrow b.$$

For instance, I accepts $aabb$ as

$$Ssaabb \;\Rightarrow\; asabb \;\Rightarrow\; aasbb \;\Rightarrow\; afb \;\Rightarrow\; f.$$

O generates this string by the following leftmost derivation:

$$Z\ {}_{lm}\!\Rightarrow \langle sSf\rangle\ {}_{lm}\!\Rightarrow a\langle saf\rangle\ {}_{lm}\!\Rightarrow aa\langle saf\rangle\langle fbf\rangle\ {}_{lm}\!\Rightarrow aab\langle faf\rangle\ {}_{lm}\!\Rightarrow aabb.$$

As a simple exercise, give a rigorous proof that $L(O) = \{a^n b^n \mid n \geq 1\}$. ■

The next crucial theorem, which says that PDAs and CFGs are equivalent, follows from Theorem 6.17, Corollary 6.4, Theorem 6.18, Algorithm 6.12, and Lemma 6.1.

Theorem 6.19. $\mathbf{L_{CFG} = L_{PDA}}$. ■

The pumping lemma for context-free languages

The pumping lemma established in this section is frequently used to disprove that a language K is context-free. The lemma says that for every CFL, L, there is a constant $k \geq 1$ such that every $z \in L$ with $|z| \geq k$ can be expressed as $z = uvwxy$, with $vx \neq \varepsilon$, so that L also contains $uv^m wx^m y$, for every $m \geq 0$. Consequently, to demonstrate the non-context freedom of a language, K, by contradiction, assume that K is a CFL, and k is its pumping lemma constant. Select a string $z \in K$, with $|z| \geq k$, consider all possible decompositions of z into $uvwxy$, and for each of these decompositions, prove that $uv^m wx^m y$ is out of K, for some $m \geq 0$, which contradicts the pumping lemma. Thus, K is not a CFL.

Lemma 6.2. *Pumping Lemma for CFLs. Let L be an infinite context-free language. Then, there exists $k \in \mathbb{N}$ such that every string $z \in L$ satisfying $|z| \geq k$ can be expressed as $z = uvwxy$, where $0 < |vx| < |vwx| \leq k$, and $uv^m wx^m y \in L$, for all $m \geq 0$.* ■

Applications of the pumping lemma

We usually use the pumping lemma in a proof by contradiction to demonstrate that a given language L is not context-free. Typically, we make a proof of this kind in the following way:

(1) Assume that L is context-free.
(2) Select a string $z \in L$ whose length depends on the pumping-lemma constant k so that $|z| \geq k$ is necessarily true.
(3) For all possible decompositions of z into $uvwxy$ satisfying the pumping lemma conditions, find $m \in {}_0\mathbb{N}$ such that $uv^m wx^m y \notin L$, which contradicts Lemma 6.2.
(4) The contradiction obtained in (3) means that the assumption in (1) is incorrect; therefore, L is not context-free.

Example 6.15. Consider $L = \{a^n b^n c^n \mid n \geq 1\}$. Next, under the guidance of the recommended proof structure preceding this example, we demonstrate that L is not context-free:

(1) Assume that L is context-free.
(2) In L, select $z = a^k b^k c^k$ with $|z| = 3k \geq k$, where k is the pumping-lemma constant.

(3) By Lemma 6.2, z can be written as $z = uvwxy$ so that this decomposition satisfies the pumping lemma conditions. As $0 < |vx| < |vwx| \leq k$, either $vwx \in \{a\}^*\{b\}^*$ or $vwx \in \{b\}^*\{c\}^*$. If $vwx \in \{a\}^*\{b\}^*, uv^0wx^0y$ has k cs but fewer than k as or bs, so $uv^0wx^0y \notin L$, but by the pumping lemma, $uv^0wx^0y \in L$. If $vwx \in \{b\}^*\{c\}^*, uv^0wx^0y$ has k as but fewer than k bs or cs, so $uv^0wx^0y \notin L$, but by the pumping lemma, $uv^0wx^0y \in L$. In either case, we obtain the contradiction that $uv^0wx^0y \notin L$ and, simultaneously, $uv^0wx^0y \in L$.

(4) By the contradiction obtained in (3), L is not context-free.

6.2 General Grammars and Turing Machines

General grammars represent grammatical counterparts to Turing machines. Next, we define these grammars, establish their normal forms, and prove their equivalence with TMs.

Definition 6.19. A *general grammar* (GG for short) is a quadruple

$$G = (V, T, P, S),$$

where V is an alphabet, $T \subseteq V, P \subseteq V^*(V - T)V^* \times V^*$ is finite, and $S \in V - T$. Set $N = V - T$. In G, V, N, and T are referred to as the *total alphabet*, the *alphabet of nonterminal symbols*, and the *alphabet of terminal symbols*, respectively. S is the *start symbol*. P is a finite *set of rules*; every rule $(u, v) \in P$ is written as $u \rightarrow v$ in what follows.

Define the relation $\Rightarrow$ over V^* so that for all $u \rightarrow v \in P$ and $x, y \in V^*, xuy \Rightarrow xvy$; in words, G *directly derives* xvy from xuy . As usual, $\Rightarrow^*$ denotes the transitive-reflexive closure of $\Rightarrow$. If $S \Rightarrow^* w$, where $w \in V^*, G$ *derives* w, and w is a *sentential form*. $F(G)$ denotes the set of all sentential forms derived by G. The *language generated by* G, symbolically denoted by $L(G)$, is defined as $L(G) = F(G) \cap T^*$; in other words, $L(G) = \{w \in T^* \mid S \Rightarrow^* w\}$. Members of $L(G)$ are called *sentences*. If $S \Rightarrow^* w$ and w is a sentence, $S \Rightarrow^* w$ is a *successful derivation* in G. ■

By $xuy \Rightarrow xvy$ $[u \rightarrow v]$, where $u, v \in V^*$ and $u \rightarrow v \in P$, we say that G directly derives xvy from xuy by using $u \rightarrow v$, or G makes a derivation step from xuy to xvy by $u \rightarrow v$. Furthermore, to express that G makes $u \Rightarrow^* w$ according to a sequence of rules, $r_1 r_2 \cdots r_n$, we write $u \Rightarrow^* v\ [r_1 r_2 \cdots r_n]$, which is read as a *derivation from* u to v by using $r_1 r_2 \cdots r_n$. On the other hand, whenever the information regarding the applied rules is immaterial, we omit these rules. In other words, we often simplify $u \Rightarrow v\ [r]$ and $u \Rightarrow^* v\ [r_1 r_2 \cdots r_n]$ to $u \Rightarrow^* v$ and $u \Rightarrow v$, respectively.

Consider Convention 6.1. We apply these conventions to GGs throughout too. We denote the set of all GGs by **GG**. We set $\mathbf{L_{GG}} = \{L(G) \mid G \in \mathbf{GG}\}$, and we refer to $\mathbf{L_{GG}}$ as the *family of languages generated by general grammars.*

Example 6.16. Consider a GG G defined by these following seven rules:

$$1 : S \to CAaD, \quad 2 : Aa \to aaA, \quad 3 : AD \to BD,$$

$$4 : aB \to Ba, \quad 5 : CB \to CA, \quad 6 : CA \to A, \quad 7 : AD \to \varepsilon$$

In G, a is the only terminal, while all the other symbols are nonterminals.

For instance, from $CAaD$, G derives $CAaaD$ by carrying out the following five derivation steps:

$$\begin{aligned} CAaD &\Rightarrow CaaAD\ [2] \\ &\Rightarrow CaaBD\ [3] \\ &\Rightarrow CaBaD\ [4] \\ &\Rightarrow CBaaD\ [4] \\ &\Rightarrow CAaaD\ [5]. \end{aligned}$$

Thus, $CAaD \Rightarrow^5 CAaaD\ [23445]$ or, briefly, $CAaD \Rightarrow^5 CAaaD$. Therefore, $CAaD \Rightarrow^+ CAaaD$ and $CAaD \Rightarrow^* CAaaD$. From a more general viewpoint, observe that every generation of a sentence in $L(G)$ has the following form:

$$\begin{aligned} &S \Rightarrow CAaD \Rightarrow^* CAa^2D \Rightarrow^* CAa^4D \Rightarrow^* \cdots \\ &\quad \Rightarrow^* CAa^{2^j}D \Rightarrow^* CAa^{2^{j+1}}D \Rightarrow^* \cdots \Rightarrow^* CAa^{2^{k-1}}D \Rightarrow^* a^{2^k}, \end{aligned}$$

for some $j \in {}_0\mathbb{N}$ and $k \in \mathbb{N}$ such that $j \leq k$. Consequently,

$$L(G) = \{a^{2^i} \mid i \geq 1\}.$$

■

Normal forms

Just like we often make use of CFGs transformed into normal forms, such as the Chomsky normal form (see Algorithm 6.8), we frequently transform GGs into some normal forms, in which their rules have a prescribed form. In the current section, we convert GGs to Kuroda normal form and its special case — Pentonnen normal form.

Definition 6.20. A GG, $G = (V, T, P, S)$, is in Kuroda normal form if every rule $r \in P$ has one of the following four forms:

$$AB \to DC, \quad A \to BC, \quad A \to a, \quad \text{or} \quad A \to \varepsilon,$$

where $A, B, C, D \in N$ and $a \in T$. ■

Basic idea. Next, we sketch how to turn any GG $I = (V_I, T_I, P_I, S_I)$ to an equivalent GG $O = (V_O, T_O, P_O, S_O)$ in Kuroda normal form.

Initially, set N_O to N_I. Move all rules satisfying Kuroda normal form from P_I to P_O. Carry out the following five steps, (1)–(5). Note that after performing (1) and (2), every $x \to y \in P_I$ satisfies $x, y \in N_O^*$ and $|x| \leq |y|$:

(1) Consider every $a \in T_I$ and every $r \in P_I$. For every occurrence of a in r, introduce a special new nonterminal, X, in N_O. In P_I, change this occurrence with X, and if, after this change, the rule satisfies Kuroda normal form, move it to P_O. Add $X \in a$ to P_O.
(2) For every $u \to v \in P_I$, with $|u| < |v|$, introduce a special new nonterminal X in N_O. In P_I, change this rule to $u \to vX^n$, and if after this change, the rule satisfies Kuroda normal form ($|u| = 2, |v| = 1$, and $n = 1$), move it to P_O. Add $X \in \varepsilon$ to P_O.
(3) For every $A \to B \in P_I$ with $A, B \in N_O$, introduce a special new nonterminal, X, in N_O. Add $A \to BX$ and $X \to \varepsilon$ to P_O. Remove $A \to B$ from P_I.

Repeat (4) and (5), given next, until $P_I = \emptyset$.

(4) Turn every $A \to X_1X_2X_3 \cdots X_{n-1}X_n \in P_I$, where $X_i \in N_O, 1 \leq i \leq n$, for some $n \geq 3$, to a set of $n - 1$ rules with two-nonterminal right-hand sides exactly like in Algorithm 6.8 *Chomsky Normal Form.* That is, introduce new nonterminals

$\langle X_2 \cdots X_n\rangle, \langle X_3 \cdots X_n\rangle, \ldots, \langle X_{n-2}X_{n-1}X_n\rangle, \langle X_{n-1}X_n\rangle$ into N_O and add the following rules:

$$\begin{aligned}
&A \rightarrow X_1\langle X_2 \cdots X_n\rangle,\\
&\langle X_2 \cdots X_n\rangle \rightarrow X_2\langle X_3 \cdots X_n\rangle,\\
&\vdots\\
&\langle X_{n-2}X_{n-1}X_n\rangle \rightarrow X_{n-2}\langle X_{n-1}X_n\rangle,\\
&\langle X_{n-1}X_n\rangle \rightarrow X_{n-1}X_n
\end{aligned}$$

to P_O. Remove $A \rightarrow X_1X_2X_3\cdots X_{n1}X_n$ from P_I.

(5) For every $A_1A_2A_3\cdots A_{n-1}A_n \rightarrow B_1B_2\cdots B_{m-1}B_m \in P_I$, where $A_i \in N_O, 1 \leq i \leq n, B_j \in N_O, 1 \leq j \leq n$, for some $n \geq 2$ and $m \geq 3$, introduce a special new nonterminal, X, into N_O. Add $A_1A_2 \rightarrow B_1X$ to P_O. If $|A_3\cdots A_{n1}A_n| = 0$ and $|B_2\cdots B_{m-1}B_m| = 2$, add $X \rightarrow B_2B_3$ to P_O; otherwise, add $XA_3\cdots A_{n-1}A_n \rightarrow B_2\cdots B_{m-1}B_m$ into P_I. Remove $A_1A_2A_3\cdots A_{n-1}A_n \rightarrow B_1B_2\cdots B_{m-1}B_m$ from P_I.

As an exercise, express the basic idea above as an algorithm, verify the algorithm formally, and prove the following theorem.

Theorem 6.20. *For every $I \in$ **GG**, there exists $O \in$ **GG** such that O is in Kuroda normal form and $L(O) = L(I)$.* ∎

Although GGs in Kuroda normal form grammars allow only four forms of rules

$$AB \rightarrow DC, \quad A \rightarrow BC, \quad A \rightarrow a, \quad \text{or} \quad A \rightarrow \varepsilon,$$

where A, B, C, D are nonterminals and a is a terminal, they are as powerful as ordinary GGs, as follows from Theorem 6.20. Note that the exclusion of any of these forms results in a decrease in their generative power, however. Indeed, without rules of the form $AB \rightarrow DC$, GGs in Kuroda normal form become CFGs, which are less powerful than GGs, as stated later in this chapter (see Theorems 6.22 and 6.26). Without rules of the form $A \rightarrow BC$, these grammars cannot actually expand the start symbol, so they can only generate terminal symbols or ε. Without rules of the form $A \rightarrow a$, they obviously generate only ε. Finally, without rules of the form $A \rightarrow \varepsilon$, GGs also decrease their power as demonstrated in Theorem 6.26.

We close this section by introducing Pentonnen normal form, which represents a more restrictive version of Kuroda normal form.

Definition 6.21. Let $G = (V, T, P, S)$ be a GG in Kuroda normal form. If every rule of the form $AB \to DC \in P$, where $A, B, C, D \in N$, satisfies $A = D, G$ is in *Pentonnen normal form.* ■

Prove the following theorem as an exercise.

Theorem 6.21. *For every $I \in \mathbf{GG}$, there exists $O \in \mathbf{GG}$ such that O is in Pentonnen normal form and $L(O) = L(I)$.* ■

Equivalence of general grammars and Turing machines

First, this section explains how to transform any GG into an equivalent TM. Then, it shows how to convert any TM into an equivalent GG. As a result, it establishes the equivalence of GGs and TMs; in symbols, $\mathbf{L_{GG}} = \mathbf{L_{TM}}$.

From general grammars to Turing machines

Any GG obviously represents a procedure, so the Church–Turing thesis immediately implies

Lemma 6.3. $\boldsymbol{L_{GG}} \subseteq \boldsymbol{L_{TM}}$. ■

Nevertheless, one might ask for an effective proof of this result. Therefore, we next sketch how to turn any $G \in \mathbf{GG}$ to an equivalent $M \in \mathbf{TM}$ algorithmically. In fact, we give two alternative methods of this conversion:

(I) Given $G \in \mathbf{GG}$, the first method constructs a TM M that explores all possible derivations in G. More precisely, M has three tracks on its tape. On the first track, M places any $w \in T_G^*$. On the second track, it systematically generates sequences of rules from P_G, according to which it simulates the derivations in G on the third track. If $w \in L(G), M$ eventually simulates the generation of w in G, accepts w, and halts. If $w \notin L(G), M$ never simulates the generation of w in G, so it runs endlessly, never accepting w. Hence, $L(G) = L(M)$.

(II) Construct M with a tape having three tracks. Initially, M writes any $w \in T_G^*$ on the first track. Then, it lists the rules of P_G on the second track. Finally, it writes S_G on the third track, which M uses to record the current sentential form of G during the upcoming simulation of a derivation in G. After this initialization phase, M works as follows (recall that for any rule $r \in P_G, lhs(r)$ and $rhs(r)$ denote its left- and right-hand sides, respectively):

(i) M non-deterministically selects any $r \in P_G$ on the second track;
(ii) if there is no occurrence of $lhs(r)$ on the third track, M rejects; otherwise, it selects any occurrence of $lhs(r)$ and replaces it with $rhs(r)$;
(iii) if the first and third tracks coincide, accept; otherwise, go to (i).

As an exercise, describe (I) and (II) above more formally.

From Turing machines to general grammars

The following algorithm converts any $I \in \mathbf{TM}$ into an equivalent $O \in \mathbf{GG}$. In essence, G simulates every accepting computation $\rhd \blacktriangleright w \lhd \Rightarrow^* \rhd \blacksquare u \lhd$ in M in reverse by a derivation of the form $S \Rightarrow^* \rhd \blacksquare u \lhd \Rightarrow^* \rhd \blacktriangleright w \lhd \Rightarrow^* w$.

Algorithm 6.13. TM-to-GG conversion.

Input: A TM $I = (Q_I, \Delta_I, \Gamma_I, R_I, s_I, F_I)$.
Output: A GG $O = (V_O, T_O, P_O, S_O)$, such that $L(O) = L(I)$.
Method:

begin
set $V_O = \Delta_I \cup \{S, 1, 2\}$, where S, 1 and 2 are new symbols, $\{S, 1, 2\} \cap \Delta_I = \emptyset, T_O = T_I, N_O = V_O - T_I$, and $P_O = \{S \rightarrow \rhd 1 \lhd, 1 \rightarrow \blacksquare, \rhd \blacktriangleright \rightarrow 2, 2\lhd \rightarrow \varepsilon\}$;

for every $a \in \Gamma_I$, add $1 \rightarrow 1a$ to P_O;
for every $y \rightarrow x \in P_I$, add $x \rightarrow y$ to P_O;
for every $a \in T_I$, add $2a \rightarrow a2$ to P_O
end

O always makes its first derivation step always by using $S \to \triangleright 1 \triangleleft$, so $S \Rightarrow^* \triangleright 1 \triangleleft$. Then, by using rules of the form $1 \to 1a$, where $a \in V_I$, O performs $\triangleright 1 \triangleleft \Rightarrow^* \triangleright 1u \triangleleft$, after which it makes $\triangleright 1u \triangleleft \Rightarrow \triangleright \blacksquare u \triangleleft$ by using $1 \to \blacksquare$, where $u \in V_I^*$. In sum, O starts every derivation as $S \Rightarrow^* \triangleright \blacksquare u \triangleleft$, for any $u \in V_I^*$. Since $y \to x \in P_I$ iff $x \to y \in P_O$, $\triangleright \blacktriangleright w \triangleleft \Rightarrow^* \triangleright \blacksquare u \triangleleft$ in I iff $\triangleright \blacksquare u \triangleleft \Rightarrow^* \triangleright \blacktriangleright w \triangleleft$ in O, where $u \in \Gamma_I^*, w \in T_I^*$. By using $\triangleright \blacktriangleright \to 2, 2a \to a2$, and $2\triangleleft \to \varepsilon$, where $a \in T_I, O$ completes the generation of w, so it crosses $\triangleright \blacktriangleright w \triangleleft$ from left to right and, simultaneously, removes $\triangleright, \blacktriangleright$, and $\triangleleft$. Thus, $L(O) = L(I)$. Based on these observations, give a formal proof of the following lemma as an exercise.

Lemma 6.4. *Algorithm 6.13 is correct. Therefore,* $\mathbf{L_{TM}} \subseteq \mathbf{L_{GG}}$. ■

Lemmas 6.3 and 6.4 imply the following theorem.

Theorem 6.22. $\mathbf{L_{TM}} = \mathbf{L_{GG}}$. ■

Context-Sensitive Grammars and Linear-Bounded Automata

Context-sensitive grammars represent special cases of GGs in which each rule has its right-hand side at least as long as its left-hand side. Linear-bounded automata (LBA) are special cases of TMs that cannot extend their tape. In this section, we demonstrate that these grammars and automata have the same power.

First, we define context-sensitive grammars. We also present their normal forms. Then, we define LBA and demonstrate that they are as powerful as context-sensitive grammars. Finally, we close this section by summarizing the relationships between the most important language families defined so far.

Throughout this section, we express only the basic ideas underlying the key results while leaving their rigorous verification as an exercise.

Context-sensitive grammars and their normal forms

Definition 6.22. Let $G = (V, T, P, S)$ be a GG. G is a *context-sensitive grammar* (CSG for short) if every $u \to v \in P$ satisfies

$|u| \leq |v|$. Otherwise, $V, N, T, S, \Rightarrow, \Rightarrow^*$, $F(G)$, and $L(G)$ are defined just like for GGs (see Section 6.2). ■

The name *context-sensitive* comes from their special variants, in which every rule is of the form $uAv \rightarrow uwv$, where A is a nonterminal and u, v, w are strings with $w \neq \varepsilon$. In essence, a rule like this permits the replacement of A with w provided that the replaced A occurs in the context of u and v. These special variants are equivalent to CSGs according to Definition 6.22; as a matter of fact, CSGs in Pentonnen normal form, given later in this section, satisfy this context-sensitive form even in a more restrictive way.

In the literature, CSGs are also referred to as length-increasing grammars, which reflect the property that in every CSG, $G = (V, T, P, S), x \Rightarrow^* y$ implies $|x| \leq |y|$ because each $u \rightarrow v \in P$ satisfies $|u| \leq |v|$. Consequently, if $S \Rightarrow^* w$, with $w \in L(G)$, then $1 \leq |w|$, so ε is necessarily out of $L(G)$.

Example 6.17. Consider a CSG G defined by the following eleven rules:

$$1: S \rightarrow ASc, \quad 2: S \rightarrow Z, \quad 3: Z \rightarrow BZ,$$

$$4: Z \rightarrow X, \quad 5: BX \rightarrow XB, \quad 6: AX \rightarrow XA, \quad 7: AX \rightarrow aY,$$

$$8: YA \rightarrow AY, \quad 9: YB \rightarrow BY, \quad 10: YB \rightarrow Xb, \quad 11: Yb \rightarrow bb.$$

In G, a, b, and c are terminals, while all the other symbols are nonterminals. In every derivation that generates a sentence from $L(G), G$ first applies rules 1–3 to produce $A^n B^m Z c^n$, for some $m, n \in \mathbb{N}$. By using the other rules, G verifies that $n = m + 1$ and, simultaneously, changes As and Bs to as and bs, respectively. As a result, the generated sentence is of the form $a^n b^n c^n$. For instance, from S, G derives $aabbcc$ as follows:

$$\begin{aligned} S \;&\Rightarrow\; ASc \;\Rightarrow\; AAScc \;\Rightarrow\; AAZcc \;\Rightarrow\; AABZcc \;\Rightarrow \\ &AABXcc \;\Rightarrow\; AAXBcc \;\Rightarrow\; AXABcc \;\Rightarrow\; aYABcc \;\Rightarrow \\ &aAYBcc \;\Rightarrow\; aAXbcc \;\Rightarrow\; aaYbcc \;\Rightarrow\; aabbcc. \end{aligned}$$

Consequently,

$$L(G) = \{a^n b^n c^n \mid n \geq 1\}.$$ ■

Normal forms

We have already introduced Kuroda and Pentonnen normal forms for GGs (see Definitions 6.20 and 6.21). We conclude the current section by adapting these forms for CSGs.

Definition 6.23. A CSG, $G = (V, T, P, S)$, is in Kuroda normal form if every rule $r \in P$ has one of the following three forms:

$$AB \to DC, \quad A \to BC, \quad \text{or} \quad A \to a,$$

where $A, B, C, D \in N$ and $a \in T$. If, in addition, every $AB \to DC \in P$ satisfies $A = D, G$ is in *Pentonnen normal form.* ■

Of course, CSGs in Pentonnen normal form are special cases of CSGs in Kuroda normal form. As an exercise, prove the following theorem by analogy with the proof of Theorem 6.20.

Theorem 6.23. *For every $I \in \boldsymbol{CSG}$, there exists an equivalent CSG O in Pentonnen normal form; thus, O is in Kuroda normal form too.* ■

Linear-bounded automata and their equivalence with context-sensitive grammars

In this section, we introduce LBA as special cases of TMs that cannot extend the tape. In other words, they contain no extending rule of the form $q\lhd \to p\blacksquare\lhd \in P$, where q and p are states (see Definition 6.24).

Definition 6.24. Let $M = (Q, \Delta, \Gamma, R, s, F)$ be a TM (see Definition 5.1). M is a linear-bounded automaton if $q\lhd \to p\blacksquare\lhd \notin R$, for any $q, p \in Q$. ■

We demonstrate that LBAs are as powerful as CSGs, except that LBAs can accept ε, while CSGs cannot generate ε, as already noted. In other words, LBAs characterize $\mathbf{L_{CSG}}$, as defined in the following.

Convention 6.3. LBA and **CSG** denote the set of all LBAs and CSGs, respectively. Set $\mathbf{L_{LBA}} = \{L(M) \mid M \in \mathbf{LBA}\}$ and $\mathbf{L_{CSG}} = \{L \mid L - \{\varepsilon\} = L(G) \text{ with } G \in \mathbf{CSG}\}$. ■

From context-sensitive grammars to linear-bounded automata

Turn any $I \in \mathbf{CSG}$ to $O \in \mathbf{LBA}$, so $L(I) = L(O)$.

Basic idea. Let $I = (V_I, T_I, P_I, S_I)$ be any CSG. By analogy with (II) in the proof of Lemma 6.3, turn I to an equivalent LBA, $O = (Q_O, \Delta_O, \Gamma_O, R_O, s_O, F_O)$. O is constructed in such a way that it has a three-track tape. Initially, O writes w on the first track. Then, it writes the rules of P_I on the second track (since P_I is finite, it can always place them all there, possibly in a condensed way). Finally, O writes S_I on the third track. Having initialized the three tracks in this way, O performs (1)–(3), described as follows (recall that for any rule $r \in P_I, lhs(r)$ and $rhs(r)$ denote its left- and right-hand sides, respectively):

(a) O non-deterministically selects $r \in P_I$ on the second track;
(b) O rejects and halts if there is no occurrence of $lhs(r)$ on the third track; otherwise, it selects any occurrence of $lhs(r)$ and replaces it with $rhs(r)$;
(c) if the first and third tracks coincide, O accepts and halts; otherwise, O goes to (1) and continues.

As an exercise, based on the idea sketched above, give an algorithm that transforms any $I \in \mathbf{CSG}$ into $O \in \mathbf{LBA}$ such that $L(I) = L(O)$. Then, prove the following lemma.

Lemma 6.5. *For every $I \in$ **CSG**, there is $O \in$ **LBA** such that $L(I) = L(O)$. Therefore, $\mathbf{L_{CSG}} \subseteq \mathbf{L_{LBA}}$.* ■

From linear-bounded automata to context-sensitive grammars

Basic idea. To convert any $I \in \mathbf{LBA}$ into $O \in \mathbf{CSG}$ so that $L(I) - \{\varepsilon\} = L(O)$, modify Algorithm 6.13 TM-to-GG Conversion, so it performs this conversion. That is, from $I \in \mathbf{LBA}$, construct $O \in \mathbf{CSG}$, so it simulates every accepting computation $\triangleright \blacktriangleright w\triangleleft \Rightarrow^* \triangleright\blacksquare u\triangleleft$ in I in reverse, where $w \in \Delta_I^*$ and $u \in \Gamma_I^*$. However, as opposed to the GG produced by Algorithm 6.13, O cannot erase any symbols. Therefore, in the string on the tape during the simulation of $\triangleright \blacktriangleright w\triangleleft \Rightarrow^* \triangleright\blacksquare u\triangleleft$, O incorporates $\triangleright$ and $\triangleleft$ into the leftmost and rightmost symbols, respectively, in order to detect that it performs rewriting at the very left or right tape end. Furthermore, O incorporates the current state of the simulated computation into the symbol that is rewritten. Otherwise, O works just like the GG constructed by Algorithm 6.13.

As an exercise, formulate the basic idea sketched above as an algorithm and prove the following lemma.

Lemma 6.6. *For every* $I \in \boldsymbol{LBA}$, *there is* $O \in \boldsymbol{CSG}$ *such that* $L(I) = L(O)$. *Therefore,* $\mathbf{L}_{\boldsymbol{LBA}} \subseteq \mathbf{L}_{\boldsymbol{CSG}}$. ■

Lemmas 6.5 and 6.6 imply

Theorem 6.24. $\mathbf{L}_{\boldsymbol{LBA}} = \mathbf{L}_{\boldsymbol{CSG}}$. ■

Theorem 6.25. $\mathbf{L}_{\boldsymbol{CFG}} \subset \mathbf{L}_{\boldsymbol{CSG}}$.

Proof. For every $L \in \mathbf{L}_{\mathbf{CFG}}$, there exists $G \in \mathbf{CFG}$ such that G is a proper CFG and $L(G) = L - \{\varepsilon\}$ (see Theorem 6.8). Every proper CFG is a special case of a CSG, so $\mathbf{L}_{\mathbf{CFG}} \subseteq \mathbf{L}_{\mathbf{CSG}}$. Consider the CSG G in Example 6.17. By Example 6.15, $L(G) \notin \mathbf{L}_{\mathbf{CFG}}$; therefore, $\mathbf{L}_{\mathbf{CFG}} \subset \mathbf{L}_{\mathbf{CSG}}$. □

Language Family Relationships

Until now, we have already introduced several types of language models and studied the language families that they define. We have also established fundamental relations between these language families. In the current section, we summarize these relationships.

Theorem 6.26. $\mathbf{L}_{\boldsymbol{FIN}} \subset \mathbf{L}_{\boldsymbol{FA}} = \mathbf{L}_{\boldsymbol{REG}} \subset \mathbf{L}_{\boldsymbol{dPDA}} \subset \mathbf{L}_{\boldsymbol{CFG}} = \mathbf{L}_{\boldsymbol{PDA}} \subset \mathbf{L}_{\boldsymbol{CSG}} = \mathbf{L}_{\boldsymbol{LBA}} \subset \mathbf{L}_{\boldsymbol{TM}} = \mathbf{L}_{\boldsymbol{GG}} \subset \mathbf{L}_{\boldsymbol{ALL}}$.

Proof. The equations of this theorem are established in Theorems 3.8, 6.19, and 6.22. Clearly, $\mathbf{L}_{\mathbf{FIN}} \subset \mathbf{L}_{\mathbf{FA}}$. By Theorem 4.5, $\mathbf{L}_{\mathbf{FA}} \subset \mathbf{L}_{\mathbf{dPDA}}$. By Theorems 4.4 and 6.19, $\mathbf{L}_{\mathbf{dPDA}} \subset \mathbf{L}_{\mathbf{CFG}}$. By Theorem 6.25, $\mathbf{L}_{\mathbf{CFG}} \subset \mathbf{L}_{\mathbf{CSG}}$. The proper inclusions $\mathbf{L}_{\mathbf{CSG}} \subset \mathbf{L}_{\mathbf{TM}}$ and $\mathbf{L}_{\mathbf{TM}} \subset \mathbf{L}_{\mathbf{ALL}}$ are established in the following chapter (see Theorem 7.17). □

Recall that $\mathbf{L_{FIN}}$ and $\mathbf{L_{INFIN}}$ denote the families of finite and infinite languages, respectively, and $\mathbf{L_{ALL}}$ denotes the family of all languages. As is obvious, every finite language is regular, but there are infinite regular languages. $\mathbf{L_{FA}}$ denotes the family of regular languages characterized by finite automata and regular expressions (see Section 3.1). CFG and pushdown automata characterize the family of context-free languages, $\mathbf{L_{CFG}}$ (see Section 6.1); deterministic pushdown automata define an important proper subfamily of $\mathbf{L_{CFG}}$, namely $\mathbf{L_{dPDA}}$, which properly contains $\mathbf{L_{FA}}$ (see Section 4.2). As demonstrated earlier in this section, $\mathbf{L_{CSG}}$ is generated by context-sensitive grammars, which are as powerful as LBA, except for the fact that the automata can accept ε, while the grammars cannot generate ε. Turing machines and general grammars characterize $\mathbf{L_{TM}}$, as proved in this section (see Theorem 6.22).

6.3 Making Context-Free Grammars Stronger

On the one hand, general grammars have the same power as Turing machines, but they are somewhat clumsy to use in practice. Indeed, they generate their languages according to context-dependent rules, whose application requires a verification of strict context-related conditions during this generation. On the other hand, CFG generate their languages according to context-independent productions, which significantly simplify the generation process. Unfortunately, they are much weaker than general grammars (see Theorem 6.26). These obvious advantages and disadvantages gave rise to the introduction of several modified versions of CFG whose power exceeds that of their ordinary versions. This section describes two versions of this kind: state grammars and scattered context grammars.

State grammars

The notion of a state grammar, G, represents, in essence, that of a CFG extended by a FA-like state mechanism. During a derivation step, G rewrites the leftmost possible occurrence of a nonterminal in the current state and, simultaneously, moves to another state. If in a derivation, consisting of a sequence of derivation steps, G always rewrites a nonterminal within the first n occurrences of nonterminals, the derivation is said to be n-limited, where $n >= 1$.

Definition 6.25. A *state grammar* (see Kasai, 1970) is a quintuple

$$G = (V, W, T, P, S),$$

where V is a *total alphabet*, W is a finite set of *states*, $T \subset V$ is an alphabet of *terminals*, $S \in V - T$ is the *start symbol*, and $P \subseteq (W \times (V - T)) \times (W \times V^+)$ is a finite relation. Instead of $(q, A, p, v) \in P$, we write $(q, A) \to (p, v) \in P$. For every $z \in V^*$, define

$$\text{states}_G(z) = \{q \in W \mid (q, A) \to (p, v) \in P, A \in \text{alph}(z)\}.$$

If $(q, A) \to (p, v) \in P$, $x, y \in V^*$, and $\text{states}_G(x) = \emptyset$, then G makes a *derivation step* from (q, xAy) to (p, xvy), symbolically written as

$$(q, xAy) \quad \Rightarrow \quad (p, xvy)\ [(q, A) \to (p, v)].$$

In addition, if n is a positive integer satisfying that $occur(xA, V - T) \leq n$, we say that $(q, xAy) \Rightarrow (p, xvy)\ [(q, A) \to (p, v)]$ is *n-limited*, symbolically written as

$$(q, xAy) \quad {}_n\!\Rightarrow \quad (p, xvy)\ [(q, A) \to (p, v)].$$

Whenever there is no danger of confusion, we simplify $(q, xAy) \Rightarrow (p, xvy)\ [(q, A) \to (p, v)]$ and $(q, xAy)\ {}_n\!\Rightarrow (p, xvy)\ [(q, A) \to (p, v)]$ to

$$(q, xAy) \quad \Rightarrow \quad (p, xvy)$$

and

$$(q, xAy) \quad {}_n\!\Rightarrow \quad (p, xvy),$$

respectively. In the standard manner, we extend $\Rightarrow$ *to* $\Rightarrow^m$, where $m \geq 0$; then, based on $\Rightarrow^m$, we define $\Rightarrow^+$ *and* $\Rightarrow^*$.

Let n be a positive integer, and let $\nu, \omega \in W \times V^+$. To express that every derivation step in $\nu \Rightarrow^m \omega, \nu \Rightarrow^+ \omega$, and $\nu \Rightarrow^* \omega$ is n-limited, we write $\nu\ {}_n\!\Rightarrow^m \omega, \nu\ {}_n\!\Rightarrow^+ \omega$, and $\nu\ {}_n\!\Rightarrow^* \omega$ instead of $\nu \Rightarrow^m \omega, \nu \Rightarrow^+ \omega$, and $\nu \Rightarrow^* \omega$, respectively.

By strings($\nu \ {}_n\!\Rightarrow^* \omega$), we denote the set of all strings occurring in the derivation of $\nu \ {}_n\!\Rightarrow^* \omega$. The *language* of G, denoted by $L(G)$, is defined as

$$L(G) = \{w \in T^* \mid (q, S) \Rightarrow^* (p, w), q, p \in W\}.$$

Furthermore, for every $n \geq 1$, define

$$L(G, n) = \{w \in T^* \mid (q, S) \ {}_n\!\Rightarrow^* (p, w), q, p \in W\}.$$

A derivation of the form $(q, S) \ {}_n\!\Rightarrow^* (p, w)$, where $q, p \in W$ and $w \in T^*$, represents a *successful n-limited generation* of w in G. ■

By $\mathbf{L_{ST}}$, we denote the family of languages generated by state grammars. For every $n \geq 1$, ${}_n\mathbf{L_{ST}}$ denotes the family of languages generated by n-limited state grammars. Set

$${}_\infty\mathbf{L_{ST}} = \bigcup_{n \geq 1} {}_n\mathbf{L_{ST}}.$$

The following theorem gives the key result concerning state grammars, originally established by Kasai (1970).

Theorem 6.27. $\boldsymbol{L_{CFG}} = {}_1\boldsymbol{L_{ST}} \subset {}_2\boldsymbol{L_{ST}} \subset \cdots \subset {}_\infty\boldsymbol{L_{ST}} \subset \boldsymbol{L_{ST}} = \boldsymbol{L_{CSG}}$. ■

Scattered context grammars

During a single derivation step, a scattered context grammar can simultaneously rewrite several nonterminals by context-free rules while keeping the rest of the rewritten sentential form unchanged.

Definition 6.26. A *scattered context grammar* is a quadruple

$$G = (V, T, P, S),$$

where V is a *total alphabet*, $T \subset V$ is an alphabet of *terminals*, P is a finite set of *rules* of the form

$$(A_1, \ldots, A_n) \rightarrow (x_1, \ldots, x_n),$$

where $n \geq 1$, $A_i \in V - T$, and $x_i \in V^*$, for all i, $1 \leq i \leq n$ (each rule may have different n), and $S \in V - T$ is the *start symbol*.

If

$$u = u_1 A_1 \cdots u_n A_n u_{n+1},$$
$$v = u_1 x_1 \cdots u_n x_n u_{n+1},$$

and $p = (A_1, \ldots, A_n) \to (x_1, \ldots, x_n) \in P$, where $u_i \in V^*$, for all i, $1 \leq i \leq n+1$, then G makes a *derivation step* from u to v according to p, symbolically written as

$$u \quad \Rightarrow \quad Gv\ [p]$$

or, simply, $u \Rightarrow_G v$. Set

$$\text{lhs}(p) = A_1 \cdots A_n,$$
$$\text{rhs}(p) = x_1 \cdots x_n,$$

and

$$\text{len}(p) = n.$$

If $\text{len}(p) \geq 2$, p is said to be a *context-sensitive rule*, while for $\text{len}(p) = 1$, p is said to be *context-free*. Define $\Rightarrow_G^k$, $\Rightarrow_G^*$, and $\Rightarrow_G^+$ in the standard way. The *language* of G is denoted by $L(G)$ and defined as

$$L(G) = \{w \in T^* \mid S \Rightarrow_G^* w\}.$$

A language L is a *scattered context language* if there exists a scattered context grammar G such that $L = L(G)$. ■

By $\mathbf{L_{SCG}}$, we denote the family of languages generated by scattered context grammars.

Chapter 7

A Metaphysics of Computation

According to the Church–Turing thesis, every procedure can be realized as a TM, so any computation beyond the power of these machines is also beyond the power of computers in general. Making use of this thesis together with the results about TMs obtained in Chapter 5, this chapter builds up a metaphysics of computation as a systematized body of unshakable and infallible knowledge concerning computer power and its limits. To do so, rather than dealing with TMs as language models, the current section considers them as computers of mathematical formulas, such as functions.

An Introductory Example

Let M be a Turing machine and f be a function over all positive integers, in which we allow leading zeros (a leading zero is any 0 that comes before the first non-zero digit in a digital string; for instance, James Bond's famous identifier, 007, has two leading zeros, and it denotes that same integer as 7). Let us state that M computes f if the following equivalence holds true:

$$f(x) = y \quad \text{if and only if } \triangleright \blacktriangleright x \triangleleft \Rightarrow^* \triangleright \blacksquare y \triangleleft \text{ in } M$$

(see Definition 5.1 for the meaning of all the symbols involved in this equivalence).

Specifically, consider the function $f(x) = 10x$, for all positive integers; for instance, $f(12) = 120$. A Turing machine M that computes

f can work in the following way. Starting from $\triangleright \blacktriangleright x\triangleleft$ with $x \in \{0,\ldots,9\}^+$, M moves across x to the left until $\triangleleft$ is encountered. Then, it change $\triangleleft$ to $0\triangleleft$, returns back to $\triangleright$, after which it enters $\blacksquare$.

From an utterly general standpoint, this chapter explores the limits of computer power in terms of computability and decidability. Regarding computability, it narrows its attention to TMs as computers of functions over non-negative integers and demonstrates the existence of functions whose computation cannot be specified by any procedure. As far as decidability is concerned, it considers TMs as computers of functions over non-negative integers.

7.1 Computability

In this section, we demonstrate the existence of functions whose computation cannot be specified by any procedure, so they can never be computed by any computer. As a matter of fact, the existence of these uncomputable functions immediately follows from the following counting argument. Consider the set of all functions that map $\mathbb{N}$ onto $\{0,1\}$ and the set of all procedures. While the former is uncountable, the latter is countable under our assumption that every procedure has a finite description. Thus, there necessarily exist functions with no procedures to compute them. In this section, based on the following TM-based formalization, we take a more specific look at functions whose computation can or, in contrast, cannot be specified by any procedure.

Definition 7.1. Let $M \in \mathbf{TM}$. The *function computed by M*, symbolically denoted by M-f, is defined over Δ^* as $M\text{-}f = \{(x,y) \mid x,y \in \Delta^*, \triangleright \blacktriangleright x\triangleleft \Rightarrow^* \triangleright\blacksquare yu\triangleleft \text{ in } M, u \in \{\square\}^*\}$. $\blacksquare$

Consider M-f, where $M \in \mathbf{TM}$, and an argument $x \in \Delta^*$. In a general case, M-f is partial, so M-$f(x)$ may or may not be defined. Clearly, if M-$f(x) = y$ is defined, M computes $\triangleright \blacktriangleright x\triangleleft \Rightarrow^* \triangleright\blacksquare yu\triangleleft$, where $u \in \{\square\}^*$. However, if M-$f(x)$ is undefined, M, starting from $\triangleright \blacktriangleright x\triangleleft$, never reaches a configuration of the form $\triangleright\blacksquare vu\triangleleft$, where

$v \in \Delta^*$ and $u \in \{\square\}^*$, so it either rejects x or loops on x (see Convention 5.3).

Definition 7.2. A function f is a *computable function* if there exists $M \in \mathbf{TM}$ such that $f = M\text{-}f$; otherwise, f is an uncomputable function. ■

Integer functions computed by TMs

By Definition 7.1, for every $M \in \mathbf{TM}$, $M\text{-}f$ is defined over Δ^*, where Δ is an alphabet. However, in mathematics, we usually study numeric functions defined over sets of infinitely many numbers. To use TMs to compute functions like these, we first need to represent these numbers by strings over Δ. In this introductory book, we restrict our attention only to integer functions over ${}_0\mathbb{N}$, so we need to represent every non-negative integer $i \in {}_0\mathbb{N}$ as a string over Δ. Traditionally, we represent i in unary as $unary(i)$, for all $i \in {}_0\mathbb{N}$, where unary is defined in Example 1.7. That is, $unary(j) = a^j$, for all $j \geq 0$; for instance, $unary(0)$, $unary(2)$, and $unary(999)$ are equal to ε, aa, and a^{999}, respectively. Under this representation used in the sequel, we obviously automatically assume that $\Delta = a$ simply because a is the only input symbol needed. Next, we formalize the computation of integer functions by TMs based on unary:

Definition 7.3.

(I) Let g be a function over ${}_0\mathbb{N}$ and $M \in \mathbf{TM}$. M computes g in *unary* or, more briefly, M *computes* g iff $unary(g) = M\text{-}f$.

(II) A function h over ${}_0\mathbb{N}$ is a *computable function* if there exists $M \in \mathbf{TM}$ such that M computes h; otherwise, h is an *uncomputable function.* ■

In greater detail, part (I) of Definition 7.3 says that M computes an integer function g over ${}_0\mathbb{N}$ if this equivalence holds:

$$g(x) = y \text{ iff } (unary(x), unary(y)) \in M\text{-}f, \quad \text{for all } x, y \in {}_0\mathbb{N}.$$

Convention 7.1. Whenever $M \in \mathbf{TM}$ works on an integer $x \in {}_0\mathbb{N}$, x is expressed as $unary(x)$. For brevity, whenever no confusion exists, instead of stating that M works on x represented as $unary(x)$, we just state that M *works* on x in what follows. ■

Example 7.1. Let g be the *successor function* defined as $g(i) = i + 1$, for all $i \geq 0$. Construct a TM M that computes $\triangleright \blacktriangleright a^i \triangleleft \Rightarrow^* \triangleright \blacksquare a^{i+1} \triangleleft$ so that it moves across a^i to the right bounder $\triangleleft$, replaces it with a $\triangleleft$, and returns to the left to finish its accepting computation in $\triangleright \blacksquare a^{i+1} \triangleleft$. As a result, M increases the number of *a*s by one on the tape. Thus, by Definition 7.3, M computes g. ■

Example 7.2. Let g be the total function defined as $g(i) = j$, for all ≥ 0, where j is the smallest prime satisfying $i \leq j$. Construct a TM M that tests whether i, represented by a^i, is a prime in the way described in Example 5.2. If i is prime, M accepts in the configuration $\triangleright \blacksquare a^i \triangleleft$. If not, M continues its computation from $\triangleright \blacktriangleright a^{i+1} \triangleleft$ and tests whether $i + 1$ is prime; if it is, it accepts in $\triangleright \blacksquare a^{i+1} \triangleleft$. In this way, it continues increasing the number of *a*s by one and testing whether the number is prime until it reaches a^j such that j is prime. As this prime j is obviously the smallest prime satisfying $i \leq j$, M accepts in $\triangleright \blacksquare a^j \triangleleft$. Thus, M computes g. ■

Both functions discussed in the previous two examples are total. However, there also exist partial integer functions which may be undefined for some arguments. Suppose that g is a function over ${}_0\mathbb{N}$, which is undefined for some arguments. Let $M \in$ **TM** compute g. According to Definition 7.3, for any $x \in {}_0\mathbb{N}, g(x)$ is undefined iff $(unary(x), unary(y)) \notin M\text{-}f$, for all $y \in {}_0\mathbb{N}$. The following example illustrates a partial integer function computed in this way.

Convention 7.2. As opposed to the previous two examples, the following function as well as all other functions discussed throughout the rest of this section are defined over the set of positive integers, $\mathbb{N}$, which excludes 0. ■

Example 7.3. In this example, we consider a partial function g over $\mathbb{N}$ that is defined for 1, 2, 4, 8, 16, ..., but it is undefined for the other positive integers. More precisely, $g(x) = 2x$ if $x = 2^n$, for some $n \in \mathbb{N}$; otherwise, $g(x)$ is undefined (see Table 7.1). ■

We construct $M \in$ **TM** that computes g as follows. Starting from $\triangleright \blacktriangleright a^i \triangleleft$, M computes $\triangleright \blacktriangleright a^i \triangleleft \Rightarrow^* \triangleright \blacktriangleright a^i A^j \triangleleft$, with j being the

Table 7.1. Partial function g discussed in Example 7.3.

x	$g(x)$
1	2
2	4
3	Undefined
4	8
5	Undefined
6	Undefined
7	Undefined
8	16
⋮	⋮

smallest natural number simultaneously satisfying $i \leq j$ and $j = 2^n$, with $n \in \mathbb{N}$. If $i = j$, then $i = 2^n$ and $g(i) = 2i = 2^{n+1}$, so M computes $\rhd \blacktriangleright a^i A^j \lhd \Rightarrow^* \rhd\blacksquare a^i a^j \lhd$ and, thereby, defines $g(i) = 2^{n+1}$. If $i = j$, then $2^{n-1} = i = 2^n$, and $g(i)$ is undefined, so M rejects a^i by $\rhd \blacktriangleright a^i A^j \lhd \Rightarrow^* \rhd\blacklozenge a^i \square^j \lhd$. In somewhat greater detail, we describe M by the following Pascal-like algorithm that explains how M changes its configurations.

Let $\rhd \blacktriangleright a^i \lhd$ be the input, for some $i \in \mathbb{N}$;
change $\rhd \blacktriangleright a^i \lhd$ *to*$\rhd \blacktriangleright a^i A \lhd$;
while the current configuration $\rhd \blacktriangleright a^i A^j \lhd$ satisfies $j \leq i$ **do**
  if $i = j$ **then**
    ACCEPT by computing $\rhd \blacktriangleright a^i A^j \lhd \Rightarrow^* \rhd\blacksquare a^i a^i \lhd$
    because $i = j = 2^m$ for some $m \in \mathbb{N}$
  else
    compute $\rhd \blacktriangleright a^i A^j \lhd \Rightarrow^* \rhd \blacktriangleright a^i A^2 j \lhd$ by changing each A to AA
REJECT by computing $\rhd \blacktriangleright a^i A^j \lhd \Rightarrow^* \rhd\blacklozenge a^i \square^j \lhd$
because $j >; i$, so $i \neq 2^m$ for any $m \in \mathbb{N}$.

The set of all TMs is countable. However, the set of all functions is uncountable (see Example 1.5). From this observation, the existence of uncomputable functions straightforwardly follows: There are just more functions than TMs. More surprisingly, however, even some simple total well-defined functions over $\mathbb{N}$ are uncomputable, as the following example illustrates.

Example 7.4. For every $k \in \mathbb{N}$, set

$$X_k = \{M \in \mathbf{TM} \mid |Q_M| = k+1, \Delta_M = \{a\}\}.$$

Informally, X_k denotes the set of all TMs in **TM**, with $k+1$ states, such that their languages are over $\{a\}$. Without any loss of generality, suppose that $Q_M = q_0, q_1, \ldots, q_k$, with $\blacktriangleright = q_0$ and $\blacksquare = q_k$. Let g be the total function over $\mathbb{N}$ defined for every $i \in \mathbb{N}$ so that $g(i)$ equals the greatest integer $j \in \mathbb{N}$ satisfying $\triangleright q_0 a \triangleleft \Rightarrow^* \triangleright q_i a^j u \triangleleft$ in M with $M \in X_i$, where $u \in \{\square\}^*$. In other words, $g(i) = j$ iff j is the greatest positive integer satisfying $M\text{-}f(1) = j$, where $M \in X_i$. Consequently, for every TM $K \in X_i$, either $K\text{-}f(1) \leq g(i)$ or $K\text{-}f(1)$ is undefined.

Observe that for every $i \in \mathbb{N}$, X_i is finite. Furthermore, X_i always contains $M \in \mathbf{TM}$ such that $\triangleright q_0 a \triangleleft \Rightarrow^* \triangleright q_i a^j u \triangleleft$ in M with $j \in \mathbb{N}$, so g is total. Finally, $g(i)$ is defined quite rigorously because each TM in X_i is deterministic (see Convention 5.3). At first glance, these favorable mathematical properties might suggest that g is computable, yet we next demonstrate that g is uncomputable by a proof based on diagonalization (see Example 1.5).

Basic idea. To demonstrate that g is uncomputable, we proceed, in essence, as follows. We assume that g is computable. Under this assumption, **TM** contains a TM M that computes g. We convert M to a TM N, which we subsequently transform into a TM O and demonstrate that O performs a computation that contradicts the definition of g, so our assumption that g is computable is incorrect. Thus, g is uncomputable.

In greater detail, let $M \in \mathbf{TM}$ be a TM that computes g. We can easily modify M to another TM $N \in \mathbf{TM}$ such that N computes $h(x) = g(2x) + 1$ for every $x \in \mathbb{N}$. Let $N \in X_m$, where $m \in \mathbb{N}$, so $Q_N = \{q_0, q_1, \ldots, q_m\}$ with $\blacktriangleright = q_0$ and $\blacksquare = q_m$. Modify N to the TM $O = (Q_O, \Delta_O, \Gamma_O, R_O, s_O, F_O), O \in \mathbf{TM}$, in the following way. Define q_m as a non-final state. Set $Q_O = \{q_0, q_1, \ldots, q_m, q_{m+1}, \ldots, q_{2m}\}$, with $\blacktriangleright = q_0$ and $\blacksquare = q_{2m}$, so that $O \in X_{2m}$. Initialize R_O with the rules of R_N. Then, extend R_O by the following new rules:

- $q_m a \rightarrow a q_m$ and $q_m \square \rightarrow a q_m$;
- $q_h \triangleleft \rightarrow q_{h+1} \square \triangleleft$ and $q_{h+1} \square \rightarrow a q_{h+1}$, for all $m \leq h \leq 2m-1$;
- $a q_2 m \rightarrow q_2 m a$.

Starting from $\triangleright q_0 a \triangleleft$, O first computes $\triangleright q_0 a \triangleleft \Rightarrow^* \triangleright q_m a^{h(m)} u \triangleleft$, with $u \in \{\square\}^*$, just like N does. Then, by the newly introduced rules, O computes $\triangleright q_m a^{h(m)} u \triangleleft \Rightarrow^* \triangleright q_{2m} a^{h(m)} a^{|u|} a^m \triangleleft$ with $q_{2m} = \blacksquare$. In brief, $\triangleright q_0 a \triangleleft \Rightarrow^* \triangleright q_{2m} a^{h(m)} a^{|u|} a^m \triangleleft$ in O, which is impossible, however. Indeed, $|a^{h(m)} a^{|u|} a^m| = |a^{g(2m)+1} a^{|u|} a^m| > g(2m)$, so O-$f(1) > g(2m)$, which contradicts K-$f(1) \leq g(2m)$, for all $K \in X_{2m}$, because $O \in X_{2m}$. From this contradiction, we conclude that g is uncomputable. ■

In what follows, we often consider an enumeration of **TM**. In essence, to enumerate **TM** means to list all TMs in **TM**. We can easily obtain a list like this, for instance, by enumerating their codes according to length and alphabetic order. If the code of $M \in \mathbf{TM}$ is the ith string in this lexicographic enumeration, we let M be the ith TM in the list.

Convention 7.3. In the sequel, ζ denotes some fixed enumeration of all possible TMs:

$$\zeta = M_1\text{-}f, M_2\text{-}f, \ldots.$$

Regarding ζ, we just require the existence of two algorithms: (1) an algorithm that translates every $i \in \mathbb{N}$ to M_i, and (2) an algorithm that translates every $M \in \mathbf{TM}$ to i so that $M = M_i$, where $i \in \mathbb{N}$. Let

$$\xi = M_1\text{-}f, M_2\text{-}f, \ldots.$$

That is, ξ corresponds to ζ, so ξ denotes the enumeration of the functions computed by the TMs listed in ζ. The positive integer i of M_i-f is referred to as the index of M_i-f; in terms of ζ, i is referred to as the index of M_i. ■

Throughout the rest of this chapter, we frequently discuss TMs that construct other TMs, represented by their codes, and the TMs constructed in this way may subsequently create some other machines, and so on. Let us note that a construction like this commonly occurs in real-world computer-science practice; for instance, a compiler produces a program that itself transforms the codes of some other programs, and so forth. Crucially, by means of universal TMs described in Section 5.3, we always know how to run any TM on any string, including a string that encodes another TM.

Recursion theorem

Consider any total computable function γ over $\mathbb{N}$ and apply γ to the indices of TMs in ζ (see Convention 7.3). The following theorem says that there necessarily exists $n \in \mathbb{N}$, customarily referred to as a fixed point of γ, such that M_n and $M_{\gamma(n)}$ compute the same function, i.e., in terms of ξ, $M_n\text{-}f = M_{\gamma(n)}\text{-}f$. As a result, this important theorem rules out the existence of a total computable function that would map each index i onto another index j, so $M_i\text{-}f \neq M_j\text{-}f$.

Theorem 7.1. Recursion Theorem. *For every total computable function γ over $\mathbb{N}$, there is $n \in \mathbb{N}$ such that $M_n\text{-}f = M_{\gamma(n)}\text{-}f$ in ξ.*

Proof. Let γ be any total computable function over $\mathbb{N}$, and let $X \in$ **TM** compute γ, i.e., $X\text{-}f = \gamma$. First, for each $i \in \mathbb{N}$, introduce a TM $N_i \in$ **TM** that works on every input $x \in \mathbb{N}$ as follows:

- N_i saves x;
- N_i runs M_i on i (according to Convention 5.5, M_i denotes the TM of index i in ζ);
- if $M_i\text{-}f(i)$ is defined; therefore, M_i actually computes $M_i\text{-}f(i)$, then N_i runs X on $M_i\text{-}f(i)$ to compute $X\text{-}f(M_i\text{-}f(i))$;
- N_i runs $M_{X\text{-}f(M_i\text{-}f(i))}$ on x to compute $M_{X\text{-}f(M_i\text{-}f(i))}\text{-}f(x)$.

Let O be a TM in ζ that computes the function $O\text{-}f$ over $\mathbb{N}$ such that for each $i \in \mathbb{N}, O\text{-}f(i)$ is equal to the index of N_i in ζ, constructed above. Note that although $M_i\text{-}f(i)$ may be undefined in (3), $O\text{-}f$ is total because N_i is defined for all $i \in \mathbb{N}$. Furthermore, $M_{O\text{-}f(i)}\text{-}f = M_{X\text{-}f(M_i\text{-}f(i))}\text{-}f$ because $O\text{-}f(i)$ is the index of N_i in ζ, and N_i computes $M_{X\text{-}f(M_i\text{-}f(i))}\text{-}f$. As $X\text{-}f = \gamma$, we have $M_{X\text{-}f(M_i\text{-}f(i))}\text{-}f = M_{\gamma(M_i\text{-}f(i))}\text{-}f$. Let $O = M_k$ in ζ, where $k \in \mathbb{N}$; in other words, k is the index of O. Set $n = O\text{-}f(k)$ to obtain $M_n\text{-}f = M_{O\text{-}f(k)}\text{-}f = M_{X\text{-}f(M_k\text{-}f(k))}\text{-}f = M_{\gamma(M_k\text{-}f(k))}\text{-}f = M_{\gamma(O\text{-}f(k))}\text{-}f = M_{\gamma(n)}\text{-}f$. Thus, n is a fixed point of γ, and Theorem 7.1 holds true. □

The recursion theorem is a powerful tool frequently applied in the theory of computation, as illustrated in the following.

Example 7.5. Consider the enumeration $\zeta = M_1, M_2, \ldots$ (see Convention 7.3). Observe that Theorem 7.1 implies the existence of $n \in \mathbb{N}$ such that $M_n\text{-}f = M_{n+1}\text{-}f$, meaning M_n and M_{n+1} compute the same function. Indeed, define the total computable function γ for each $i \in \mathbb{N}$ as $\gamma(i) = i + 1$. By Theorem 7.1, there is $n \in \mathbb{N}$ such that M_n-$f = M_{\gamma(n)}$-f in ξ, and by the definition of γ, $\gamma(n) = n+1$. Thus, M_n-$f = M_{n+1}$-f. ∎

From a broader perspective, this result holds in terms of any enumeration of **TM**, which may differ from ζ, provided that it satisfies the simple requirements stated in Convention 7.3. That is, the enumeration can be based on any representation whatsoever, provided that there exists an algorithm that translates each representation to the corresponding machine in **TM**, and vice versa. As an exercise, consider an alternative enumeration of this kind, and prove that it necessarily contains two consecutive TMs that compute the same function. To rephrase this generalized result in terms of the Turing–Church thesis, any enumeration of procedures contains two consecutive procedures that compute the same function.

Before closing this section, we generalize functions in such a way that they map multiple arguments onto a set, and we briefly discuss their computation by TMs. For k elements, $a_1, \ldots, a_k$, where $k \in \mathbb{N}$, $(a_1, \ldots, a_k)$ denotes the ordered k-tuple consisting of a_1 through a_k in this order. Let $A_1, \ldots, A_k$ be k sets. The *Cartesian product* of $A_1, \ldots, A_k$ is denoted by $A_1 \times \cdots \times A_k$ and defined as

$$A_1 \times \cdots \times A_k = \{(a_1, \ldots, a_k) \mid a_i \in A_i, 1 \leq i \leq k\}.$$

Let $m \in \mathbb{N}$ and B be a set. Roughly, an *m-argument function* from $A_1 \times \cdots \times A_m$ to B maps each $(a_1, \ldots, a_m) \in A_1 \times \cdots \times A_m$ onto no more than one $b \in B$. To express that a function f represents an m-argument function, we write $f^{\underline{m}}$ (carefully distinguish $f^{\underline{m}}$ from f^m, which denotes the m-fold product of f, defined in the conclusion of Section 1.3). If $f^{\underline{m}}$ maps $(a_1, \ldots, a_m) \in A_1 \times \cdots \times A_m$ onto $b \in B$, then $f^{\underline{m}}(a_1, \ldots, a_m)$ is defined as b, written as $f^{\underline{m}}(a_1, \ldots, a_m) = b$, where b is the value of $f^{\underline{m}}$ for arguments $a_1, \ldots, a_m$. If $f^{\underline{m}}$ maps $(a_1, \ldots, a_m)$ onto no member of B, $f^{\underline{m}}(a_1, \ldots, a_m)$ is undefined. If $f^{\underline{m}}(a_1, \ldots, a_m)$ is defined for all $(a_1, \ldots, a_m) \in A_1 \times \cdots \times A_m$, $f^{\underline{m}}$ is total. If we want to emphasize that $f^{\underline{m}}$ may not be total, we say that $f^{\underline{m}}$ is partial.

Next, we generalize Definition 7.1 to the m-argument function M-$f^{\underline{m}}$ computed by $M \in$ **TM**. For the sake of this generalization, we assume that Δ contains #, used in the following definition to separate the m arguments of $M\text{-}f^{\underline{m}}$.

Definition 7.4. Let $M \in \mathbf{TM}$. The m-argument function computed by M is denoted by $M\text{-}f^{\underline{m}}$ and defined as

$$M\text{-}f^{\underline{m}} = \{(x, y) \mid x \in \Delta^*, occur(x, \#) = m - 1, \\ y \in (\Delta - \{\#\})^*, \rhd \blacktriangleright x \lhd \quad \Rightarrow^* \quad \rhd \blacksquare yu \lhd \text{ in } M, u \in \{\square\}^*\}.$$

■

That is, $f^{\underline{m}}(x_1, x_2, \ldots, x_m) = y$ iff $\rhd \blacktriangleright x_1\#x_2\# \ldots \#x_m\lhd \Rightarrow^* \rhd\blacksquare yu\lhd$ in M, with $u \in \{\square\}^*$, and $f^{\underline{m}}(x_1, x_2, \ldots, x_m)$ is undefined iff M loops on $x_1\#x_2\# \ldots \#x_m$ or rejects $x_1\#x_2\# \ldots \#x_m$. Note that $M\text{-}f^1$ coincides with $M\text{-}f$ (see Definition 7.1).

According to Definition 7.4, for every $M \in \mathbf{TM}$ and every $m \in \mathbb{N}$, there exists $M\text{-}f^{\underline{m}}$. At a glance, it is hardly credible that every $M \in \mathbf{TM}$ defines $M\text{-}f^{\underline{m}}$ because **TM** obviously contains TMs that never perform a computation that defines any member of $M\text{-}f^{\underline{m}}$. However, if we realize that we might have $M\text{-}f^{\underline{m}}$ completely undefined, i.e., $M\text{-}f^{\underline{m}} = \emptyset$, which is perfectly legal from a mathematical point of view, then the existence of $M\text{-}f^{\underline{m}}$ corresponding to every $M \in \mathbf{TM}$ comes as no surprise.

Definition 7.5. Let $m \in \mathbb{N}$. A function $f^{\underline{m}}$ is a *computable function* if there exists $M \in \mathbf{TM}$ such that $f^{\underline{m}} = M\text{-}f^{\underline{m}}$; otherwise, $f^{\underline{m}}$ is *uncomputable*. ■

To use TMs as computers of m-argument integer functions, we automatically assume that TMs work with the unary-based representation of integers by analogy with one-argument integer functions computed by TMs (see Definition 7.3 and Convention 7.1).

Definition 7.6. Let $M \in \mathbf{TM}$, $m \in \mathbb{N}$, and $f^{\underline{m}}$ be an m-argument function from $A_1 \times \cdots \times A_m$ to $\mathbb{N}$, where $A_i = \mathbb{N}$, for all $1 \leq i \leq m$. M computes $f^{\underline{m}}$ iff the following equivalence holds:

$$f^{\underline{m}}(x_1, \ldots, x_m) \\ = y \text{ iff } (unary(x_1)\# \cdots \#unary(x_m), unary(y)) \in M\text{-}f^{\underline{m}}.$$

■

Kleene's *s-m-n* Theorem

The following theorem says that for all $m, n \in \mathbb{N}$, there is a total computable function s of $m+1$ arguments such that $M_i\text{-}f^{\underline{m+n}}(x_1, \ldots, x_m, y_1, \ldots, y_n) = M_{s(i,x_1,\ldots,x_m)}\text{-}f^{\underline{n}}(y_1, \ldots, y_n)$, for all $i, x_1, \ldots, x_m, y_1, \ldots, y_n$. In other words, considering the Turing–Church thesis, there is an algorithm such that from M_i and $x_1, \ldots, x_m$, it determines another TM that computes $M_i\text{-}f^{\underline{m+n}}(x_1, \ldots, x_m, y_1, \ldots, y_n)$, with only n arguments $y_1, \ldots, y_n$. In this way, the number of arguments is lowered, yet the same function is computed.

Convention 7.4. In this chapter, we often construct $M \in \mathbf{TM}$ from a finite sequence of strings, $z_1, \ldots, z_n$ (see, for instance, Theorem 7.2 and Example 7.6), and in order to express this clearly and explicitly, we denote M constructed in this way by $M_{[z_1,\ldots,z_n]}$. Specifically, in the proof of the following theorem, $M_{[i,x_1,\ldots,x_m]}$ is constructed from $i, x_1, \ldots, x_m$, which are unary strings representing integers (see Convention 7.1). ■

Theorem 7.2. *Kleene's s-m-n Theorem. For all $i, m, n \in \mathbb{N}$, there is a total computable $(m+1)$-argument function $s^{\underline{m+1}}$ such that*

$$M_i\text{-}f^{\underline{m+n}}(x_1, \ldots, x_m, y_1, \ldots, y_n) = M_{s^{\underline{m+1}}(i,x_1,\ldots,x_m)}\text{-}f^{\underline{n}}(y_1, \ldots, y_n).$$

Proof. We first construct a TM $S \in \mathbf{TM}$. Then, we demonstrate that $S\text{-}f^{\underline{m+1}}$ satisfies the properties of $s^{\underline{m+1}}$ stated in Theorem 7.2, so we just take $s^{\underline{m+1}} = S\text{-}f^{\underline{m+1}}$ to complete the proof.

Construction of S. Let $m, n \in \mathbb{N}$. We construct a TM $S \in \mathbf{TM}$ so that S itself constructs another machine in **TM** and produces its index in ζ as the resulting output value. More precisely, given input $i\#x_1\#\ldots\#x_m$, S constructs a TM, denoted by $M_{[i,x_1,\ldots,x_m]}$, for $i = 1, 2, \ldots$, and produces the index of $M_{[i,x_1,\ldots,x_m]}$, i.e., j satisfying $M_{[i,x_1,\ldots,x_m]} = M_j$ in ζ, as the resulting output value. $M_{[i,x_1,\ldots,x_m]}$ constructed by S works as follows:

(1) when given input $y_1\#\ldots\#y_n$, $M_{[i,x_1,\ldots,x_m]}$ shifts $y_1\#\ldots\#y_n$ to the right, writes $x_1\#\ldots\#x_m$ to its left, so it actually changes $y_1\#\ldots\#y_n$ to $x_1\#\ldots\#x_m\#y_1\#\ldots\#y_n$;
(2) $M_{[i,x_1,\ldots,x_m]}$ runs M_i on $x_1\#\ldots\#x_m\#y_1\#\ldots\#y_n$.

Properties of $S\text{-}\underline{f^{(m+1)}}$. Consider the $(m+1)$-argument function $S\text{-}\underline{f^{(m+1)}}$ computed by S constructed above. Recall that $S\text{-}\underline{f^{(m+1)}}$ maps $(i, x_1, \ldots, x_m)$ to the resulting output value equal to the index of $M_{[i,x_1,\ldots,x_m]}$ in ζ. More briefly, $S\text{-}\underline{f^{(m+1)}}(i, x_1, \ldots, x_m) = j$, with j satisfying $M_{[i,x_1,\ldots,x_m]} = M_j$ in ζ. Observe that $M_{[i,x_1,\ldots,x_m]}$ computes $M_i\text{-}\underline{f^{m+n}}(x_1, \ldots, x_m, y_1, \ldots, y_n)$ on every input $(y_1, \ldots, y_n)$, where $M_i\text{-}\underline{f^{m+n}}$ denotes the $(m+n)$-argument computable function. By these properties,

$$M_i\text{-}\underline{f^{m+n}}(x_1, \ldots, x_m, y_1, \ldots, y_n) = M_j\text{-}f^n(y_1, \ldots, y_n)$$
$$= M_{S\text{-}\underline{f^{m+1}}(i,x_1,\ldots,x_m)}\text{-}\underline{f^n}(y_1, \ldots, y_n).$$

Therefore, to obtain the total computable $(m+1)$-argument function $\underline{s^{m+1}}$ satisfying Theorem 7.2, set $\underline{s^{m+1}} = S\text{-}\underline{f^{m+1}}$. □

Theorem 7.2 represents a powerful tool for demonstrating closure properties concerning computable functions. To illustrate, by using this theorem, we prove that the set of computable one-argument functions is closed with respect to composition in the following example.

Example 7.6. There is a total computable 2-argument function g^2 such that $M_i\text{-}f(M_j\text{-}f(x)) = M_{g^{\underline{2}}(i,j)}\text{-}f(x)$, for all $i, j, x \in \mathbb{N}$. We define the 3-argument function h^3 as $h^3(i, j, x) = M_i\text{-}f(M_j\text{-}f(x))$, for all $i, j, x \in \mathbb{N}$. First, we demonstrate that h^3 is computable. Given $i, j, x \in \mathbb{N}$, we introduce a TM H that computes h^3 so that it works on every input x as follows:

(1) H runs M_j on x;
(2) if $M_j\text{-}f(x)$ is defined and, therefore, produced by H in (1), H runs M_i on $M_j\text{-}f(x)$;
(3) if $M_i\text{-}f(M_j\text{-}f(x))$ is defined, H produces $M_i\text{-}f(M_j\text{-}f(x))$, so H computes $M_i\text{-}f(M_j\text{-}f(x))$.

Thus, $h^{\underline{3}}$ is computable. Let $h^{\underline{3}}$ be computed by M_k in ζ. That is, $M_k\text{-}f^{\underline{3}} = h^{\underline{3}}$. By Theorem 7.2, there is a total computable function s such that $M_{s^{\underline{3}}(k,i,j)}\text{-}f(x) = M_k\text{-}f^{\underline{3}}(i, j, x)$, for all $i, j, x \in \mathbb{N}$. Set $g^{\underline{2}}(i, j) = s^{\underline{3}}(k, i, j)$, for all $i, j \in \mathbb{N}$ (of course, k has the same meaning as above). Thus, $M_i\text{-}f(M_j\text{-}f(x)) = M_{s^{\underline{3}}(k,i,j)}\text{-}f(x) = M_{g^{\underline{2}}(i,j)}\text{-}f(x)$, for all $i, j, x \in \mathbb{N}$.

As already noted, from a broader perspective, we have actually proved that the composition of two computable functions is again computable, so the set of computable one-argument functions is closed with respect to composition. Establishing more closure properties concerning other common operations, such as addition and product, is left to the reader. ■

Most topics concerning the computability of multi-argument functions are far beyond this introductory text. Therefore, we narrow our attention to one-argument functions. In fact, we just consider total functions from Δ^* to $\{\varepsilon\}$, upon which we base the following section, which discusses another crucially important topic of computation theory — decidability.

7.2 Decidability

In this section, we formally explore the power of algorithms that decide problems. Since problem-deciding algorithms are used in most computer science areas, it is unfeasible to examine them all. Therefore, we only consider the algorithmic decidability concerning problems related to the language models, such as automata and grammars, discussed earlier in this book.

Turing deciders

In essence, we express every problem by a language in this book. More specifically, a problem P is associated with the set of all its instances Π and with a property π that each instance either satisfies or, in contrast, does not satisfy. Given a particular instance $i \in \Pi$, P asks whether or not i satisfies π. To decide P by means of Turing deciders, which work on strings like any TMs, we represent P by an encoding language as

$$_PL = \{\langle i \rangle \mid i \in \Pi, i \text{ satisfies } \pi\},$$

where $\langle i \rangle$ is a string representing instance i (see Convention 5.4). A Turing decider M, which halts on all inputs, decides P if: (1) M rejects every input that represents no instance from Π; and (2) for every $\langle i \rangle$ with $i \in \Pi$, M accepts $\langle i \rangle$ iff i satisfies π, so M rejects $\langle i \rangle$ iff

i does not satisfy π. More formally, $L(M) = {}_{P}L$, and $\Delta^* - L(M) = (\Delta^* - \{\langle i \rangle \mid i \in \Pi\}) \cup \{\langle i \rangle \mid i \in \Pi, i \textit{ does not satisfy } \pi\}$.

In brief, we state P as follows.

Problem 7.1. *P*.
Question: a formulation of P.
Language: ${}_{P}L$.

To illustrate our approach to decidability, we consider the problem referred to as *FA-Emptiness*. For any FA M, *FA-Emptiness* asks whether the language accepted by M is empty. **FA** is thus the set of its instances. The language encoding *FA-Emptiness* is defined as

$${}_{FA\text{-}Emptiness}L = \{\langle M \rangle \mid M \in \mathbf{FA}, L(M) = \emptyset\}.$$

Formally, *FA-Emptiness* is specified as follows.

Problem 7.2. *FA-Emptiness*.
Question: Let $M \in \mathbf{FA}$. Is $L(M)$ empty?
Language: ${}_{FA\text{-}Emptiness}L = \{\langle M \rangle \mid M \in \mathbf{FA}, L(M) = \emptyset\}$.

We can construct a Turing decider for ${}_{FA\text{-}Emptiness}L$ in a trivial way, as demonstrated shortly (see Theorem 7.5).

In general, a problem that can be decided by a Turing decider is referred to as a decidable problem, while an undecidable problem cannot be decided by any Turing decider. For instance, the above problem of emptiness reformulated in terms of TMs, symbolically referred to as *TM-Emptiness*, is undecidable, as we demonstrate later in this chapter. That is, we encode this important undecidable problem by its encoding language

$${}_{TM\text{-}Emptiness}L = \{\langle M \rangle \mid M \in \mathbf{TM}, L(M) = \emptyset\}$$

and prove that no Turing decider accepts ${}_{TM\text{-}Emptiness}L$ (see Theorem 7.21).

Next, we define Turing deciders rigorously. As already pointed out, any Turing decider M halts on every input string. In addition, we require that M always halt with its tape completely blank. The following definition makes use of M-f (see Definition 7.1) and the domain of a function (see Section 1.3), so recall both notions before reading it.

Definition 7.7.

(I) Let $M \in \mathbf{TM}$. If M always halts and M-f is a function from Δ^* to $\{\varepsilon\}$, then M is a Turing decider (TD for short).

(II) Let L be a language and M be a TD. M is a TD for L if $domain(M\text{-}f) = L$.

(III) A language is *decidable* if there is a TD for it; otherwise, the language is *undecidable*. $\blacksquare$

By (I) in Definition 7.7, $M \in \mathbf{TM}$ is a TD if it never loops, and for every $x \in \Delta^*, \rhd \blacktriangleright x\lhd \Rightarrow^* \rhd iu\lhd$ in M, with $i \in \{\blacksquare, \blacklozenge\}$ and $u \in \{\square\}^*$. By (II), a TD M for a language L satisfies $\rhd \blacktriangleright x\lhd \Rightarrow^* \rhd\blacksquare u\lhd$ in M, for every $x \in L$ and $\rhd \blacktriangleright y\lhd \Rightarrow^* \rhd\blacklozenge v\lhd$ in M, for every $y \in \Delta_M^* - L$, where $u, v \in \{\square\}^*$.

Convention 7.5. **TD** denotes the set of all TDs, and $\mathbf{L_{TD}} = \{L(M) \mid M \in \mathbf{TD}\}$; in other words, $\mathbf{L_{TD}}$ denotes the family of all decidable languages. $\blacksquare$

Example 7.7. Return to Example 5.1, in which we have designed a TM M that accepts $L = \{x \mid x \in a, b, c^*, occur(x, a) = occur(x, b) = occur(x, c)\}$. As is obvious, M does not satisfy Convention 5.3; in fact, it is not even deterministic. As a result, it is out of **TM**, so it is definitely out of **TD** as well.

In the current example, we design another TM D such that $D \in \mathbf{TD}$ and D accepts L. D repeatedly scans across the tape in a left-to-right way, erasing the leftmost occurrences of a, b, and c during every single scan. When it reaches $\lhd$ after erasing all these three occurrences, it moves left to $\rhd$ and makes another scan like this. However, when D reaches $\lhd$ while some of the three symbols is missing on the tape, it can decide whether the input string is accepted. Indeed, if all three symbols are missing, D accepts; otherwise, it rejects. Therefore, D performs its final return to $\rhd$ in either of the following two ways:

- If the tape is completely blank and, therefore, all as, bs, and cs have been erased during the previous scans, D moves its head left to $\rhd$ and accepts in a configuration of the form $\rhd\blacksquare\square\cdots\square\lhd$.
- If the tape is not blank and, therefore, contains some occurrences of symbols from X, where $\emptyset \subset X \subset a, b, c$, then during its return to $\rhd$,

Table 7.2. Acceptance of *babcca*.

Scan	Tape
0	*babcca*
1	$\Box\Box b\Box ca$
2	$\Box\Box\Box\Box\Box\Box$
ACCEPT	

D changes all these occurrences to $\Box$ and rejects in a configuration of the form $\rhd\blacklozenge\Box\ldots\Box\lhd$.

Omitting the state specification, Table 7.2 schematically describes the acceptance of *babcca* by D. Clearly, D is a TD for L, so L is a decidable language. Symbolically and briefly, $D \in \mathbf{TD}$ and $L \in \mathbf{L_{TD}}$. ■

As $\mathbf{L_{TD}}$ represents another important language family, we revisit the hierarchy from Theorem 6.26 and demonstrate that $\mathbf{L_{CSG}} \subset \mathbf{L_{TD}}$. That is, we first prove that every language in $\mathbf{L_{CSG}}$ is decidable, so $\mathbf{L_{CSG}} \subseteq \mathbf{L_{TD}}$. Then, we demonstrate that $\mathbf{L_{TD}} - \mathbf{L_{CSG}} \neq \emptyset$; consequently, $\mathbf{L_{CSG}} \subset \mathbf{L_{TD}}$.

To prove that every language in $\mathbf{L_{CSG}}$ belongs to $\mathbf{L_{TD}}$, we next give an algorithm that decides whether or not $w \in L(G)$, for every $G \in \mathbf{CSG}$, so $\mathbf{L_{CSG}} \subseteq \mathbf{L_{TD}}$.

Convention 7.6. We suppose there exists a fixed encoding and decoding of CSGs obtained by analogy with obtaining the encoding and decoding of the members of $\mathbf{TM} \times T^*$ (see Convention 5.4). We denote the codes of $(G, w) \in \mathbf{CSG} \times T^*$ by $\langle G, w\rangle$. ■

Problem 7.3. *CSG-Membership*.
Question: Let $G = (V, T, P, S)$ be a CSG and $w \in T_G^*$. Is w a member of $L(G)$?
Language: ${}_{CSG\text{-}Membership}L = \{\langle G, w\rangle \mid G \in \mathbf{CSG}, w \in T_G^*, w \in L(G)\}$.

Basic idea. We sketch an algorithm that decides Problem 7.3 *CSG-Membership* based upon the property that any CSG generates every sentence by a derivation that contains no sentential form longer than the sentence. That is, given $\langle G, w\rangle$, where $G = (V, T, P, S)$ is a

CSG and $w \in T^*$, we consider the set of all derivations of the form $S \Rightarrow^* x$, where $x \in V^*$, such that $|x| \leq |w|$ and $S \Rightarrow^* x$ does not contain two identical sentential forms. As the length of any string cannot decrease at any step of the derivations satisfying the properties stated above, we can obviously design a Turing machine D that constructs this set of derivations in such a way that it does not loop forever. If during this construction, D produces a derivation that generates w, it accepts $\langle G, w\rangle$ and halts. If D completes this construction without ever producing a derivation that generates w, it rejects $\langle G, w\rangle$ and halts (note that $\langle G, \varepsilon\rangle$ is always rejected). Therefore, D represents a Turing decider, so ${}_{CSG\text{-}Membership}L \in \mathbf{L_{TD}}$.

As an exercise, formalize the design of D and prove the following theorem.

Theorem 7.3. ${}_{CSG\text{-}Membership}L \in \boldsymbol{L_{TD}}$. *Therefore,* $\boldsymbol{L_{CSG}} \subseteq \boldsymbol{L_{TD}}$. ■

Some decidable languages, however, are out of $\mathbf{L_{CSG}}$, as demonstrated in the following by using diagonalization (see Example 1.5).

Lemma 7.1. $\boldsymbol{L_{TD}} - \boldsymbol{L_{CSG}} \neq \emptyset$.

Proof. Consider a fixed enumeration of all possible CSGs:

$$G_1, G_2, \ldots,$$

(this enumeration can be obtained similarly to obtaining the enumeration of all TMs, in Convention 7.3). Let

$$x_1, x_2, \ldots$$

be all the strings in $\{0,1\}^+$ listed in a canonical order, such as the lexicographic order, i.e., $0, 1, 00, 01, \ldots$. Consider the binary language

$$L = \{x_i \mid x_i \notin L(G_i), i \in \mathbb{N}\}.$$

Given $x \in \{0,1\}^+$, we can obviously find the x_i satisfying $x_i = x$ in $x_1, x_2, \ldots$, and by Theorem 7.3, we can decide whether $x \in L(G_i)$, so $L \in \mathbf{L_{TD}}$.

By contradiction, we prove that $L \notin \mathbf{L_{CSG}}$. Suppose that $L \in \mathbf{L_{CSG}}$. Thus, there is G_j in $G_1, G_2, \ldots$ such that $L = L(G_j)$. Consider x_j. If $x_j \in L$, then the definition of L implies $x_j \notin L(G_j)$, and if

$x_j \notin L$, then $x_j \in L(G_j)$. This statement contradicts $L = L(G_j)$, so $L \notin \mathbf{L_{CSG}}$.

Therefore, $L \in \mathbf{L_{TD}} - \mathbf{L_{CSG}}$, and Lemma 7.1 holds true. □

Theorem 7.4. $\boldsymbol{L_{CSG} \subset L_{TD}}$.

Proof. By Theorem 7.3 and Lemma 7.1, $\mathbf{L_{CSG}} \subset \mathbf{L_{TD}}$. □

Decidable problems for finite automata

Let M be any FA (see Definition 3.1). We give algorithms for deciding the following three problems:

- Is the language accepted by M empty?
- Is the language accepted by M finite?
- Is the language accepted by M infinite?

In addition, for any input word w, we decide the following problem.

- Is w a member of the language accepted by M?

Strictly, we decide all these four problems for CSDFAs (as stated in Section 3.3, a CSDFA stands for a completely specified deterministic finite automaton) because deciding them in this way is somewhat simpler than deciding these problems for the general versions of FAs, contained in **FA** (see Definitions 3.1 and 3.6). However, since any FA can be algorithmically converted into an equivalent CSDFA, all four problems are decidable for the general versions of FAs in **FA** too.

Convention 7.7. CSDFA denotes the set of all CSDFAs. We suppose there exist a fixed encoding and decoding of automata in **CSDFA** by analogy with the encoding and decoding of TMs (see Convention 5.4). That is, $\langle M \rangle$ represents the code of $M \in$ **CSDFA**. Similarly, we suppose there exist an analogical encoding and decoding of the members of **CSDFA** $\times \Delta^*$ and **CSDFA** $\times$ **CSDFA**. For brevity, we denote the codes of $(M, w) \in$ **CSDFA** $\times \Delta^*$ and $(M, N) \in$ **CSDFA** $\times$ **CSDFA** by $\langle M, w \rangle$ and $\langle M, N \rangle$, respectively. ■

As already stated at the beginning of this section, the *FA-Emptiness* problem asks whether the language accepted by an FA is empty.

Next, we prove that $FA\text{-}Emptiness$ is decidable by demonstrating that its encoding language ${}_{FA\text{-}Emptiness}L$ belongs to $\mathbf{L_{TD}}$.

Problem 7.4. $FA\text{-}Emptiness$.
Question: Let $M \in \mathbf{CSDFA}$. Is $L(M)$ empty?
Language: ${}_{FA\text{-}Emptiness}L = \{\langle M \rangle \mid M \in \mathbf{CSDFA}, L(M) = \emptyset\}$.

Theorem 7.5. ${}_{FA\text{-}Emptiness}L \in \boldsymbol{L_{TD}}$.

Proof. As M is a $CSDFA$, each of its states is reachable (see Definition 3.6). Thus, $L(M) = \emptyset$ iff $F_M = \emptyset$, which says that M has no final state. Design a TD D that works on every $\langle M \rangle$, where $M \in \mathbf{CSDFA}$, so D accepts $\langle M \rangle$ iff $F_M = \emptyset$, and D rejects $\langle M \rangle$ iff $F_M \neq \emptyset$. □

The $FA\text{-}Membership$ problem asks whether a string $w \in \Delta_M^*$ is a member of the language accepted by $M \in \mathbf{CSDFA}$. Like $FA\text{-}Emptiness$, $FA\text{-}Membership$ is decidable.

Problem 7.5. $FA\text{-}Membership$.
Question: Let $M \in \mathbf{CSDFA}$ and $w \in \Delta_M^*$. Is w a member of $L(M)$?
Language: ${}_{FA\text{-}Membership}L = \{\langle M, w \rangle \mid M \in \mathbf{CSDFA}, w \in \Delta_M^*, w \in L(M)\}$.

Theorem 7.6. ${}_{FA\text{-}Membership}L \in \boldsymbol{L_{TD}}$.

Proof. Recall that any $M \in \mathbf{CSDFA}$ reads an input symbol during every move. Thus, after making precisely $|w|$ moves on $w \in \Delta^*$, M either accepts or rejects w. Therefore, construct a TD D that works on every $\langle M, w \rangle$ as follows:

(1) D runs M on w until M accepts or rejects w (after $|w|$ moves);
(2) D accepts $\langle M, w \rangle$ iff M accepts w, and D rejects $\langle M, w \rangle$ iff M rejects w.

□

$FA\text{-}Infiniteness$ is a problem that asks whether the language accepted by $M \in \mathbf{CSDFA}$ is infinite. To demonstrate that the decidability of the same problem can often be proved in several different ways, we next give two alternative proofs that $FA\text{-}Infiniteness$ is decidable. We only sketch the first proof while describing the other in detail.

Problem 7.6. $FA\text{-}Infiniteness$.
Question: Let $M \in \mathbf{CSDFA}$. Is $L(M)$ infinite?
Language: ${}_{FA\text{-}Infiniteness}L = \{\langle M\rangle \mid M \in \mathbf{CSDFA}, L(M) \text{ is infinite}\}$.

Under our assumption that M is from **CSDFA**, we obviously see that $L(M)$ is infinite iff its state diagram contains a cycle, so we can easily reformulate and decide this problem in terms of graph theory in this way. Alternatively, we can prove this by using the pumping lemma for regular languages (see Lemma 3.8) in the following way. For every $M \in \mathbf{CSDFA}$, let ${}_{\infty?}L(M)$ denote the finite language

$${}_{\infty?}L(M) = \{x \mid x \in L(M), |Q_M| \leq |x| < 2|Q_M|\} \subseteq L(M).$$

Lemma 7.2. *For every $M \in$ **CSDFA**, $L(M)$ is infinite iff ${}_{\infty?}L(M) \neq \emptyset$.*

Proof. To prove the if part of the equivalence, suppose that ${}_{\infty?}L(M) \neq \emptyset$. Take any $z \in {}_{\infty?}L(M)$. Recall that the pumping lemma constant k equals $|Q_M|$ in the proof of Lemma 3.8. As $|Q_M| \leq |z|$ by the definition of ${}_{\infty?}L(M)$, Lemma 3.8 implies that $z = uvw$, where $0 < |v| \leq |uv| \leq |Q_M|$, and most importantly, $uv^m w \in L$, for all $m \geq 0$. Hence, $L(M)$ is infinite.

To prove the only if part, assume that L is infinite. Let z be the shortest string such that $z \in L(M)$ and $|z| \geq 2|Q_M|$. As $|z| \geq 2|Q_M| \geq |Q_M|$, Lemma 3.8 implies that $z = uvw$, where $0 < |v| \leq |uv| \leq |Q_M|$, and $uv^m w \in L$, for all $m \geq 0$. Take $uv^0 w = uw \in L(M)$. Observe that $uw \in L(M)$ and $0 < |v|$ imply $2|Q_M| > |uw|$; indeed, if $2|Q_M| \leq |uw| = |z|$, then z would not be the shortest string satisfying $z \in L(M)$ and $|z| \geq 2|Q_M|$ — a contradiction. As $0 < |v| \leq |Q_M|, |Q_M| \leq |uw| = 2|Q_M| \leq |z|$, so $uw \in {}_{\infty?}L(M)$; therefore, ${}_{\infty?}L(M) \neq \emptyset$. □

Theorem 7.7. ${}_{FA\text{-}Infiniteness}L \in \boldsymbol{L_{TD}}$.

Proof. Construct a TD D that works on every $\langle M\rangle \in {}_{FA\text{-}Infiniteness}L$, so it first constructs ${}_{\infty?}L(M)$. After the construction of this finite language, D accepts $\langle M\rangle$ iff ${}_{\infty?}L(M) \neq \emptyset$, and D rejects $\langle M\rangle$ iff ${}_{\infty?}L(M) = \emptyset$. □

Consequently, the following problem is decidable as well.

Problem 7.7. *FA-Finiteness.*
Question: Let $M \in \mathbf{CSDFA}$. Is $L(M)$ finite?
Language: ${}_{FA\text{-}Finiteness}L = \{\langle M\rangle \mid M \in \mathbf{CSDFA}, L(M) \text{ is finite}\}$.

Corollary 7.1. ${}_{FA\text{-}Finiteness}L \in \boldsymbol{L_{TD}}$. ■

The *FA-Equivalence* problem asks whether two *CSDFA*s are equivalent; in other words, it asks whether both automata accept the same language. We decide this problem by using some elementary results of the set theory.

Problem 7.8. *FA-Equivalence.*
Question: Let $M, N \in \mathbf{CSDFA}$. Are M and N equivalent?
Language: ${}_{FA\text{-}Equivalence}L = \{\langle M, N\rangle \mid M, N \in \mathbf{CSDFA}, L(M) = L(N)\}$.

Theorem 7.8. ${}_{FA\text{-}Equivalence}L \in \boldsymbol{L_{TD}}$.

Proof. Let M and N be in **CSDFA**. As an exercise, prove that $L(M) = L(N)$ iff $\emptyset = (L(M) \cap \sim L(N)) \cup (L(N) \cap \sim L(M))$. Construct a TD D that works on every $\langle M, N\rangle \in {}_{FA\text{-}Equivalence}L$ as follows:

(1) from M and N, D constructs an FA O such that $L(O) = (L(M) \cap \sim L(N)) \cup (L(N) \cap \sim L(M))$;
(2) from O, D constructs an equivalent ${}_{C}sDFAP$;
(3) D decides whether $L(P) = \emptyset$ (see Theorem 7.5 and its proof);
(4) if $L(P) = \emptyset, L(M) = L(N)$ and D accepts $\langle M, N\rangle$, while if $L(P) \neq \emptyset$, D rejects $\langle M, N\rangle$.

Consider the constructions in (1) and (2); as an exercise, describe them in detail. □

Decidable problems for context-free grammars

Let G be any CFG (see Definition 6.1). We give algorithms for deciding the following three problems:

- Is the language generated by G finite?
- Is the language accepted by G infinite?

In addition, for any input word w, we decide the following problem:

- Is w a member of the language generated by G?

Rather than discuss these problems in general terms of CFGs in **CFG**, we decide them for CFGs in the Chomsky normal form, in which every rule has either a single terminal or two nonterminals on its right-hand side (see Definition 6.14). Making use of this form, we find it easier to decide two of them: *CFG-Membership* and *CFG-Infiniteness*. As any grammar in **CFG** can be turned to an equivalent grammar in the Chomsky normal form (see Algorithm 6.8), deciding these problems for grammars satisfying the Chomsky normal form obviously implies their decidability for grammars in **CFG** as well.

Convention 7.8. **CNF-CFG** denotes the set of all CFGs in Chomsky normal form. We suppose there exist a fixed encoding and decoding of the grammars in **CNF-CFG**. Similarly to TMs and FAs (see Conventions 5.4 and 7.7), $\langle G \rangle$ represents the code of $G \in$ **CNF-CFG**. Similarly, we suppose there exists an analogical encoding and decoding of the members of **CNF-CFG** $\times \Delta^*$ and **CNF-CFG** $\times$ **CNF-CFG**. Again, for brevity, we denote the codes of $(G, w) \in$ **CNF-CFG** $\times \Delta^*$ and $(G, H) \in$ **CNF-CFG** $\times$ **CNF-CFG** by $\langle G, w \rangle$ and $\langle G, H \rangle$, respectively. ■

Problem 7.9. *CFG-Emptiness.*
Question: Let $G \in$ **CNF-CFG**. Is $L(G)$ empty?
Language: ${}_{\textit{CFG-Emptiness}}L = \{\langle G \rangle \mid G \in \textbf{CNF-CFG}, L(G) = \emptyset\}$.

Theorem 7.9. ${}_{\textit{CFG-Emptiness}}L \in \boldsymbol{L_{TD}}$.

Proof. Let $G \in$ **CNF-CFG**. Recall that a symbol in G is terminating if it derives a string of terminals (see Definition 6.7). As a result, $L(G)$ is non-empty iff S_G is terminating, where S_G denotes the start symbol of G (see Convention 6.1). Therefore, construct a TD D that works on $\langle G \rangle$ as follows:

- D decides whether S_G is terminating by Algorithm 6.1;
- D rejects $\langle G \rangle$ if S_G is terminating; otherwise, D accepts $\langle G \rangle$.

□

Note that the decision of *CFG-Emptiness* described in the proof of 7.9 is straightforwardly applicable to any CFG in **CFG** because this decision does not actually make use of the Chomsky normal form. During the decision of the next two problems, however, we make use of this form significantly. Given a string w in Δ^* and a grammar G in **CNF-CFG**, the *CFG-Membership* problem asks whether w is a member of $L(G)$. Of course, we can easily decide this problem by any of the general parsing algorithms discussed in Chapter (see Algorithms 11.3 and 11.4). Next, we add yet another algorithm that decides this problem based on the Chomsky normal form.

Problem 7.10. *CFG-Membership.*
Question: Let $G \in$ **CNF-CFG** and $w \in \Delta^*$. Is w a member of $L(G)$?
Language: ${}_{\textit{CFG-Membership}}L = \{\langle G, w\rangle \mid G \in \textbf{CNF-CFG}, w \in \Delta^*, w \in L(G)\}$.

The proof of the following lemma is simple and left as an exercise.

Lemma 7.3. *Let $G \in$ **CNF-CFG**. Then, G generates every $w \in L(G)$ by making no more than $2|w| - 1$ derivation steps.* ■

Theorem 7.10. ${}_{\textit{CFG-Membership}}L \in \boldsymbol{L_{TD}}$.

Proof. As follows from the Chomsky normal form, **CNF-CFG** contains no grammar that generates ε. Therefore, we construct the following TD D that works on every $\langle G, w\rangle$ in either of the following two ways, (A) and (B), depending on whether $w = \varepsilon$ or not.

(1) Let $w = \varepsilon$. Clearly, $\varepsilon \in L(G)$ iff S_G is an ε-nonterminal, i.e., S_G derives ε (see Definition 6.11). Thus, D decides whether S_G is an ε-nonterminal by Algorithm 6.4, and if so, D accepts $\langle G, w\rangle$; otherwise, D rejects $\langle G, w\rangle$.
(2) Let $w \neq \varepsilon$. Then, D works on $\langle G, w\rangle$ as follows:
    (a) D constructs the set of all sentences that G generates by making no more than $2|w| - 1$ derivation steps;
    (b) if this set contains w, D accepts $\langle G, w\rangle$; otherwise, it rejects $\langle G, w\rangle$. □

The *CFG-Infiniteness* problem asks whether the language generated by a CFG is infinite.

Problem 7.11. *CFG-Infiniteness.*
Question: Let $G \in$ **CNF-CFG**. Is $L(G)$ infinite?

Language: ${}_{CFG\text{-}Infiniteness}L = \{\langle G\rangle \mid G \in \textbf{CNF-CFG}, L(G) \text{ is infinite}\}$.

As an exercise, prove the following lemma.

Lemma 7.4. *Let $G \in$ **CNF-CFG**. $L(G)$ is infinite iff $L(G)$ contains a sentence x such that $k \leq |x| < 2k$ with $k = 2^{|N_G|}$.* ■

Theorem 7.11. ${}_{CFG\text{-}Infiniteness}L \in \mathbf{L_{TD}}$.

Proof. Construct a TD D that works on every $G \in$ **CNF-CFG** as follows:

(1) D constructs the set of all sentences x in $L(G)$ satisfying $k \leq |x| < 2k$ with $k = 2^{card(N_G)}$;
(2) if this set is empty, D rejects $\langle G\rangle$; otherwise, D accepts $\langle G\rangle$. □

Theorem 7.11 implies that we can also decide the following problem.

Problem 7.12. *CFG-Finiteness.*
Question: Let $G \in$ **CNF-CFG**. Is $L(G)$ finite?
Language: ${}_{CFG\text{-}Finiteness}L = \{\langle G\rangle \mid G \in \textbf{CNF-CFG}, L(G) \text{ is finite }\}$.

Corollary 7.2. ${}_{CFG\text{-}Finiteness}L \in \mathbf{L_{TD}}$. ■

Recall that for FAs, we have formulated the problem *FA-Equivalence* and proved that it is decidable for them (see Theorem 7.8). However, we have not reformulated this problem for CFGs in this section. The reason for this is that this problem is undecidable for these grammars, which brings us to the topic of the next section — undecidable problems.

Undecidable Problems

As the central topic of this section, we consider several problems concerning TMs and demonstrate that they are undecidable. In addition, without any rigorous proofs, we briefly describe some undecidable problems not concerning TMs in the conclusion of this section.

Let P be a problem concerning TMs, and let P be encoded by a language ${}_PL$. Demonstrating that P is undecidable consists in proving that ${}_PL$ is an undecidable language. Like every rigorous proof in mathematics, a proof like this requires some ingenuity. Nevertheless, it is usually achieved by contradiction based on either of two proof techniques: diagonalization and reduction.

Diagonalization

As a rule, a diagonalization-based proof is schematically performed in the following way:

(1) Assume that ${}_PL$ is decidable, and consider a TD D such that $L(D) = {}_PL$.
(2) From D, construct another TD O; then, by using the diagonalization technique (see Example 1.5), apply O on its own description $\langle O \rangle$ so that this application results in a contradiction.
(3) The contradiction obtained in (2) implies that the assumption in (1) is incorrect, so ${}_PL$ is undecidable.

Following this proof scheme almost literally, we next demonstrate the undecidability of the famous halting problem that asks whether $M \in \mathbf{TM}$ halts on input x. Observe that the following formulation of the *TM-Halting* problem makes use of the encoding language ${}_{TM\text{-}Halting}L$, introduced in the conclusion of Section 5.3.

Problem 7.13. *TM-Halting.*
Question: Let $M \in \mathbf{TM}$ and $w \in \Delta^*$. Does M halt on w?
Language: ${}_{TM\text{-}Halting}L = \{\langle M, w \rangle \mid M \in \mathbf{TM}, w \in \Delta^*, M$ halts on $w\}$.

Theorem 7.12. ${}_{TM\text{-}Halting}L \notin \boldsymbol{L_{TD}}$.

Proof. Assume that ${}_{TM\text{-}Halting}L$ is decidable. Then, there exists a TD D such that $L(D) = {}_{TM\text{-}Halting}L$. That is, for any $\langle M, w \rangle \in {}_{TM\text{-}Halting}L$, D accepts $\langle M, x \rangle$ iff M halts on x, and D rejects $\langle M, x \rangle$ iff M loops on x. From D, construct another TD O that works on every input w, where $w = \langle M \rangle$, with $M \in \mathbf{TM}$ (recall that according to Convention 5.4, every input string encodes a TM in **TM**, so the case when w encodes no TM is ruled out), as follows:

- O replaces w with $\langle M, M\rangle$, where $w = \langle M\rangle$;
- O runs D on $\langle M, M\rangle$;
- O accepts iff D rejects, and O rejects iff D accepts.

That is, for every $w = \langle M\rangle$, O accepts $\langle M\rangle$ iff D rejects $\langle M, M\rangle$, and since $L(D) = {}_{TM\text{-}Halting}L$, D rejects $\langle M, M\rangle$ iff M loops on $\langle M\rangle$. Thus, O accepts $\langle M\rangle$ iff M loops on $\langle M\rangle$. Now, we apply the diagonalization technique in this proof: As O works on every input w, it also works on $w = \langle O\rangle$. Consider this case. Since O accepts $\langle M\rangle$ iff M loops on $\langle M\rangle$ for every $w = \langle M\rangle$, this equivalence holds for $w = \langle O\rangle$ as well, so O accepts $\langle O\rangle$ iff O loops on $\langle O\rangle$. Thus, $\langle O\rangle \in L(O)$ iff $\langle O\rangle \notin L(O)$ — a contradiction. Therefore, ${}_{TM\text{-}Halting}L$ is undecidable; in symbols, ${}_{TM\text{-}Halting}L \notin \mathbf{L_{TD}}$. □

This theorem straightforwardly implies the following proper inclusion.

Theorem 7.13. $\boldsymbol{L_{TD} \subset L_{TM}}$. ■

Observe that Theorem 7.12 has its important consequences in practice. Indeed, considering the Turing–Church thesis, it rules out the existence of a universal algorithm that would decide, for any procedure, whether the procedure is an algorithm, which halts on all inputs (see Section 2.1). Thus, no operation system can ever provide us with a universal software tool that decides whether any implemented procedure or, to put it simply and plainly, any program, written in a programming language, always stops. From a more general and, in fact, ethical standpoint, as far as computational power is concerned, we see that neither humans nor their computers are omnipotent, and they never will be. Thus, the metaphysics of computation presented in this chapter, in which Theorem 7.12 fulfills a crucially important role, leads us to a certain humility, not pride: Before starting to write a program to decide a problem, we should ask whether the problem under consideration is decidable at all. The reader is more than encouraged to derive similar consequences from all the upcoming results concerning undecidability.

As *TM-Halting* is undecidable, it comes as no surprise that the problem whether $M \in \mathbf{TM}$ loops on $w \in \Delta^*$ is not decidable either.

Problem 7.14. *TM-Looping.*
Question: Let $M \in \mathbf{TM}$ and $w \in \Delta^*$. Does M loop on x?
Language: ${}_{TM\text{-}Looping}L = \{\langle M, w\rangle \mid M \in \mathbf{TM}, x \in \Delta^*, M$ loops on $w\}$.

To prove the undecidability of *TM-Looping*, we establish the following two theorems. The first is obvious.

Theorem 7.14. *${}_{TM\text{-}Looping}L$ is the complement of ${}_{TM\text{-}Halting}L$.* ■

Next, we prove that a language L is decidable iff $\mathbf{L_{TM}}$ contains both L and its complement $\sim L$.

Theorem 7.15. *Let $L \subseteq \Delta^*$. $L \in \mathbf{L_{TD}}$ iff both L and $\sim L$ are in $\mathbf{L_{TM}}$.*

Proof. To prove the only if part of the equivalence, suppose that L is any decidable language, symbolically written $L \in \mathbf{L_{TD}}$. By Theorem 7.13, $L \in \mathbf{L_{TM}}$. By Definition 7.7, there is $M \in \mathbf{TD}$ such that $L = L(M)$. Change M to a TM $N \in \mathbf{TM}$ so that N enters a non-final state in which it keeps looping exactly when M enters the final state ■ (see Convention 5.3). As a result, $L(N) = \sim L(M) = \sim L$, so $\sim L \in \mathbf{L_{TM}}$. Thus, L and $\sim L$ are in $\mathbf{L_{TM}}$.

To prove the if part of the equivalence, suppose that $L \in \mathbf{L_{TM}}$ and $\sim L \in \mathbf{L_{TM}}$. That is, there exist $N \in \mathbf{TM}$ and $O \in \mathbf{TM}$ such that $L(N) = L$ and $L(O) = \sim L$. Observe that N and O cannot accept the same string because $L \cap \sim L = \emptyset$. On the other hand, every input w is accepted by either N or O because $L \cup \sim L = \Delta^*$. These properties underlie the next construction of a TD M for L from N and O. M works on every input w in the following way:

(1) M simultaneously runs N and O on w so M executes by turns one move in N and O, i.e., step by step, M computes the first move in N, the first move in O, the second move in N, the second move in O, and so forth.
(2) M continues the simulation described in (1) until a move that would take N or O to an accepting configuration, and in this way, M finds out whether $w \in L(N)$ or $w \in L(O)$.

(3) Instead of entering the accepting configuration in N or O, M halts and either accepts if $w \in L(N)$ or rejects if $w \in L(O)$ — in greater detail, M changes the current configuration to a halting configuration of the form $\triangleright iu \triangleleft$, where $u \in \{\square\}^*, i \in \blacksquare, \blacklozenge, i = \blacksquare$ iff $w \in L(N)$, and $i = \blacklozenge$ iff $w \in L(O)$.

Observe that $L(M) = L$. Furthermore, M always halts, so $M \in \mathbf{TD}$ and $L \in \mathbf{L_{TD}}$. $\square$

Making use of Theorems 5.7 and 7.15, we easily demonstrate *TM-Looping* as an undecidable problem. In fact, we prove a much stronger result stating that ${}_{TM\text{-}Looping}L$ is not even in $\mathbf{L_{TM}}$.

Theorem 7.16. ${}_{TM\text{-}Looping}L \notin \mathbf{L_{TM}}$.

Proof. Assume ${}_{TM\text{-}Looping}L \in \mathbf{L_{TM}}$. Recall that ${}_{TM\text{-}Looping}L$ is the complement of ${}_{TM\text{-}Halting}L$ (see Theorem 7.14). Furthermore, ${}_{TM\text{-}Halting}L \in \mathbf{L_{TM}}$ (see Theorem 5.7). Thus, by Theorem 7.15, ${}_{TM\text{-}Halting}L$ would be decidable, which contradicts Theorem 7.12. Thus, ${}_{TM\text{-}Looping}L \notin \mathbf{L_{TM}}$. $\square$

From this theorem, we obtain the following proper inclusion.

Corollary 7.3. $\mathbf{L_{TM}} \subset \mathbf{L_{ALL}}$. $\blacksquare$

Therefore, putting Theorems 6.26, 7.13 and Corollary 7.3 together, we obtain the following language hierarchy, which includes $\mathbf{L_{TD}}$ as well.

Theorem 7.17. $\mathbf{L_{FIN}} \subset \mathbf{L_{FA}} \subset \mathbf{L_{dPDA}} \subset \mathbf{L_{CFG}} \subset \mathbf{L_{CSG}} \subset \mathbf{L_{TD}} \subset \mathbf{L_{TM}} \subset \mathbf{L_{ALL}}$. $\blacksquare$

Reduction

Apart from diagonalization, we often establish the undecidability of a problem P so the decidability of P would imply the decidability of a well-known undecidable problem U, and from this contradiction, we conclude that P is undecidable. In other words, from the well-known undecidability of U, we actually derive the undecidability of P; hence, we usually say that we reduce U to P when demonstrating that P is undecidable in this way. In terms of problem-encoding languages, to prove that a language ${}_PL$, encoding P, is undecidable, we usually follow the proof scheme given in the following:

(1) Assume that ${}_PL$ is decidable, and consider a TD D such that $L(D) = {}_PL$.
(2) Modify D to another TD that would decide a well-known undecidable language L_U — a contradiction.
(3) The contradiction obtained in (2) implies that the assumption in (1) is incorrect, so ${}_PL$ is undecidable.

Based on this reduction-proof scheme, we next demonstrate the undecidability of the *TM-Membership* problem that asks whether input w is a member of $L(M)$, where $M \in \mathbf{TM}$ and $w \in \Delta^*$. It is worth noting that the following formulation of this problem makes use of the encoding language ${}_{\textit{TM-Membership}}L$ that coincides with ${}_{\textit{TM-Acceptance}}L$ defined in Section 5.3.

Problem 7.15. *TM-Membership.*
Question: Let $M \in \mathbf{TM}$ and $w \in \Delta^*$. Is w a member of $L(M)$?
Language: ${}_{\textit{TM-Membership}}L = \{\langle M, w\rangle \mid M \in \mathbf{TM}, w \in \Delta^*, w \in L(M)\}$.

We prove the undecidability of this problem by reducing Problem 7.13 *TM-Halting* to it. That is, we show that if there were a way of deciding the *TM-Membership* problem, we could decide Problem 7.13 *TM-Halting*, which contradicts Theorem 7.12.

Theorem 7.18. ${}_{\textit{TM-Membership}}L \notin \mathbf{L_{TD}}$.

Proof. Given $\langle M, x\rangle$, construct a TM N that coincides with M except that N accepts x iff M halts on x (recall that M halts on x iff M either accepts or rejects x according to Convention 5.2). In other words, $x \in L(N)$ iff M halts on x. If there were a TD D for ${}_{\textit{TM-Membership}}L$, we could use D and this equivalence to decide ${}_{\textit{TM-Halting}}L$. Indeed, we could decide ${}_{\textit{TM-Halting}}L$ by transforming M into N, as described above, and asking whether $x \in L(N)$; from $x \in L(N)$, we would conclude that M halts on x, while from $x \notin L(N)$, we would conclude that M loops on x. However, Problem 7.13 *TM-Halting* is undecidable (see Theorem 7.12), which rules out the existence of D. Thus, there is no TD for ${}_{\textit{TM-Membership}}L$, so ${}_{\textit{TM-Membership}}L \notin \mathbf{L_{TD}}$. □

Next, we formulate the *Non-TM-Membership* problem, and based upon Theorems 5.6 and 7.18, we prove that it is not decidable either.

Problem 7.16. *Non-TM-Membership.*
Question: Let $M \in \mathbf{TM}$ and $w \in \Delta^*$. Is w out of $L(M)$?
Language: ${}_{Non\text{-}TM\text{-}Membership}L = \{\langle M, w\rangle \mid M \in \mathbf{TM}, w \in \Delta^*, w \notin L(M)\}$.

By analogy with the proof of Theorem 7.18, we prove that ${}_{Non\text{-}TM\text{-}Membership}L$ is even out of $\mathbf{L_{TM}}$.

Theorem 7.19. ${}_{Non\text{-}TM\text{-}Membership}L \notin \boldsymbol{L_{TM}}$.

Proof. For the sake of obtaining a contradiction, suppose that ${}_{Non\text{-}TM\text{-}Membership}L \in \mathbf{L_{TM}}$. As already pointed out,

$$ {}_{TM\text{-}Membership}L = {}_{TM\text{-}Acceptance}L, \text{ so } {}_{TM\text{-}Membership}L \in \mathbf{L_{TM}}. $$

As is obvious, ${}_{Non\text{-}TM\text{-}Membership}L$ is the complement of ${}_{TM\text{-}Membership}L$. Thus, by Theorem 7.15, ${}_{TM\text{-}Membership}L$ would belong to $\mathbf{L_{TD}}$, which contradicts Theorem 7.18. Thus, ${}_{Non\text{-}TM\text{-}Membership}L \notin \mathbf{L_{TM}}$. □

From Theorems 7.13 and 7.19, we obtain the following corollary, saying that *Non-TM-Membership* is an undecidable problem.

Corollary 7.4. ${}_{Non\text{-}TM\text{-}Membership}L \notin \boldsymbol{L_{TD}}$. ■

The following problem asks whether $L(M)$ is regular, where $M \in \mathbf{TM}$. By reducing *TM-Halting* to it, we prove its undecidability.

Problem 7.17. *TM-Regularness.*
Question: Let $M \in \mathbf{TM}$. Is $L(M)$ regular?
Language: ${}_{TM\text{-}Regularness}L = \{\langle M\rangle \mid M \in \mathbf{TM}, L(M) \in \mathbf{L_{REG}}\}$.

Theorem 7.20. ${}_{TM\text{-}Regularness}L \notin \boldsymbol{L_{TD}}$.

Proof. Consider ${}_{TM\text{-}Halting}L = \{\langle M, w\rangle \mid M \in \mathbf{TM}, w \in \Delta^*, M$ halts on $w\}$. Recall that ${}_{TM\text{-}Halting}L \in \mathbf{L_{TM}} - \mathbf{L_{TD}}$ (see Theorems 7.12 and 7.13). Take any TM O such that $L(O) = {}_{TM\text{-}Halting}L$; for instance, in the proof of Theorem 5.7, ${}_{TM\text{-}Halting}U$ satisfies this requirement because $L({}_{TM\text{-}Halting}U) = {}_{TM\text{-}Halting}L$. Next, we construct a TM $W \in \mathbf{TM}$ so that W converts every input $\langle M, w\rangle$, where

$M \in \mathbf{TM}$ and $w \in \Delta^*$, to a new TM, denoted by $N_{[M,w]}$. Next, we describe this conversion in a greater detail. Given $\langle M, w\rangle, W$ constructs a TM $N_{[M,w]}$ that works on every input $y \in \Delta^*$ as follows:

(1) $N_{[M,w]}$ places w somewhere behind y on its tape;
(2) $N_{[M,w]}$ runs M on w;
(3) if M halts on $w, N_{[M,w]}$ runs O on y and accepts if and when O accepts.

If M loops on $w, N_{[M,w]}$ never gets behind (2), so the language accepted by $N_{[M,w]}$ equals $\emptyset$ in this case. If M halts on $w, N_{[M,w]}$ accepts y in (3) if and when O accepts y, so the language accepted by $N_{[M,w]}$ coincides with $L(O)$ in this case. Thus, the language accepted by $N_{[M,w]}$ equals $L(O)$ iff M halts on w, and the language accepted by $N_{[M,w]}$ equals $\emptyset$ iff M loops on w. By Theorem 7.17, $L(O)$ is not regular because $L(O) \in \mathbf{L_{TM}} - \mathbf{L_{TD}}$ and $\mathbf{L_{REG}} \subset \mathbf{L_{TD}}$. By Definition 3.8, $\emptyset$ is regular. Thus, the language accepted by $N_{[M,w]}$ is regular iff M loops on w, and the language accepted by $N_{[M,w]}$ is non-regular iff M halts on w. Hence, if ${}_{\textit{TM-Regularness}}L$ were decidable by a TD $V \in \mathbf{TD}, W$ could make use of V and these equivalences to decide ${}_{\textit{TM-Halting}}L$. Simply put, W would represent a TD for ${}_{\textit{TM-Halting}}L$, which contradicts Theorem 7.12. Thus, ${}_{\textit{TM-Regularness}}L$ is undecidable. □

As an exercise, demonstrate the undecidability of the following three problems by analogy with the proof of Theorem 7.20.

Problem 7.18. *TM-Emptiness.*
Question: Let $M \in \mathbf{TM}$. Is $L(M)$ empty?
Language: ${}_{\textit{TM-Emptiness}}L = \{\langle M\rangle \mid M \in \mathbf{TM}, L(M) = \emptyset\}$.

Theorem 7.21. ${}_{\textit{TM-Emptiness}}L \notin \mathbf{L_{TD}}$. ■

Problem 7.19. *TM-Finiteness.*
Question: Let $M \in \mathbf{TM}$. Is $L(M)$ finite?
Language: ${}_{\textit{TM-Finiteness}}L = \{\langle M\rangle \mid M \in \mathbf{TM}, L(M) \text{ is finite}\}$.

Theorem 7.22. ${}_{\textit{TM-Finiteness}}L \notin \mathbf{L_{TD}}$. ■

Problem 7.20. *TM-Contextfreeness.*
Question: Let $M \in \mathbf{TM}$. Is $L(M)$ context-free?
Language: ${}_{\textit{TM-Contextfreeness}}L = \{\langle M\rangle \mid M \in \mathbf{TM}, L(M) \in \mathbf{L_{CF}}\}$.

Theorem 7.23. ${}_{TM\text{-}Contextfreeness}L \notin \mathbf{L_{TD}}$. ■

Consider the following problem that asks whether $L(M) = \Delta^*$, where $M \in \mathbf{TM}$. We again prove its undecidability by reducing Problem 7.13 *TM-Halting* to it.

Problem 7.21. *TM-Universality*.
Question: Let $M \in \mathbf{TM}$. Is $L(M)$ equal to Δ^*?
Language: ${}_{TM\text{-}Universality}L = \{\langle M\rangle \mid M \in \mathbf{TM}, L(M) = \Delta^*\}$.

Theorem 7.24. ${}_{TM\text{-}Universality}L \notin \mathbf{L_{TD}}$.

Proof. We reduce Problem 7.13 *TM-Halting* to Problem 7.21 *TM-Universality*. Once again, recall that ${}_{TM\text{-}Halting}L = \{\langle M, w\rangle \mid M \in \mathbf{TM}, w \in \Delta^*, M \text{ halts on } w\}$. We introduce a TM $W \in \mathbf{TM}$ so that W constructs the following TM $N_{[M,w]}$ from every input $\langle M, w\rangle$, where $M \in \mathbf{TM}$ and $w \in \Delta^*$. That is, given $\langle M, w\rangle$, W makes $N_{[M,w]}$ that works on every input $y \in \Delta^*$ as follows:

(1) $N_{[M,w]}$ replaces y with w;
(2) $N_{[M,w]}$ runs M on w and halts if and when M halts.

As $N_{[M,w]}$ works on every y in this way, its language equals Δ^* iff M halts on w, while its language is empty iff M loops on w. Assume that ${}_{TM\text{-}Universality}L \in \mathbf{L_{TD}}$, so there is a TD V for ${}_{TM\text{-}Universality}L$. Thus, W could use V and these equivalences to decide ${}_{TM\text{-}Halting}L$, which contradicts Theorem 7.12. Hence, ${}_{TM\text{-}Universality}L \notin \mathbf{L_{TD}}$. □

Undecidable problems not concerning Turing machines

We have concentrated our attention on the undecidability concerning TMs and their languages so far. However, undecidable problems arise in a large variety of areas in the formal language theory as well as outside of this theory. Therefore, before concluding this section, we present some of them, but we completely omit proofs that rigorously demonstrate their undecidability.

For CFGs, the following problems are undecidable.

Problem 7.22. *CFG-Equivalence*.
Question: Let $G, H \in \mathbf{CFG}$. Are G and H equivalent?
Language: ${}_{CFG\text{-}Equivalence}L = \{\langle G, H\rangle \mid G, H \in \mathbf{CFG}, L(G) = L(H)\}$.

Problem 7.23. *CFG-Containment.*
Question: Let $G, H \in \mathbf{CFG}$. Does $L(G)$ contain $L(H)$?
Language: ${}_{\textit{CFG-Containment}}L = \{\langle G, H\rangle \mid G, H \in \mathbf{CFG}, L(H) \subseteq L(G)\}$.

Problem 7.24. *CFG-Intersection.*
Question: Let $G, H \in \mathbf{CFG}$. Is the intersection of $L(G)$ and $L(H)$ empty?
Language: ${}_{\textit{CFG-Intersection}}L = \{\langle G, H\rangle \mid G, H \in \mathbf{CFG}, L(H) \cap L(G) = \emptyset\}$.

Problem 7.25. *CFG-Universality.*
Question: Let $G \in \mathbf{CFG}$. Is $L(G)$ equal to T_G^*?
Language: ${}_{\textit{CFG-Universality}}L = \{\langle G\rangle \mid G \in \mathbf{CFG}, L(G) = T_G^*\}$.

Problem 7.26. *CFG-Ambiguity.*
Question: Let $G \in \mathbf{CFG}$. Is G ambiguous?
Language: ${}_{\textit{CFG-Ambiguity}}L = \{\langle G\rangle \mid G \in \mathbf{CFG}, G \text{ is ambiguous}\}$.

Within the formal language theory, however, there exist many undecidable problems concerning languages without involving their models, and some of them were introduced a long time ago. To illustrate, in 1946, Post introduced a famous problem, which we here formulate in terms of ε-free homomorphisms, defined in Section 1.3. Let X, Y be two alphabets and g, h be two ε-free homomorphisms from X^* to Y^*; *Post's Correspondence Problem* is to determine whether there is $w \in X^+$ such that $g(w) = h(w)$. For example, consider $X = \{1, 2, 3\}, Y = \{a, b, c\}, g(1) = abbb, g(2) = a, g(3) = ba, h(1) = b, h(2) = aab$, and $h(3) = b$, and observe that $2231 \in X^+$ satisfies $g(2231) = h(2231)$. Consider a procedure that systematically produce all possible $w \in X^+$, makes $g(w)$ and $h(w)$, and tests whether $g(w) = h(w)$. If and when the procedure finds out that $g(w) = h(w)$, it halts and answers yes; otherwise, it continues to operate endlessly. Although there is a procedure like this, there is no algorithm, which halts on every input, to decide this problem. Simply put, Post's Correspondence Problem is undecidable.

Of course, outside formal language theory, there exist many undecidable problems as well. To illustrate, mathematics will never have a general algorithm that decides whether statements in number theory with the plus and times are true or false. Although these results are obviously more than significant from a purely mathematical point of view, they are somewhat out of the scope of this book, which primarily concentrates its attention to formal languages and their models; therefore, we leave their discussion as an exercise.

A general approach to undecidability

As demonstrated in the previous section, many reduction-based proofs of undecidability are very similar. This similarity has inspired the theory of computation to undertake a more general approach to reduction, based on the following definition, which makes use of the notion of a computable function (see Definition 7.2).

Definition 7.8. Let $K, L \subseteq \Delta^*$ be two languages. A total computable function f over Δ^* is a reduction of K to L, symbolically written as $K \angle_f L$, if for all $w \in \Delta^*, w \in K$ iff $f(w) \in L$. ■

Convention 7.9. Let $K, L \subseteq \Delta^*$. We write $K \angle L$ to express that there exists a reduction of K to L. Let us note that instead of $\angle$, $\leq$ is also used in the literature. ■

First, we establish a general theorem concerning $\angle$ in terms of $\mathbf{L_{TM}}$.

Theorem 7.25. *Let $K, L \subseteq \Delta^*$. If $K \angle L$ and $L \in \boldsymbol{L_{TM}}$, then $K \in \boldsymbol{L_{TM}}$.*

Proof. Let $K, L \subseteq \Delta^*, K \angle L$, and $L \in \mathbf{L_{TM}}$. Recall that $K \angle L$ means that there exists a reduction f of K to L, written as $K \angle_f L$ (see Definition 7.8 and Convention 7.9). As $L \in \mathbf{L_{TM}}$, there is a TM M satisfying $L = L(M)$. Construct a new TM N that works on every input $w \in \Delta^*$ as follows:

(1) N computes $f(w)$ (according to Definition 7.2, f is computable);
(2) N runs M on $f(w)$;
(3) if M accepts, then N accepts, and if M rejects, then N rejects.

Note that N accepts w iff M accepts $f(w)$. As $L = L(M), M$ accepts $f(w)$ iff $f(w) \in L$. As $K \angle L$ (see Definition 7.8), $w \in K$ iff $f(w) \in L$. Thus, $K = L(N)$, so $K \in \mathbf{L_{TM}}$. □

Corollary 7.5. *Let $K, L \subseteq \Delta^*$. If $K \angle L$ and $K \notin \boldsymbol{L_{TM}}$, then $L \notin \boldsymbol{L_{TM}}$.* ■

By Theorem 7.25, we can easily prove that a language K belongs to $\mathbf{L_{TM}}$. Indeed, we take a language $L \in \mathbf{L_{TM}}$ and construct a TM M that computes a reduction of K to L, so $K \angle L$. Then, by Theorem 7.25, $K \in \mathbf{L_{TM}}$. For instance, from Theorem 5.6 (recall that ${}_{\textit{TM-Membership}}L = {}_{\textit{TM-Acceptance}}L$), it follows that ${}_{\textit{TM-Membership}}L \in \mathbf{L_{TM}}$. Take this language. Demonstrate that ${}_{\textit{TM-Halting}}L \angle {}_{\textit{TM-Membership}}L$ to prove ${}_{\textit{TM-Halting}}L \in \mathbf{L_{TM}}$. As a result, we have obtained an alternative proof that ${}_{\textit{TM-Halting}}L \in \mathbf{L_{TM}}$, which also follows from Theorem 5.7.

Perhaps even more importantly, Corollary 7.5 saves us much work to prove that a language L is out of $\mathbf{L_{TM}}$. Typically, a proof like this is made in one of the following two ways.

(I) Take a well-known language $K \notin \mathbf{L_{TM}}$, and construct a TM M that computes a reduction of K to $L, K \angle L$. As a result, Corollary 7.5 implies $L \notin \mathbf{L_{TM}}$.

(II) By Definition 7.8, if f is a reduction of K to L, then f is a reduction of $\sim K$ to $\sim L$ as well. Therefore, to prove that $L \notin \mathbf{L_{TM}}$, take a language K with its complement $\sim K \notin \mathbf{L_{TM}}$ and construct a TM that computes a reduction of K to $\sim L$. As $K \angle \sim L$, we have $\sim K \angle \sim\sim L$. That is, $\sim K \angle L$, and by Corollary 7.5, $L \notin \mathbf{L_{TM}}$.

In fact, by a clever use of Corollary 7.5, we can sometimes demonstrate that both $L \notin \mathbf{L_{TM}}$ and $\sim L \notin \mathbf{L_{TM}}$, and both proofs frequently resemble each other very much. To illustrate, in this way, we next prove that ${}_{\textit{TM-Equivalence}}L \notin \mathbf{L_{TM}}$ and its complement ${}_{\textit{Non-TM-Equivalence}}L \notin \mathbf{L_{TM}}$, where

$$ {}_{\textit{TM-Equivalence}}L = \{\langle M, N \rangle \mid M, N \in \mathbf{TM}, L(M) = L(N)\}, \text{ and} $$

$$ {}_{\textit{Non-TM-Equivalence}}L = \{\langle M, N \rangle \mid M, N \in \mathbf{TM}, L(M) \neq L(N)\}. $$

Theorem 7.26. ${}_{\textit{TM-Equivalence}}L \notin \mathbf{L_{TM}}$.

Proof. To demonstrate ${}_{\textit{TM-Equivalence}}L \notin \mathbf{L_{TM}}$, we follow proof method (II) above. More specifically, we prove that ${}_{\textit{TM-Membership}}L \angle {}_{\textit{Non-TM-Equivalence}}L$ (see Problem 7.15 *TM-Membership* for ${}_{\textit{TM-Membership}}L$); therefore, ${}_{\textit{TM-Equivalence}}L \notin \mathbf{L_{TM}}$ because ${}_{\textit{Non-TM-Membership}}L \notin \mathbf{L_{TM}}$ (see (II) above). To establish ${}_{\textit{TM-Membership}}L \angle {}_{\textit{Non-TM-Equivalence}}L$, we construct a TM X that computes a reduction of ${}_{TM\text{-}Membership}L$ to ${}_{Non\text{-}TM\text{-}Equivalence}L$. Specifically, X transforms every $\langle O, w\rangle$, where $O \in \mathbf{TM}$ and $w \in \Delta^*$, to the following two TMs, M and $N_{[O,w]}$, and produces $\langle M, N_{[O,w]}\rangle$ as output (we denote M without any information concerning $\langle O, w\rangle$ because its construction is completely independent of it; that is, X produces the same M for every $\langle O, w\rangle$). M and $N_{[O,w]}$ work as follows:

(1) M rejects every input;
(2) on every input $x \in \Delta^*, N_{[O,w]}$ works so it runs O on w and accepts x if and when O accepts w.

As obvious, $L(M) = \emptyset$. Since $N_{[O,w]}$ works on every input $x \in \Delta^*$ in the way described above, these two implications hold:

- if $w \in L(O)$, then $L(N_{[O,w]}) = \Delta^*$, which implies $L(M) \neq L(N_{[O,w]})$;
- if $w \notin L(O)$, then $L(N_{[O,w]}) = \emptyset$, which means $L(M) = L(N_{[O,w]})$.

Thus, X computes a reduction of ${}_{TM\text{-}Membership}L$ to ${}_{Non\text{-}TM\text{-}Equivalence}L$, so ${}_{TM\text{-}Equivalence}L \notin \mathbf{L_{TM}}$. □

Observe that the proof of the following theorem, which says that the complement of ${}_{TM\text{-}Equivalence}L$ is out of $\mathbf{L_{TM}}$ as well, parallels the previous proof significantly. As a matter of fact, while in the previous proof, M always rejects, in the following proof, M always accepts; otherwise, both proofs coincide with each other.

Theorem 7.27. ${}_{Non\text{-}TM\text{-}Equivalence}L \notin \boldsymbol{L_{TM}}$.

Proof. To show that

$$ {}_{Non\text{-}TM\text{-}Equivalence}L \notin \mathbf{L_{TM}}, $$

we prove that ${}_{TM\text{-}Membership}L \angle {}_{TM\text{-}Equivalence}L$. We define a reduction of ${}_{TM\text{-}Membership}L$ to ${}_{TM\text{-}Equivalence}L$ by a TM X that transforms every $\langle O, w\rangle$, where $O \in \mathbf{TM}$ and $w \in \Delta^*$, to the following two TMs $M, N_{[O,w]} \in \mathbf{TM}$ and produces $\langle M, N_{[O,w]}\rangle$ as output. M and $N_{[O,w]}$ are defined as follows:

(1) M accepts every input string;
(2) on every input string x, $N_{[O,w]}$ runs O on w and accepts x if and when O accepts w.

As is obvious, $L(M) = \Delta^*$. If $w \in L(O), L(N_{[O,w]}) = \Delta^*$ and $L(M) = L(N_{[O,w]})$; otherwise, $L(M) \neq L(N_{[O,w]})$. Hence, ${}_{TM\text{-}Membership}L \angle {}_{TM\text{-}Equivalence}L$. Therefore, by using proof method (II) above, we obtain ${}_{Non\text{-}TM\text{-}Equivalence}L \notin \mathbf{L_{TM}}$. □

Returning to the key topic of this section, we see that such results as Theorem 7.25 and Corollary 7.5 often have significant consequences in terms of undecidability. Indeed, if $L \notin \mathbf{L_{TM}}$, then a problem encoded by L is undecidable because $\mathbf{L_{TD}} \subset \mathbf{L_{TM}}$ (see Theorem 7.13). Specifically, in this way, Theorems 7.26 and 7.27 imply the undecidability of the following two problems encoded by languages ${}_{TM\text{-}Equivalence}L$ and ${}_{Non\text{-}TM\text{-}Equivalence}L$, introduced above; for convenience, we repeat the definition of ${}_{TM\text{-}Equivalence}L$ and ${}_{Non\text{-}TM\text{-}Equivalence}L$ in the following problems again.

Problem 7.27. *TM-Equivalence.*
Question: Are M and N equivalent, where $M, N \in \mathbf{TM}$?
Language: ${}_{TM\text{-}Equivalence}L = \{\langle M, N\rangle \mid M, N \in \mathbf{TM}, L(M) = L(N)\}$.

Problem 7.28. *Non-TM-Equivalence.*
Question: Are M and N non-equivalent, where $M, N \in \mathbf{TM}$?
Language: ${}_{Non\text{-}TM\text{-}Equivalence}L = \{\langle M, N\rangle \mid M, N \in \mathbf{TM}, L(M) \neq L(N)\}$.

Corollary 7.6. *${}_{TM\text{-}Equivalence}L \notin \boldsymbol{L_{TD}}$ and ${}_{Non\text{-}TM\text{-}Equivalence}L \notin \boldsymbol{L_{TD}}$.* ■

Next, we state results analogical to Theorem 7.25 and Corollary 7.5 in terms of $\mathbf{L_{TD}}$.

Theorem 7.28. *Let $K, L \subseteq \Delta^*$. If $K \angle L$ and $L \in \boldsymbol{L_{TD}}$, then $K \in \boldsymbol{L_{TD}}$.*

Proof. Let $K, L \subseteq \Delta^*, K \angle L$, and $L \in \mathbf{L_{TD}}$. Let f be a reduction of K to L. As already pointed out, by Definition 7.8, f is a reduction of $\sim K$ to $\sim L$ too. By Theorem 7.15, $L \in \mathbf{L_{TD}}$ iff $L \in \mathbf{L_{TM}}$ and $\sim L \in \mathbf{L_{TM}}$. By Theorem 7.15, $K \in \mathbf{L_{TM}}$ and $\sim K \in \mathbf{L_{TM}}$. Thus, $K \in \mathbf{L_{TD}}$ by Theorem 7.28. □

Corollary 7.7. *Let $K, L \subseteq \Delta^*$. If $K \angle L$ and $K \notin \boldsymbol{L_{TD}}$, then $L \notin \boldsymbol{L_{TD}}$.* ■

Theorem 7.28 and Corollary 7.7 often save us much work when we demonstrate undecidability. In the following examples, we revisit some of our earlier results concerning undecidability to see how they follow from Corollary 7.7.

Example 7.8. Reconsider Problem 7.18 *TM-Emptiness* and Theorem 7.21, stating that this problem is undecidable. In essence, this undecidability is established so that from any TM M and any string x, we algorithmically construct a TM N such that $L(N) = \emptyset$ iff M halts on x. To rephrase this in terms of languages, we define a reduction of ${}_{TM\text{-}Halting}L$ to ${}_{TM\text{-}Emptiness}L$, so ${}_{TM\text{-}Halting}L \angle {}_{TM\text{-}Emptiness}L$. As ${}_{TM\text{-}Halting}L \notin \mathbf{L_{TD}}$ (see Theorem 7.12), ${}_{TM\text{-}Emptiness}L \notin \mathbf{L_{TD}}$ by Corollary 7.7, so Problem 7.18 *TM-Emptiness* is undecidable. ■

Example 7.9. Earlier in this section, by using diagonalization, we proved that Problem 7.13 *TM-Halting* is undecidable, after which we demonstrated that Problem 7.15 *TM-Membership* is undecidable, so we reduced *TM-Halting* to *TM-Membership*. As the current example demonstrates, we could proceed the other way around. That is, first, by using diagonalization, we could prove that Problem 7.15 *TM-Membership* is undecidable; a proof like this is similar to the proof of Theorem 7.12 and, therefore, left as an exercise. Next, we show ${}_{TM\text{-}Membership}L \angle {}_{TM\text{-}Halting}L$, so the undecidability of *TM-Membership* implies the undecidability of *TM-Halting* by Corollary 7.7.

We construct a TD D that computes a total function f over Δ^* that maps every $\langle M, w\rangle$ to $\langle N, v\rangle$, so

$$\langle M, w\rangle \in {}_{TM\text{-}Membership}L \text{ iff } \langle N, v\rangle \in {}_{TM\text{-}Halting}L.$$

Therefore, ${}_{TM\text{-}Membership}L \angle_f {}_{TM\text{-}Halting}L$. D is defined as follows:

(1) on every input $\langle M, w\rangle, D$ constructs a TM $W_{[M]}$ that works on every input in this way:
   (a) $W_{[M]}$ runs M on the input;
   (b) $W_{[M]}$ accepts if M accepts, and $W_{[M]}$ loops if M rejects;
(2) write $\langle W_{[M]}, w\rangle$ as output.

Observe that $W_{[M]}$ loops on w iff M rejects w or loops on w; thus, in terms of the above equivalence, $W_{[M]}$ fulfills the role of N with v equal to w. Clearly, ${}_{TM\text{-}Membership}L \angle {}_{TM\text{-}Halting}L$. As ${}_{TM\text{-}Membership}L \notin \mathbf{L_{TD}}$, ${}_{TM\text{-}Halting}L \notin \mathbf{L_{TD}}$ by Corollary 7.7. ■

Rice's theorem

Next, we discuss the undecidability concerning the properties of Turing languages in $\mathbf{L_{TM}}$ rather than TMs in **TM**. More specifically, we identify a property of Turing languages, π, with the subfamily of $\mathbf{L_{TM}}$ defined by this property; that is, this subfamily contains precisely the Turing languages that satisfy π. For instance, the property of being finite equals $\{L \in \mathbf{L_{TM}} \mid L \text{ is finite}\}$. In this way, we consider π as a decidable property if there exists a TD $D \in \mathbf{TD}$ such that $L(D)$ consists of all descriptions of TMs whose languages are in the subfamily defined by π.

Definition 7.9. Let $\pi \subseteq \mathbf{L_{TM}}$. Then, π is said to be a property of Turing languages.

(1) A language $L \in \mathbf{L_{TM}}$ satisfies π if $L \in \pi$.
(2) Set $L_\pi = \{\langle M \rangle \mid M \in \mathbf{TM}, L(M) \in \pi\}$. We say that π is decidable if $L_\pi \in \mathbf{L_{TD}}$; otherwise, π is undecidable.
(3) We say that π is trivial if $\pi = \mathbf{L_{TM}}$ or $\pi = \emptyset$; otherwise, π is non-trivial. ■

For instance, the property of being finite is non-trivial because $\{L \in \mathbf{L_{TM}} \mid L \text{ is finite}\}$ is a non-empty proper subfamily of $\mathbf{L_{TM}}$. As a matter of fact, there are only two trivial properties: $\mathbf{L_{TM}}$ and $\emptyset$, and both are trivially decidable because they are true either for all members of $\mathbf{L_{TM}}$ or for no member of $\mathbf{L_{TM}}$. As a result, we concentrate our attention on the non-trivial properties in what follows. Surprisingly, Rice's theorem, as follows, states that all non-trivial properties are undecidable.

Theorem 7.29. Rice's Theorem. *Every non-trivial property of Turing languages is undecidable.*

Proof. Let π be a non-trivial property. Without any loss of generality, suppose that $\emptyset \notin \pi$ (as an exercise, reformulate this proof in terms of $\sim \pi$ if $\emptyset \in \pi$). As π is non-trivial, π is non-empty, so there exists a Turing language $K \in \pi$. Let $N \in \mathbf{TM}$ be a TM such that $K = L(N)$.

For the sake of obtaining a contradiction, assume that π is decidable. In other words, there exists a TD $D \in \mathbf{TD}$ that decides L_π. Next, we demonstrate that under this assumption, ${}_{TM\text{-}Halting}L$ would belong to $\mathbf{L_{TD}}$, which contradicts Theorem 7.12. Indeed, we construct an algorithm that takes any $\langle M, x\rangle$, where $M \in \mathbf{TM}$ and $x \in \Delta^*$, and produces $\langle O\rangle$ as output, where $O \in \mathbf{TM}$, so $\langle M, x\rangle \in {}_{TM\text{-}Halting}L$ iff $\langle O\rangle \in L_\pi$, and by using this equivalence and D, we would decide ${}_{TM\text{-}Halting}L$. O is designed in such a way that it works on every input string y as follows:

(1) O saves y and runs M on x;
(2) if M halts on x, O runs N on y and accepts iff N accepts y.

If M loops on x, so does O. As O works on every y in this way, $L(O) = \emptyset$ iff M loops on x. If M halts on x, O runs N on y, and O accepts y iff N accepts y, so $L(O) = L(N) = K$ in this case (recall that the case when $K = \emptyset$ is ruled out because $\emptyset \notin \pi$). Thus, $\langle M, x\rangle \in {}_{TM\text{-}Halting}L$ iff $\langle O\rangle \in L_\pi$. Apply D to decide whether $\langle O\rangle \in L_\pi$. If so, $\langle M, x\rangle \in {}_{TM\text{-}Halting}L$, and if not, $\langle M, x\rangle \notin {}_{TM\text{-}Halting}L$, so ${}_{TM\text{-}Halting}L$ would be in $\mathbf{L_{TD}}$, which contradicts ${}_{TM\text{-}Halting}L \notin \mathbf{L_{TD}}$ (see Theorem 7.12). Therefore, L_π is undecidable. □

Rice's theorem is a powerful result that has a great variety of consequences. For instance, consider the properties of being finite, regular, and context-free as properties of Turing languages. Rice's theorem straightforwardly implies that all these properties are undecidable.

7.3 Complexity

This section takes a finer look at TDs by discussing their computational complexity. This complexity is measured according to their time and space computational requirements. The time complexity equals the number of moves they need to make a decision, while the space complexity is defined as the number of visited tape squares. Perhaps most importantly, this section points out that some problems are tractable for their reasonable computational requirements, while others are intractable for their unmanageably high computational requirements to decide them. Simply put, there exist problems that are decidable in theory, but their decision is intractable in practice.

As most topics concerning complexity are too complicated to discuss them in this introductory text, this section differs from the previous sections of this chapter, which have discussed their material in the form of mathematical formulas and proofs. Rather than give a fully rigorous presentation of computational complexity, this section only explains the basic ideas underlying it. Indeed, it restricts its attention to the very fundamental concepts and results, which are usually described informally. The section omits mathematically precise proofs. On the other hand, it points out some important open problems concerning computational complexity.

We begin with an explanation of time complexity, after which we briefly conceptualize space complexity.

Time complexity

Observe that the following definition that formalizes the time complexity of a TD considers the worst-case scenario concerning this complexity.

Definition 7.10. Let M be a TD. The *time-complexity* function of M, denoted by $time_M$, is defined over ${}_0\mathbb{N}$, so for all $n \in {}_0\mathbb{N}$, $time_M(n)$ is the maximal number of moves M makes on an input string of length n before halting. ■

Example 7.10. Return to the TD D in Example 5.1 such that $L(D) = \{x \mid x \in \{a, b, c\}^*, occur(x, a) = occur(x, b) = occur(x, c)\}$. Recall that D scans across the tape in a left-to-right way while erasing the leftmost occurrences of a, b, and c. When it reaches $\lhd$ after erasing all three occurrences, it moves left to $\rhd$ and makes another scan of this kind. However, when D reaches $\lhd$, while some of the three symbols are missing on the tape, D makes its final return to $\rhd$ and halts by making one more move during which it accepts or rejects as described in Example 5.1. Let g be the integer function over ${}_0\mathbb{N}$ defined for all ${}_0\mathbb{N}$ so that if $n \in {}_0\mathbb{N}$ is divisible by $3, g(n) = n(2(n/3)) + 1$, and if $n \in \mathbb{N}$ is indivisible by $3, g(n) = g(m) + 2n$, where m is the smallest $m \in {}_0\mathbb{N}$ such that $m \leq n$ and m is divisible by 3. Observe that $time_D(n) = g(n)$.

As an exercise, design another TD E such that $L(D) = L(E)$ and $time_E(n) = time_D(n)$, for all $n \in \mathbb{N}$. ■

As a general rule, for $M \in \mathbf{L_{TD}}$, $time_M$ is a complicated polynomial, whose determination represents a tedious and difficult task. Besides this difficulty, we are usually interested in the time complexity of M only when it is run on large inputs. As a result, rather than determine $time_M$ rigorously, we often consider the highest-order term of $time_M$; on the other hand, we disregard the coefficient of this term as well as any lower terms. The elegant *big-O notation*, defined in the following, is customarily used for this purpose.

Definition 7.11.

(I) Let f and g be two functions over ${}_0\mathbb{N}$. If there exist $c, d \in \mathbb{N}$ such that for every $n \geq d$, $f(n)$ and $g(n)$ are defined and $f(n) \leq cg(n)$, then g is an upper bound for f, written as $f = O(g)$.

(II) If $f = O(g)$ and g is of the form n^m, where $m \in \mathbb{N}$, then g is a polynomial bound for f.

(III) Let $M \in \mathbf{L_{TD}}$. M is polynomially bounded if there is a polynomial bound for $time_M$. ■

Let f and g be two polynomials. In essence, according to (I) and (II) in Definition 7.11, $f = O(g)$ says that f is less than or equal to g if we disregard differences regarding multiplicative constants and lower-order terms. Indeed, $f = O(g)$ implies $kf = O(g)$, for any $k \in \mathbb{N}$, so the multiplication constants are ignored. As $f(n) = cg(n)$ holds for all $n \geq d$, the values of any $n \leq d$ are also completely ignored. In practice, to obtain $g = n^m$, as described in (II), we simply take n^m as the highest-order term of f without its coefficient; for instance, if $f(n) = 918273645n^5 + 999n^4 + 1111n^3 + 71178n^2 + 98765431n + 1298726$, then $f = O(n^5)$. On the other hand, if $f \neq O(g)$, then there exist infinitely many values of n satisfying $f(n) > cg(n)$.

Based upon (III) of Definition 7.11, from a more practical point of view, we next distinguish the decidable problems that are possible to compute from the decidable problems that are not.

Definition 7.12. Let P be a decidable problem. If P is decided by a polynomially bounded TD, P is *tractable*; otherwise, P is *intractable*. ■

Informally, this definition says that although intractable problems are decidable in principle, they can hardly be decided in reality as no decision maker can decide them in polynomial time. On the other hand, tractable problems can be decided in polynomial time, so they are central to practically oriented computer science. Besides their practical significance, however, tractable problems lead to some crucial topics of theoretical computer science, as demonstrated next.

According to Convention 5.3, until now, we have automatically assumed that the TMs work deterministically. We also know that deterministic TMs are as powerful as their non-deterministic versions (see Definition 5.2 and Theorem 5.1). In terms of their time complexity, however, their relationship remains open, as pointed out shortly. Before this, we reformulate some of the previous notions in terms of non-deterministic TMs.

Definition 7.13.

(1) Let M be a TM according to Definition 5.1 (thus, M may not be deterministic). M is a non-deterministic TD if M halts on every input string.

(2) Let M be a non-deterministic TD. The time complexity of $M, time_M$, is defined by analogy with Definition 7.10; that is, for all $n \in {}_0\mathbb{N}, time_M(n)$ is the maximal number of moves M makes on an input string of length n before halting.

(3) Like in Definition 7.11, M is polynomially bounded if there is a polynomial bound for $time_M$. ∎

Convention 7.10. $\mathbf{L_P}$ denotes the family of languages accepted by polynomially bounded (deterministic) TDs, and $\mathbf{L_{NP}}$ denotes the family of languages accepted by polynomially bounded non-deterministic TDs. ∎

Note that any TD represents a special case of a non-deterministic TD, so $\mathbf{L_P} \subseteq \mathbf{L_{NP}}$. However, it is a longstanding open problem whether $\mathbf{L_P} = \mathbf{L_{NP}}$, referred to as the $P = NP$ problem. By using various methods, theoretical computer science has intensively attempted to decide this problem. One of the most important approaches to this problem is based on ordering the languages in $\mathbf{L_{NP}}$. The equivalence classes defined by this ordering consist of languages coding equally difficult problems. Considering the class corresponding to the most difficult problems, any problem coded by a language from this family is as difficult as any other problem coded by a language from $\mathbf{L_{NP}}$. Consequently, if we prove that this class contains a language that also belongs to $\mathbf{L_P}$, then $\mathbf{L_P} = \mathbf{L_{NP}}$; on the other hand, if we demonstrate that this class contains a language that does not belong to $\mathbf{L_P}$, then $\mathbf{L_P} \subset \mathbf{L_{NP}}$. Next, we describe this approach to the $P = NP$ problem in somewhat greater detail.

Definition 7.14. Let Δ and ς be two alphabets, $J \subseteq \Delta^*$, and $K \subseteq \varsigma^*$. Then, J is polynomially transformable into K, symbolically written as $J \propto K$, if there is a polynomially bounded TD M such that M-f (see Definition 7.1) is a total function from Δ^* to ς^* satisfying $x \in J$ iff M-$f(x) \in K$. ■

In other words, $J \propto K$ means that the difficulty of deciding J is no greater than the difficulty of deciding K, so the problem encoded by J is no more difficult than the problem encoded by K.

Definition 7.15. Let $L \in \mathbf{L_{NP}}$. If $J \propto L$, for every $J \in \mathbf{L_{NP}}$, then L is NP-complete. ■

A decision problem coded by an NP-complete language is an NP-complete problem. There exist a number of well-known NP-complete problems, such as the following problem.

Problem 7.29. *Time-BoundedAcceptance.*
Question: Let M be a non-deterministic TM, $w \in \Delta_M^*$, and $i \in \mathbb{N}$. Does M accept w by computing no more than i moves?
Language: ${}_{TBA}L = \{\langle M, w, i\rangle \mid M$ is a non-deterministic TM, $w \in \Delta_M^*, i \in \mathbb{N}, M$ accepts w by computing i or fewer moves$\}$.

Once again, by finding an NP-complete language L and proving either $L \in \mathbf{L_P}$ or $L \notin \mathbf{L_P}$, we would decide the $P = NP$ problem. Indeed, if $L \in \mathbf{L_P}$, then $\mathbf{L_P} = \mathbf{L_{NP}}$, and if $L \notin \mathbf{L_P}$, then $\mathbf{L_P} \subset \mathbf{L_{NP}}$. So far, however, a proof like this has not been achieved yet, and the $P = NP$ problem remains open.

Space complexity

We close this section by a remark about the space complexity of TDs.

Definition 7.16. Let $M = (Q, \Delta, \Gamma, R, s, F)$ be a TD. A function g over ${}_0\mathbb{N}$ represents the space complexity of M, denoted by $space_M$, if $space_M(i)$ equals the minimal number $j \in {}_0\mathbb{N}$ such that for all $x \in \Delta^i, y, v \in \Gamma^*, \triangleright s_M x \triangleleft \Rightarrow^* \triangleright yqv \triangleleft$ in M implies $|yv| \leq j$. ■

Thus, starting with an input string of length i, M always occurs in a configuration with no more than $space_M(i)$ symbols on the tape. As an exercise, define polynomially space-bounded (deterministic) TDs and polynomially space-bounded non-deterministic TDs by analogy with the corresponding deciders in terms of time complexity (see Definitions 7.11 and 7.13).

Convention 7.11. $\mathbf{L_{PS}}$ denotes the family of languages accepted by polynomially space-bounded (deterministic) TDs, and $\mathbf{L_{NPS}}$ denotes the family of languages accepted by polynomially space-bounded non-deterministic TDs. ■

As opposed to the unknown relationship between $\mathbf{L_P}$ and $\mathbf{L_{NP}}$, we know more about $\mathbf{L_{PS}}$ and $\mathbf{L_{NPS}}$. Indeed, it holds that $\mathbf{L_{PS}} = \mathbf{L_{NPS}} \subset \mathbf{L_{TD}}$. It is also well known that $\mathbf{L_{NP}} \subseteq \mathbf{L_{PS}}$, but it is not known whether this inclusion is proper — another important long-standing open problem in the theory of computation.

Part 3

Trends

This part, consisting of Chapters 8–10, describes three vivid trends in automata theory today. Chapter 8 covers regulated automata, whose computation is controlled by an additional mathematical mechanism. Chapter 9 modifies classical automata into jumping automata as formal models of discontinuous computation. Chapter 10 studies deep pushdown automata, which are based on stacks that can be modified deeper than at their tops.

Chapter 8

Regulated Automata

All the automata covered in Part 2 represent unregulated versions of formal models because they are free to use any applicable rules during their computation at will. This chapter places a restriction on this use. That is, it introduces finite and pushdown automata in which the use of these rules is somehow regulated. The chapter undertakes two essential approaches to this topic: self-regulation (Section 8.1) and external regulation (Section 8.2).

Section 8.1 discusses regulated automata without extending them by any additional regulating means. In these automata, the selection of a rule according to which the current move is made follows from the rule applied during the previous move. In this sense, they actually regulate themselves, hence their name — self-regulating automata.

Section 8.2 deals with automata extended by additional external regulating means. Specifically, these means consist of control languages, which prescribe the use of rules during computation. Working in this way, they represent automata externally regulated by their control languages.

8.1 Self-Regulating Automata

This chapter defines and investigates *self-regulating automata.* They regulate the selection of a rule according to which the current move is made by a rule according to which a previous move was made. Both finite and pushdown versions of these automata are investigated. The chapter is divided into two sections. Section 8.1 discusses *self-regulating finite automata.* It establishes two infinite hierarchies of language families resulting from them. Both hierarchies lie between the family of regular languages and the family of context-sensitive languages. Section 8.1 studies *self-regulating pushdown automata.* Based on them, this section characterizes the families of context-free and recursively enumerable languages. However, as opposed to the results about self-regulating finite automata, many questions concerning their pushdown versions remain open; indeed, Section 8.1 formulates several specific open problem areas, including questions concerning infinite language-family hierarchies resulting from them.

In this chapter, as the first type of regulated automata discussed in Part 4, we define and investigate *self-regulating finite automata.* In essence, self-regulating finite automata regulate the selection of a rule according to which the current move is made by a rule according to which a previous move was made.

To give a more precise insight into self-regulating automata, consider a finite automaton M with a finite binary relation R over the set of rules in M. Furthermore, suppose that M makes a sequence of moves ρ that leads to the acceptance of a string, so ρ can be expressed as a concatenation of $n+1$ consecutive subsequences, $\rho = \rho_0\rho_1 \cdots \rho_n$, where $|\rho_i| = |\rho_j|$, $0 \leq i, j \leq n$, in which r_i^j denotes the rule according to which the ith move in ρ_j is made, for all $0 \leq j \leq n$ and $1 \leq i \leq |\rho_j|$ (as usual, $|\rho_j|$ denotes the length of ρ_j). If for all $0 \leq j < n$, $(r_1^j, r_1^{j+1}) \in R$, then M represents an *n-turn first-move self-regulating finite automaton with respect to R.* If for all $0 \leq j < n$ and all $1 \leq i \leq |\rho_i|$, $(r_i^j, r_i^{j+1}) \in R$, then M represents an *n-turn all-move self-regulating finite automaton with respect to R.*

First, we demonstrate that n-turn first-move self-regulating finite automata give rise to an infinite hierarchy of language families coinciding with the hierarchy resulting from $(n+1)$-parallel right-linear grammars (see Rosebrugh and Wood, 1973, 1975; Wood, 1973, 1975). Recall that n-parallel right-linear grammars generate a proper language subfamily of the language family generated by $(n+1)$-parallel right-linear grammars (see Rosebrugh and Wood, 1975, Theorem 5). As a result, n-turn first-move self-regulating finite automata accept a proper language subfamily of the language family accepted by $(n+1)$-turn first-move self-regulating finite automata for all $n \geq 0$. Similarly, we prove that n-turn all-move self-regulating finite automata give rise to an infinite hierarchy of language families coinciding with the hierarchy resulting from $(n+1)$-right-linear simple matrix grammars (see Dassow and Păun, 1989; Ibarra, 1970; Wood, 1975). As n-right-linear simple matrix grammars generate a proper subfamily of the language family generated by $(n+1)$-right-linear simple matrix grammars (see Dassow and Păun, 1989, Theorem 1.5.4), n-turn all-move self-regulating finite automata accept a proper language subfamily of the language family accepted by $(n+1)$-turn all-move self-regulating finite automata. Furthermore, since the families of right-linear simple matrix languages coincide with the language families accepted by multi-tape non-writing automata (see Fischer and Rosenberg, 1968) and by finite-turn checking automata (see Siromoney, 1971), all-move self-regulating finite automata characterize these families, too. Finally, we summarize the results for both infinite hierarchies.

By analogy with self-regulating finite automata, we introduce and discuss *self-regulating pushdown automata.* Regarding self-regulating all-move pushdown automata, we prove that they do not give rise to any infinite hierarchy analogical to the achieved hierarchies resulting from the self-regulating finite automata. Indeed, zero-turn all-move self-regulating pushdown automata define the family of context-free languages, while one-turn all-move self-regulating pushdown automata define the family of recursively enumerable languages. On the other hand, as far as self-regulating first-move pushdown automata are concerned, the question of whether they define an infinite hierarchy is open.

Self-regulating finite automata

Next, we define and illustrate n-turn first-move self-regulating finite automata and n-turn all-move self-regulating finite automata.

Definition 8.1. A *self-regulating finite automaton* (SFA for short) is a septuple

$$M = (Q, \Delta, \delta, q_0, q_t, F, R),$$

where $(Q, \Delta, \delta, q_0, F)$ is a finite automaton, $q_t \in Q$ is a *turn state*, and $R \subseteq \Psi \times \Psi$ is a finite relation on the alphabet of rule labels. ■

We consider two ways of self-regulation: first-move and all-move. According to these two types of self-regulation, two types of n-turn self-regulating finite automata are defined.

Definition 8.2. Let $n \geq 0$ and $M = (Q, \Delta, \delta, q_0, q_t, F, R)$ be a self-regulating finite automaton. M is said to be an *n-turn first-move self-regulating finite automaton* (n-first-SFA for short) if every $w \in L(M)$ is accepted by M in the following way:

$$q_0 w \vdash_M^* f\,[\mu]$$

such that

$$\mu = r_1^0 \cdots r_k^0 r_1^1 \cdots r_k^1 \cdots r_1^n \cdots r_k^n,$$

where $k \geq 1$, r_k^0 is the first rule of the form $qx \to q_t$, for some $q \in Q$, $x \in \Delta^*$, and

$$(r_1^j, r_1^{j+1}) \in R,$$

for all $j = 0, 1, \ldots, n$. ■

The family of languages accepted by n-first-SFAs is denoted by ${}_n\mathbf{L}_{\mathbf{FSFA}}$.

Example 8.1. Consider a 1-first-SFA

$$M = (\{s, t, f\}, \{a, b\}, \delta, s, t, \{f\}, \{(1, 3)\}),$$

with δ containing the rules (see Figure 8.1)

$$\begin{aligned} 1&: sa \to s, \\ 2&: sa \to t, \\ 3&: tb \to f, \\ 4&: fb \to f. \end{aligned}$$

With $aabb$, M makes

$$saabb \vdash_M sabb\,[1] \vdash_M tbb\,[2] \vdash_M fb\,[3] \vdash_M f\,[4].$$

In brief, $saabb \vdash_M^* f\,[1234]$. Observe that $L(M) = \{a^n b^n \mid n \geq 1\}$, which belongs to $\mathbf{L}_{\mathbf{CF}} - \mathbf{L}_{\mathbf{REG}}$. ■

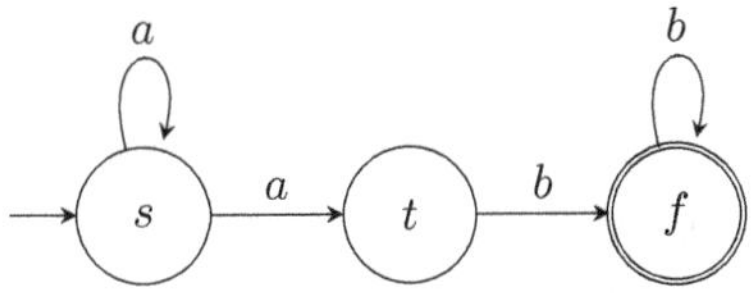

Figure 8.1. 1-turn first-move self-regulating finite automaton M.

Definition 8.3. Let $n \geq 0$ and $M = (Q,\Delta,\delta, q_0, q_t, F, R)$ be a self-regulating finite automaton. M is said to be an *n-turn all-move self-regulating finite automaton* (n-all-SFA for short) if every $w \in L(M)$ is accepted by M in the following way:

$$q_0 w \vdash_M^* f\,[\mu]$$

such that

$$\mu = r_1^0 \cdots r_k^0 r_1^1 \cdots r_k^1 \cdots r_1^n \cdots r_k^n,$$

where $k \geq 1$, r_k^0 is the first rule of the form $qx \to q_t$, for some $q \in Q$, $x \in \Delta^*$, and

$$(r_i^j, r_i^{j+1}) \in R,$$

for all $i = 1, 2, \ldots, k$ and $j = 0, 1, \ldots, n-1$. ■

The family of languages accepted by n-all-SFAs is denoted by ${}_n\mathbf{L}_{\mathbf{ASFA}}$.

Example 8.2. Consider a 1-all-SFA

$$M = \big(\{s,t,f\}, \{a,b\}, \delta, s, t, \{f\}, \{(1,4),(2,5),(3,6)\}\big),$$

with δ containing the following rules (see Figure 8.2):

$$1\colon sa \to s,$$
$$2\colon sb \to s,$$
$$3\colon s \to t,$$
$$4\colon ta \to t,$$
$$5\colon tb \to t,$$
$$6\colon t \to f.$$

With $abab$, M makes

$$sabab \vdash_M sbab\,[1] \vdash_M sab\,[2] \vdash_M tab\,[3] \vdash_M tb\,[4] \vdash_M t\,[5] \vdash_M f\,[6].$$

In brief, $sabab \vdash_M^* f\,[123456]$. Observe that $L(M) = \{ww \mid w \in \{a,b\}^*\}$, which belongs to $\mathbf{L_{CS}} - \mathbf{L_{CF}}$. ■

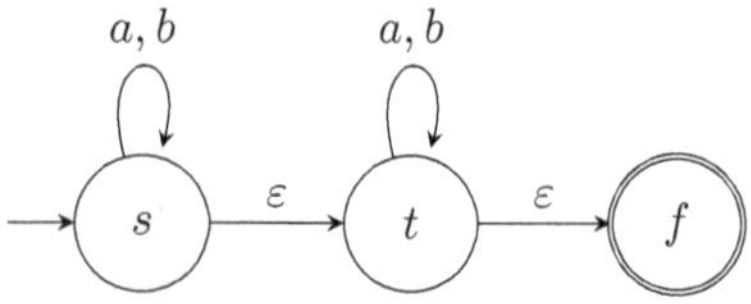

Figure 8.2. 1-turn all-move self-regulating finite automaton M.

Accepting power

In this section, we discuss the accepting power of n-first-SFAs and n-all-SFAs.

n-turn first-move self-regulating finite automata

We prove that the family of languages accepted by n-first-SFAs coincides with the family of languages generated by the so-called $(n+1)$-parallel right-linear grammars (see Rosebrugh and Wood, 1973, 1975; Wood, 1973, 1975). First, however, we define these grammars formally.

Definition 8.4. For $n \geq 1$, an *n-parallel right-linear grammar* (see Rosebrugh and Wood, 1973, 1975; Wood, 1973, 1975) (n-PRLG for short) is an $(n+3)$-tuple

$$G = (N_1, \ldots, N_n, T, S, P),$$

where N_i, $1 \leq i \leq n$, are pairwise disjoint *nonterminal alphabets*, T is a *terminal alphabet*, $S \notin N$ is an *initial symbol*, where $N = N_1 \cup \cdots \cup N_n$, and P is a finite set of *rules* that contains the following three kinds of rules:

1. $S \to X_1 \cdots X_n$, $\quad X_i \in N_i$, $1 \leq i \leq n$;
2. $X \to wY$, $\quad X, Y \in N_i$ for some i, $1 \leq i \leq n$, $w \in T^*$;
3. $X \to w$, $\quad X \in N$, $w \in T^*$.

For $x, y \in (N \cup T \cup \{S\})^*$,

$$x \Rightarrow_G y$$

if and only if:

(1) either $x = S$ and $S \to y \in P$,
(2) or $x = y_1 X_1 \cdots y_n X_n$, $y = y_1 x_1 \cdots y_n x_n$, where $y_i \in T^*$, $x_i \in T^* N \cup T^*$, $X_i \in N_i$, and $X_i \to x_i \in P$, $1 \le i \le n$.

Let $x, y \in (N \cup T \cup \{S\})^*$ and $\ell > 0$. Then, $x \Rightarrow_G^\ell y$ if and only if there exists a sequence

$$x_0 \Rightarrow_G x_1 \Rightarrow_G \cdots \Rightarrow_G x_\ell,$$

where $x_0 = x$ and $x_\ell = y$. As usual, $x \Rightarrow_G^+ y$ if and only if there exists $\ell > 0$ such that $x \Rightarrow_G^\ell y$, and $x \Rightarrow_G^* y$ if and only if $x = y$ or $x \Rightarrow_G^+ y$.

The *language* of G is defined as

$$L(G) = \{w \in T^* \mid S \Rightarrow_G^+ w\}.$$

A language $K \subseteq T^*$ is an *n-parallel right-linear language* (n-PRLL for short) if there is an n-PRLG G such that $K = L(G)$. ■

The family of n-PRLLs is denoted by ${}_n\mathbf{L}_{\mathbf{PRL}}$.

Definition 8.5. Let $G = (N_1, \ldots, N_n, T, S, P)$ be an n-PRLG, for some $n \ge 1$, and $1 \le i \le n$. By the *ith component* of G, we understand the 1-PRLG

$$G = (N_i, T, S', P'),$$

where P' contains rules of the following forms:

1. $S' \to X_i$ if $S \to X_1 \cdots X_n \in P$, $X_i \in N_i$;
2. $X \to wY$ if $X \to wY \in P$ and $X, Y \in N_i$;
3. $X \to w$ if $X \to w \in P$ and $X \in N_i$.

■

To prove that the family of languages accepted by n-first-SFAs coincides with the family of languages generated by $(n+1)$-PRLGs, we need the following normal form of PRLGs.

Lemma 8.1. *For every n-PRLG $G = (N_1, \ldots, N_n, T, S, P)$, there is an equivalent n-PRLG $G' = (N'_1, \ldots, N'_n, T, S, P')$ that satisfies:*

(i) *if $S \to X_1 \cdots X_n \in P'$, then X_i does not occur on the right-hand side of any rule, for $i = 1, 2, \ldots, n$;*
(ii) *if $S \to \alpha$, $S \to \beta \in P'$, and $\alpha \ne \beta$, then* $\mathrm{alph}(\alpha) \cap \mathrm{alph}(\beta) = \emptyset$.

Proof. If G does not satisfy the conditions from the lemma, then we construct a new n-PRLG:

$$G' = (N'_1, \dots, N'_n, T, S, P'),$$

where P' contains all rules of the form $X \to \beta \in P$, $X \neq S$, and $N_j \subseteq N'_j$, $1 \leq j \leq n$. For each rule $S \to X_1 \cdots X_n \in P$, we add new nonterminals $Y_j \notin N'_j$ into N'_j, and the rules include $S \to Y_1 \cdots Y_n$ and $Y_j \to X_j$ in P', $1 \leq j \leq n$. Clearly,

$$S \Rightarrow_G X_1 \cdots X_n \text{ if and only if } S \Rightarrow_{G'} Y_1 \cdots Y_n \Rightarrow_{G'} X_1 \cdots X_n.$$

Thus, $L(G) = L(G')$. □

Lemma 8.2. *Let G be an n-PRLG. Then, there is an $(n-1)$-first-SFA M such that $L(G) = L(M)$.*

Proof. Informally, M is divided into n parts (see Figure 8.3). The ith part represents a finite automaton accepting the language of the ith component of G, and R also connects the ith part to the $(i+1)$st part, as depicted in Figure 8.3.

Formally, without loss of generality, we assume $G = (N_1, \dots, N_n, T, S, P)$ to be in the form from Lemma 8.1. We construct an $(n-1)$-first-SFA

$$M = (Q, T, \delta, q_0, q_t, F, R),$$

where

$$\begin{aligned}
Q &= \{q_0, \dots, q_n\} \cup N, N = N_1 \cup \cdots \cup N_n, \{q_0, q_1, \dots, q_n\} \cap N = \emptyset,\\
&\quad F = \{q_n\},\\
\delta &= \{q_i \to X_{i+1} \mid S \to X_1 \cdots X_n \in P, 0 \leq i < n\} \cup\\
&\quad \{Xw \to Y \mid X \to wY \in P\} \cup\\
&\quad \{Xw \to q_i \mid X \to w \in P, w \in T^*, X \in N_i, i \in \{1, \dots, n\}\},\\
q_t &= q_1,\\
\Psi &= \delta,\\
R &= \{(q_i \to X_{i+1}, q_{i+1} \to X_{i+2}) \mid S \to X_1 \cdots X_n \in P, 0 \leq i \leq n-2\}.
\end{aligned}$$

Next, we prove that $L(G) = L(M)$. To prove that $L(G) \subseteq L(M)$, consider any derivation of w in G, and construct an acceptance of w in M depicted in Figure 8.3.

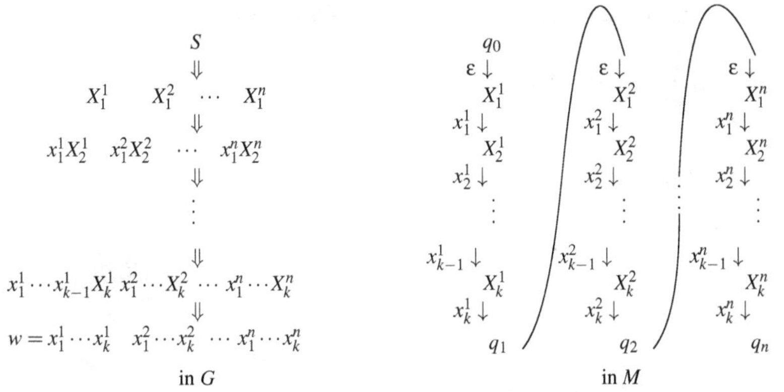

Figure 8.3. A derivation of w in G and the corresponding acceptance of w in M.

This figure clearly demonstrates the fundamental idea behind this part of the proof; its complete and rigorous version is left to the reader. Thus, M accepts every $w \in T^*$ such that $S \Rightarrow_G^* w$.

To prove that $L(M) \subseteq L(G)$, consider any $w \in L(M)$ and any acceptance of w in M. Observe that the acceptance is of the form depicted on the right-hand side of Figure 8.3. It means that the number of steps M made from q_{i-1} to q_i is the same as from q_i to q_{i+1} since the only rule in relation with $q_{i-1} \to X_1^i$ is the rule $q_i \to X_1^{i+1}$. Moreover, M can never come back to a state corresponding to a previous component. (By a component of M, we mean the finite automaton

$$M_i = \big(Q, \Delta, \delta, q_{i-1}, \{q_i\}\big),$$

for $1 \le i \le n$.) Next, construct a derivation of w in G. By Lemma 8.1, we have

$$\big|\{X \mid (q_i \to X_1^{i+1}, q_{i+1} \to X) \in R\}\big| = 1,$$

for all $0 \le i < n-1$. Thus, $S \to X_1^1 X_1^2 \cdots X_1^n \in P$. Moreover, if $X_j^i x_j^i \to X_{j+1}^i$, we apply $X_j^i \to x_j^i X_{j+1}^i \in P$, and if $X_k^i x_k^i \to q_i$, we apply $X_k^i \to x_k^i \in P$, $1 \le i \le n$, $1 \le j < k$.

Hence, Lemma 8.2 holds. □

Lemma 8.3. *Let M be an n-first-SFA. Then, there is an $(n+1)$-PRLG G such that $L(G) = L(M)$.*

Proof. Let $M = (Q,\Delta,\delta, q_0, q_t, F, R)$. Consider

$$G = \big(N_0, \ldots, N_n,\Delta, S, P\big),$$

where

$$\begin{aligned}
N_i &= (Q\Delta^l \times Q \times \{i\} \times Q) \cup (Q \times \{i\} \times Q),\\
l \; &= \max(\{|w| \mid qw \to p \in \delta\}), 0 \le i \le n,\\
P \; &= \{S \to \; [q_0x_0, q^0, 0, q_t][q_tx_1, q^1, 1, q_{i_1}][q_{i_1}x_2, q^2, 2, q_{i_2}] \cdots\\
&\quad [q_{i_{n-1}}x_n, q^n, n, q_{i_n}] \mid\\
&\quad\; r_0\colon q_0x_0 \to q^0, r_1\colon q_tx_1 \to q^1, r_2\colon q_{i_1}x_2 \to q^2, \ldots, r_n\colon\\
&\quad q_{i_{n-1}}x_n \to q^n \in \delta,\}\\
&\quad\; (r_0, r_1), (r_1, r_2), \ldots, (r_{n-1}, r_n) \in R,\ q_{i_n} \in F\} \cup\\
&\quad \{[px, q, i, r] \to x[q, i, r]\} \cup\\
&\quad \{[q, i, q] \to \varepsilon \mid q \in Q\} \cup\\
&\quad \{[q, i, p] \to w[q', i, p] \mid qw \to q' \in \delta\}.
\end{aligned}$$

Next, we prove that $L(G) = L(M)$. To prove that $L(G) \subseteq L(M)$, observe that we make $n{+}1$ copies of M and go through them similarly to Figure 8.3. Consider a derivation of w in G. Then, in a greater detail, this derivation is of the form

$$\begin{aligned}
S &\Rightarrow_G [q_0x_0^0, q_1^0, 0, q_t][q_tx_0^1, q_1^1, 1, q_{i_1}] \cdots [q_{i_{n-1}}x_0^n, q_1^n, n, q_{i_n}]\\
&\Rightarrow_G x_0^0[q_1^0, 0, q_t]x_0^1[q_1^1, 1, q_{i_1}] \cdots x_0^n[q_1^n, n, q_{i_n}]\\
&\Rightarrow_G x_0^0x_1^0[q_2^0, 0, q_t]x_0^1x_1^1[q_2^1, 1, q_{i_1}] \cdots x_0^nx_1^n[q_2^n, n, q_{i_n}] \qquad (8.1)\\
&\;\;\vdots\\
&\Rightarrow_G x_0^0x_1^0 \cdots x_k^0[q_t, 0, q_t]x_0^1x_1^1 \cdots x_k^1[q_{i_1}, 1, q_{i_1}] \cdots x_0^nx_1^n \cdots x_k^n[q_{i_n}, n, q_{i_n}]\\
&\Rightarrow_G x_0^0x_1^0 \cdots x_k^0x_0^1x_1^1 \cdots x_k^1 \cdots x_0^nx_1^n \cdots x_k^n,
\end{aligned}$$

and

$$r_0\colon q_0x_0^0 \to q_1^0, r_1\colon q_tx_0^1 \to q_1^1, r_2\colon q_{i_1}x_0^2 \to q_1^2, \ldots, r_n\colon q_{i_{n-1}}x_0^n \to q_1^n \in \delta,$$

$$(r_0, r_1), (r_1, r_2), \ldots, (r_{n-1}, r_n) \in R,$$

and $q_{i_n} \in F$.

Thus, the sequence of rules used in the acceptance of w in M is

$$\begin{aligned}\mu = & (q_0 x_0^0 \rightarrow q_1^0)(q_1^0 x_1^0 \rightarrow q_2^0) \cdots (q_k^0 x_k^0 \rightarrow q_t) \\ & (q_t x_0^1 \rightarrow q_1^1)(q_1^1 x_1^1 \rightarrow q_2^1) \cdots (q_k^1 x_k^1 \rightarrow q_{i_1}) \\ & (q_{i_1} x_0^2 \rightarrow q_1^2)(q_1^2 x_1^2 \rightarrow q_2^2) \cdots (q_k^2 x_k^2 \rightarrow q_{i_2}) \\ & \vdots \\ & (q_{i_{n-1}} x_0^n \rightarrow q_1^n)(q_1^n x_1^n \rightarrow q_2^n) \cdots (q_k^n x_k^n \rightarrow q_{i_n}).\end{aligned} \tag{8.2}$$

Next, we prove that $L(M) \subseteq L(G)$. Informally, the acceptance is divided into $n+1$ parts of the same length. The grammar G generates the ith part by the ith component and records the state from which the next component starts.

Let μ be a sequence of rules used in an acceptance of

$$w = x_0^0 x_1^0 \cdots x_k^0 x_0^1 x_1^1 \cdots x_k^1 \cdots x_0^n x_1^n \cdots x_k^n$$

in M of the form (8.2). Then, the derivation of the form (8.1) is the corresponding derivation of w in G since $[q_j^i, i, p] \rightarrow x_j^i [q_{j+1}^i, i, p] \in P$ and $[q, i, q] \rightarrow \varepsilon$, for all $0 \leq i \leq n$, $1 \leq j < k$.

Hence, Lemma 8.3 holds. □

Following is the first main result of this chapter.

Theorem 8.1. *For all* $n \geq 0$, ${}_n\mathbf{L}_{\mathbf{FSFA}} = {}_{n+1}\mathbf{L}_{\mathbf{PRL}}$.

Proof. This proof follows from Lemmas 8.2 and 8.3. □

Corollary 8.1. *The following statements hold true:*

(i) $\mathbf{L}_{\mathbf{REG}} = \mathbf{L}_{\mathbf{FSFA}} - 0 \subset \mathbf{L}_{\mathbf{FSFA}} - 1 \subset \mathbf{L}_{\mathbf{FSFA}} - 2 \subset \cdots \subset \mathbf{L}_{\mathbf{CS}}$.
(ii) $\mathbf{L}_{\mathbf{FSFA}} - 1 \subset \mathbf{L}_{\mathbf{CF}}$.
(iii) $\mathbf{L}_{\mathbf{FSFA}} - 2 \not\subseteq \mathbf{L}_{\mathbf{CF}}$.
(iv) $\mathbf{L}_{\mathbf{CF}} \not\subseteq {}_n\mathbf{L}_{\mathbf{FSFA}}$ *for any* $n \geq 0$.
(v) *For all* $n \geq 0$, ${}_n\mathbf{L}_{\mathbf{FSFA}}$ *is closed under union, finite substitution, homomorphism, intersection with a regular language, and right quotient with a regular language.*
(vi) *For all* $n \geq 1$, ${}_n\mathbf{L}_{\mathbf{FSFA}}$ *is not closed under intersection and complement.*

Proof. Recall the following statements that were proved by Rosebrugh and Wood (1975):

- $\mathbf{L_{REG}} = \mathbf{L_{PRL}} - 1 \subset \mathbf{L_{PRL}} - 2 \subset \mathbf{L_{PRL}} - 3 \subset \cdots \subset \mathbf{L_{CS}}$.
- $\mathbf{L_{PRL}} - 2 \subset \mathbf{L_{CF}}$.
- $\mathbf{L_{CF}} \not\subseteq {}_n\mathbf{L_{PRL}}$, $n \geq 1$.
- For all $n \geq 1$, ${}_n\mathbf{L_{PRL}}$ is closed under union, finite substitution, homomorphism, intersection with a regular language, and right quotient with a regular language.
- For all $n \geq 2$, ${}_n\mathbf{L_{PRL}}$ is not closed under intersection and complement.

These statements and Theorem 8.1 imply statements (i), (ii), (iv), (v), and (vi) in Corollary 8.1. Moreover, observe that

$$\{a^n b^n c^{2n} \mid n \geq 0\} \in \mathbf{L_{FSFA}} - 2 - \mathbf{L_{CF}},$$

which proves (iii). □

Theorem 8.2. *For all $n \geq 1$, ${}_n\mathbf{L_{FSFA}}$ is not closed under inverse homomorphism.*

Proof. For $n = 1$, let $L = \{a^k b^k \mid k \geq 1\}$, and let the homomorphism $h : \{a, b, c\}^* \to \{a, b\}^*$ be defined as $h(a) = a$, $h(b) = b$, and $h(c) = \varepsilon$. Then, $L \in \mathbf{L_{FSFA}} - 1$, but

$$L' = h^{-1}(L) \cap c^*a^*b^* = \{c^* a^k b^k \mid k \geq 1\} \notin \mathbf{L_{FSFA}} - 1.$$

Assume that L' is in $\mathbf{L_{FSFA}} - 1$. Then, by Theorem 8.1, there is a 2-PRLG

$$G = (N_1, N_2, T, S, P)$$

such that $L(G) = L'$. Let

$$k > |P| \cdot \max(\{|w| \mid X \to wY \in P\}).$$

Consider a derivation of $c^k a^k b^k \in L'$. The second component can generate only finitely many as; otherwise, it derives $\{a^k b^n \mid k < n\}$, which is not regular. Analogously, the first component generates only finitely many bs. Therefore, the first component generates any number of as, and the second component generates any number of bs.

Moreover, there is a derivation of the form $X \Rightarrow_G^m X$, for some $X \in N_2$ and $m \geq 1$, used in the derivation of the second component. In the first component, there is a derivation $A \Rightarrow_G^l a^s A$, for some $A \in N_1$ and $s, l \geq 1$. Then, we can modify the derivation of $c^k a^k b^k$ so that in the first component, we repeat the cycle $A \Rightarrow_G^l a^s A$ $(m+1)$-times, and in the second component, we repeat the cycle $X \Rightarrow_G^m X$ $(l+1)$-times. The derivations of both components have the same length: The added cycles are of length ml, and the rest is of the same length as in the derivation of $c^k a^k b^k$. Therefore, we have derived $c^k a^r b^k$, where $r > k$, which is not in L' — a contradiction.

For $n > 1$, the proof is analogous and left to the reader. □

Corollary 8.2. *For all $n \geq 1$, ${}_n\mathbf{L}_{\mathbf{FSFA}}$ is not closed under concatenation. Therefore, it is not closed under Kleene closure either.*

Proof. For $n = 1$, let $L_1 = \{c\}^*$ and $L_2 = \{a^k b^k \mid k \geq 1\}$. Then,

$$L_1 L_2 = \{c^j a^k b^k \mid k \geq 1, j \geq 0\}.$$

Analogously, prove this corollary for $n > 1$. □

n-turn all-move self-regulating finite automata

We next turn our attention to n-all-SFAs. We prove that the family of languages accepted by n-all-SFAs coincides with the family of languages generated by the so-called n-right-linear simple matrix grammars (see Dassow and Păun, 1989; Ibarra, 1970; Wood, 1975). First, however, we define these grammars formally.

Definition 8.6. For $n \geq 1$, an *n-right-linear simple matrix grammar* (see Dassow and Păun, 1989; Ibarra, 1970; Wood, 1975), n-RLSMG for short, is an $(n+3)$-tuple

$$G = (N_1, \ldots, N_n, T, S, P),$$

where N_i, $1 \leq i \leq n$, are pairwise disjoint *nonterminal alphabets*, T is a *terminal alphabet*, $S \notin N$ is an *initial symbol*, where $N = N_1 \cup \cdots \cup N_n$, and P is a finite set of *matrix rules*. A matrix rule can be in one of the following three forms:

1. $[S \to X_1 \cdots X_n]$, $\quad X_i \in N_i$, $1 \le i \le n$;
2. $[X_1 \to w_1Y_1, \cdots, X_n \to w_nY_n]$, $\quad w_i \in T^*$, $X_i, Y_i \in N_i$, $1 \le i \le n$;
3. $[X_1 \to w_1, \cdots, X_n \to w_n]$, $\quad X_i \in N_i$, $w_i \in T^*$, $1 \le i \le n$.

Let m be a matrix. Then, $m[i]$ denotes the ith rule of m. For $x, y \in (N \cup T \cup \{S\})^*$,

$$x \Rightarrow_G y$$

if and only if

(1) either $x = S$ and $[S \to y] \in P$,
(2) or $x = y_1X_1 \cdots y_nX_n$, $y = y_1x_1 \cdots y_nx_n$, where $y_i \in T^*$, $x_i \in T^*N \cup T^*$, $X_i \in N_i$, $1 \le i \le n$, and $[X_1 \to x_1, \cdots, X_n \to x_n] \in P$.

We define $x \Rightarrow_G^+ y$ and $x \Rightarrow_G^* y$ as in Definition 8.4.

The *language* of G is defined as

$$L(G) = \{w \in T^* \mid S \Rightarrow_G^* w\}.$$

A language $K \subseteq T^*$ is an *n-right linear simple matrix language* (n-RLSML for short) if there is an n-RLSMG G such that $K = L(G)$. ■

The family of n-RLSMLs is denoted by ${}_n\mathbf{L}_{\mathbf{RLSM}}$. Furthermore, the ith component of an n-RLSMG is defined analogously to the ith component of an n-PRLG (see Definition 8.5).

To prove that the family of languages accepted by n-all-SFAs coincides with the family of languages generated by n-RLSMGs, the following lemma is needed.

Lemma 8.4. *For every n-RLSMG, $G = (N_1, \ldots, N_n, T, S, P)$, there is an equivalent n-RLSMG G' that satisfies (i)– (iii), given as follows:*

(i) *If $[S \to X_1 \cdots X_n]$, then X_i does not occur on the right-hand side of any rule, $1 \le i \le n$.*
(ii) *If $[S \to \alpha]$, $[S \to \beta] \in P$ and $\alpha \ne \beta$, then* $\mathrm{alph}(\alpha) \cap \mathrm{alph}(\beta) = \emptyset$.
(iii) *For any two matrices $m_1, m_2 \in P$, if $m_1[i] = m_2[i]$, for some $1 \le i \le n$, then $m_1 = m_2$.*

Proof. The first two conditions can be proved analogously to Lemma 8.1. Suppose that there are matrices m and m' such that

$m[i] = m'[i]$, for some $1 \leq i \leq n$. Let

$$\begin{aligned} m \; &= [X_1 \to x_1, \ldots, X_n \to x_n], \\ m' &= [Y_1 \to y_1, \ldots, Y_n \to y_n]. \end{aligned}$$

Replace these matrices with the matrices

$$\begin{aligned} m_1 &= [X_1 \to X_1', \ldots, X_n \to X_n'], \\ m_2 &= [X_1' \to x_1, \ldots, X_n' \to x_n], \\ m_1' &= [Y_1 \to Y_1'', \ldots, Y_n \to Y_n''], \\ m_2' &= [Y_1'' \to y_1, \ldots, Y_n'' \to y_n], \end{aligned}$$

where X_i' and Y_i'' are new nonterminals for all i. These new matrices satisfy condition (iii). Repeat this replacement until the resulting grammar satisfies the properties of G' given in this lemma. □

Lemma 8.5. *Let G be an n-RLSMG. There is an $(n-1)$-all-SFA M such that $L(G) = L(M)$.*

Proof. Without loss of generality, we assume that $G = (N_1, \ldots, N_n, T, S, P)$ is in the form described in Lemma 8.4. We construct an $(n-1)$-all-SFA

$$M = \big(Q, T, \delta, q_0, q_t, F, R\big),$$

where

$$\begin{aligned}
&Q = \{q_0, \ldots, q_n\} \cup N, N = N_1 \cup \cdots \cup N_n, \{q_0, q_1, \ldots, q_n\} \cap N = \emptyset, \\
&F = \{q_n\}, \\
&\delta \; = \{q_i \to X_{i+1} \mid [S \to X_1 \cdots X_n] \in P, 0 \leq i < n\} \cup \\
&\qquad \{X_i w_i \to Y_i \mid [X_1 \to w_1 Y_1, \ldots, X_n \to w_n Y_n] \in P, 1 \leq i \leq n\} \cup \\
&\qquad \{X_i w_i \to q_i \mid [X_1 \to w_1, \ldots, X_n \to w_n] \in P, w_i \in T^*, 1 \leq i \leq n\}, \\
&q_t = q_1, \\
&\Psi = \delta, \\
&R = \{(q_i \to X_{i+1}, q_{i+1} \to X_{i+2}) \mid \\
&\qquad\qquad [S \to X_1 \cdots X_n] \in P, 0 \leq i \leq n-2\} \cup \\
&\qquad \{(X_i w_i \to Y_i, X_{i+1} w_{i+1} \to Y_{i+1}) \mid \\
&\qquad\qquad [X_1 \to w_1 Y_1, \ldots, X_n \to w_n Y_n] \in P,\; 1 \leq i < n\} \cup \\
&\qquad \{(X_i w_i \to q_i, X_{i+1} w_{i+1} \to q_{i+1}) \mid \\
&\qquad\qquad [X_1 \to w_1, \ldots, X_n \to w_n] \in P, w_i \in T^*, 1 \leq i < n\}.
\end{aligned}$$

Next, we prove that $L(G) = L(M)$. A proof of $L(G) \subseteq L(M)$ can be made by analogy with the proof of the same inclusion of Lemma 8.2, which is left to the reader.

To prove that $L(M) \subseteq L(G)$, consider $w \in L(M)$ and an acceptance of w in M. As in Lemma 8.2, the derivation looks like the one depicted on the right-hand side of Figure 8.3. Next, we describe how G generates w. By Lemma 8.4, there is the matrix

$$[S \rightarrow X_1^1 X_1^2 \cdots X_1^n] \in P.$$

Moreover, if $X_j^i x_j^i \rightarrow X_{j+1}^i$, $1 \leq i \leq n$, then

$$(X_j^i \rightarrow x_j^i X_{j+1}^i, X_j^{i+1} \rightarrow x_j^{i+1} X_{j+1}^{i+1}) \in R,$$

for $1 \leq i < n$, $1 \leq j < k$. We apply

$$[X_j^1 \rightarrow x_j^1 X_{j+1}^1, \ldots, X_j^n \rightarrow x_j^n X_{j+1}^n] \in P.$$

If $X_k^i x_k^i \rightarrow q_i$, $1 \leq i \leq n$, then

$$(X_k^i \rightarrow x_k^i, X_k^{i+1} \rightarrow x_k^{i+1}) \in R,$$

for $1 \leq i < n$, and we apply

$$[X_k^1 \rightarrow x_k^1, \ldots, X_k^n \rightarrow x_k^n] \in P.$$

Thus, $w \in L(G)$.

Hence, Lemma 8.5 holds. □

Lemma 8.6. *Let M be an n-all-SFA. There is an $(n+1)$-RLSMG G such that $L(G) = L(M)$.*

Proof. Let $M = (Q,\Delta,\delta, q_0, q_t, F, R)$. Consider

$$G = \big(N_0, \dots, N_n, \Delta, S, P\big),$$

where

$$\begin{aligned}
N_i &= (Q\Delta^l \times Q \times \{i\} \times Q) \cup (Q \times \{i\} \times Q),\\
l \ &= \max(\{|w| \mid qw \to p \in \delta\}), 0 \le i \le n,\\
P \ &= \{[S \to [q_0x_0, q^0, 0, q_t][q_tx_1, q^1, 1, q_{i_1}] \cdots [q_{i_{n-1}}x_n, q^n, n, q_{i_n}]] \mid \\
&\qquad r_0\colon q_0x_0 \to q^0, r_1\colon q_tx_1 \to q^1, \dots, r_n\colon q_{i_{n-1}}x_n \to q^n \in \delta\\
&\qquad (r_0, r_1), \dots, (r_{n-1}, r_n) \in R, q_{i_n} \in F\} \cup\\
&\quad \{[[p_0x_0, q_0, 0, r_0] \to x_0[q_0, 0, r_0], \dots, [p_nx_n, q_n, n, r_n] \to\\
&\qquad x_n[q_n, n, r_n]]\} \cup\\
&\quad \{[[q_0, 0, q_0] \to \varepsilon, \dots, [q_n, n, q_n] \to \varepsilon] : q_i \in Q, 0 \le i \le n\} \cup\\
&\quad \{[[q_0, 0, p_0] \to w_0[q_0', 0, p_0], \dots, [q_n, n, p_n] \to w_n[q_n', n, p_n]] \mid\\
&\qquad r_j\colon q_jw_j \to q_j' \in \delta, 0 \le j \le n, (r_i, r_{i+1}) \in R, 0 \le i < n\}.
\end{aligned}$$

Next, we prove that $L(G) = L(M)$. To prove that $L(G) \subseteq L(M)$, consider a derivation of w in G. Then, the derivation is of the form (8.1), and there are the rules

$$r_0\colon q_0x_0^0 \to q_1^0, r_1\colon q_tx_0^1 \to q_1^1, \dots, r_n\colon q_{i_{n-1}}x_0^n \to q_1^n \in \delta$$

such that $(r_0, r_1), \dots, (r_{n-1}, r_n) \in R$. Moreover, $(r_j^l, r_j^{l+1}) \in R$, where $r_j^l\colon q_j^lx_j^l \to q_{j+1}^l \in \delta$, and $(r_k^l, r_k^{l+1}) \in R$, where $r_k^l\colon q_k^lx_k^l \to q_{i_l} \in \delta$, $0 \le l < n$, $1 \le j < k$, q_{i_0} denotes q_t, and $q_{i_n} \in F$. Thus, M accepts w with the sequence of rules μ of the form (8.2).

To prove that $L(M) \subseteq L(G)$, let μ be a sequence of rules used in an acceptance of

$$w = x_0^0x_1^0 \cdots x_k^0x_0^1x_1^1 \cdots x_k^1 \cdots x_0^nx_1^n \cdots x_k^n$$

in M of the form (8.2). Then, the derivation is of the form (8.1) because

$$[[q_j^0, 0, q_t] \to x_j^0[q_{j+1}^0, 0, q_t], \dots, [q_j^n, n, q_{i_n}] \to x_j^n[q_{j+1}^n, n, q_{i_n}]] \in P,$$

for all $q_j^i \in Q$, $1 \le i \le n$, $1 \le j < k$, and $[[q_t, 0, q_t] \to \varepsilon, \dots, [q_{i_n}, n, q_{i_n}] \to \varepsilon] \in P$.

Hence, Lemma 8.6 holds. □

Next, we establish another important result of this chapter.

Theorem 8.3. *For all* $n \geq 0$, ${}_n\mathbf{L}_{\mathbf{ASFA}} = {}_{n+1}\mathbf{L}_{\mathbf{RLSM}}$.

Proof. This proof follows from Lemmas 8.5 and 8.6. □

Corollary 8.3. *The following statements hold true*:

(i) $\mathbf{L}_{\mathbf{REG}} = \mathbf{L}_{\mathbf{ASFA}} - 0 \subset \mathbf{L}_{\mathbf{ASFA}} - 1 \subset \mathbf{L}_{\mathbf{ASFA}} - 2 \subset \cdots \subset \mathbf{L}_{\mathbf{CS}}$.
(ii) $\mathbf{L}_{\mathbf{ASFA}} - 1 \not\subseteq \mathbf{L}_{\mathbf{CF}}$.
(iii) $\mathbf{L}_{\mathbf{CF}} \not\subseteq {}_n\mathbf{L}_{\mathbf{ASFA}}$, *for every* $n \geq 0$.
(iv) *For all* $n \geq 0$, ${}_n\mathbf{L}_{\mathbf{ASFA}}$ *is closed under union, concatenation, finite substitution, homomorphism, intersection with a regular language, and right quotient with a regular language.*
(v) *For all* $n \geq 1$, ${}_n\mathbf{L}_{\mathbf{ASFA}}$ *is not closed under intersection, complement, and Kleene closure.*

Proof. Recall the following statements that were proved by Wood (1975):

- $\mathbf{L}_{\mathbf{REG}} = \mathbf{L}_{\mathbf{RLSM}} - 1 \subset \mathbf{L}_{\mathbf{RLSM}} - 2 \subset \mathbf{L}_{\mathbf{RLSM}} - 3 \subset \cdots \subset \mathbf{L}_{\mathbf{CS}}$.
- For all $n \geq 1$, ${}_n\mathbf{L}_{\mathbf{RLSM}}$ is closed under union, finite substitution, homomorphism, intersection with a regular language, and right quotient with a regular language.
- For all $n \geq 2$, ${}_n\mathbf{L}_{\mathbf{RLSM}}$ is not closed under intersection and complement.

Furthermore, recall these statements proved by Siromoney (1969) and Siromoney (1971).

- For all $n \geq 1$, ${}_n\mathbf{L}_{\mathbf{RLSM}}$ is closed under concatenation.
- For all $n \geq 2$, ${}_n\mathbf{L}_{\mathbf{RLSM}}$ is not closed under Kleene closure.

These statements and Theorem 8.3 imply statements (i), (iv), and (v) of Corollary 8.3. Moreover, observe that

$$\{ww \mid w \in \{a,b\}^*\} \in \mathbf{L}_{\mathbf{ASFA}} - 1 - \mathbf{L}_{\mathbf{CF}}$$

(see Example 8.2), which proves (ii). Finally, let

$$L = \{wc\,reversal(w) \mid w \in \{a,b\}^*\}.$$

By Theorem 1.5.2 given by Dassow and Păun (1989), $L \notin {}_n\mathbf{L}_{\mathbf{RLSM}}$, for any $n \geq 1$. Thus, (iii) follows from Theorem 8.3. □

Theorem 8.4, given next, follows from Theorem 8.3 and from Corollary 3.3.3 in [Siromoney (1971)]. However, Corollary 3.3.3 given by Siromoney (1971) is not proved effectively. We next prove Theorem 8.4 effectively.

Theorem 8.4. ${}_n\mathbf{L_{ASFA}}$ *is closed under inverse homomorphism, for all* $n \geq 0$.

Proof. For $n = 1$, let $M = (Q, \Delta, \delta, q_0, q_t, F, R)$ be a 1-all-SFA, and let $h : \Delta^* \to \Delta^*$ be a homomorphism. Next, we construct a 1-all-SFA

$$M' = \left(Q', \Delta, \delta', q_0', q_t', \{q_f'\}, R'\right),$$

accepting $h^{-1}(L(M))$ as follows. Set

$$k = \max\left(\{|w| \mid qw \to p \in \delta\}\right) + \max\left(\{|h(a)| \mid a \in \Delta\}\right)$$

and

$$Q' = \{q_0'\} \cup \{[x, q, y] \mid x, y \in \Delta^*, |x|, |y| \leq k, q \in Q\}.$$

Initially, set δ' and R' to $\emptyset$. Then, extend δ' and R' by performing (1)–(5), given as follows:

(1) For $y \in \Delta^*$, $|y| \leq k$, add
$(q_0' \to [\varepsilon, q_0, y], q_t' \to [y, q_t, \varepsilon])$ to R'.
(2) For $A \in Q'$, $q \neq q_t$, add
$([x, q, y]a \to [xh(a), q, y], A \to A)$ to R'.
(3) For $A \in Q'$, add
$(A \to A, [x, q, \varepsilon]a \to [xh(a), q, \varepsilon])$ to R'.
(4) For $(qx \to p, q'x' \to p') \in R$, $q \neq q_t$, add
$([xw, q, y] \to [w, p, y], [x'w', q', \varepsilon] \to [w', p', \varepsilon])$ to R'.
(5) For $q_f \in F$, add
$([y, q_t, y] \to q_t', [\varepsilon, q_f, \varepsilon] \to q_f')$ to R'.

In essence, M' simulates M in the following way. In a state of the form $[x, q, y]$, the three components have the following meaning:

- $x = h(a_1 \cdots a_n)$, where $a_1 \cdots a_n$ is the input string that M' has already read;
- q is the current state of M;

- y is the suffix remaining as the first component of the state that M' enters during a turn, y is thus obtained when M' reads the last symbol right before the turn occurs in M, and M reads y after the turn.

More precisely, $h(w) = w_1 y w_2$, where w is an input string, w_1 is accepted by M before making the turn, i.e., from q_0 to q_t, and yw_2 is accepted by M after making the turn, i.e., from q_t to $q_f \in F$. A rigorous version of this proof is left to the reader.

For $n > 1$, the proof is analogous and left to the reader. □

Language families accepted by *n*-first-SFAs and *n*-all-SFAs

Next, we compare the family of languages accepted by n-first-SFAs with the family of languages accepted by n-all-SFAs.

Theorem 8.5. *For all* $n \geq 1$, ${}_n\mathbf{L}_{\mathbf{FSFA}} \subset {}_n\mathbf{L}_{\mathbf{ASFA}}$.

Proof. Rosebrugh and Wood (1975) and Wood (1975) proved that for all $n > 1$, ${}_n\mathbf{L}_{\mathbf{PRL}} \subset {}_n\mathbf{L}_{\mathbf{RLSM}}$. The proof of Theorem 8.5 thus follows from Theorems 8.1 and 8.3. □

Theorem 8.6. ${}_n\mathbf{L}_{\mathbf{FSFA}} \not\subseteq \mathbf{L}_{\mathbf{ASFA}} - n - 1$, $n \geq 1$.

Proof. Recall that ${}_n\mathbf{L}_{\mathbf{FSFA}} = \mathbf{L}_{\mathbf{PRL}} - n + 1$ (see Theorem 8.1) and $\mathbf{L}_{\mathbf{ASFA}} - n - 1 = {}_n\mathbf{L}_{\mathbf{RLSM}}$ (see Theorem 8.3). It is easy to see that

$$L = \left\{a_1^k a_2^k \cdots a_{n+1}^k \mid k \geq 1\right\} \in \mathbf{L}_{\mathbf{PRL}} - n + 1.$$

However, Lemma 1.5.6 by Dassow and Păun (1989) implies that

$$L \notin {}_n\mathbf{L}_{\mathbf{RLSM}}.$$

Hence, the theorem holds. □

Lemma 8.7. *For each regular language* L, $\{w^n \mid w \in L\} \in \mathbf{L}_{\mathbf{ASFA}} - n - 1$.

Proof. Let $L = L(M)$, where M is a finite automaton. Make n copies of M. Rename their states so that all the sets of states are pairwise disjoint. In this way, also rename the states in the rules of each of these n automata; however, keep the labels of the rules unchanged. For each rule label r, include (r, r) into R. As a result, we obtain an n-all-SFA that accepts $\{w^n \mid w \in L\}$. A rigorous version of this proof is left to the reader. □

Theorem 8.7. ${}_n\mathbf{L}_{\mathbf{ASFA}} - \mathbf{L}_{\mathbf{FSFA}} \neq \emptyset$, *for all* $n \geq 1$, *where* $\mathbf{L}_{\mathbf{FSFA}} = \bigcup_{m=1}^{\infty} {}_m\mathbf{L}_{\mathbf{FSFA}}$.

Proof. By induction on $n \geq 1$, we prove that

$$L = \{(cw)^{n+1} \mid w \in \{a, b\}^*\} \notin \mathbf{L}_{\mathbf{FSFA}}.$$

From Lemma 8.7, it follows that $L \in {}_n\mathbf{L}_{\mathbf{ASFA}}$.

Basis. For $n = 1$, let G be an m-PRLG generating L, for some positive integer m. Consider a sufficiently large string $cw_1cw_2 \in L$ such that $w_1 = w_2 = a^{n_1}b^{n_2}$, $n_2 > n_1 > 1$. Then, there is a derivation of the form

$$\begin{aligned} S &\Rightarrow^p_G x_1A_1x_2A_2 \cdots x_mA_m \\ &\Rightarrow^k_G x_1y_1A_1x_2y_2A_2 \cdots x_my_mA_m \end{aligned} \tag{8.3}$$

in G, where cycle (8.3) generates more than one a in w_1. The derivation continues as

$$\begin{aligned} &x_1y_1A_1 \cdots x_my_mA_m \Rightarrow^r_G \\ &x_1y_1z_1B_1 \cdots x_my_mz_mB_m \Rightarrow^l_G x_1y_1z_1u_1B_1 \cdots x_my_mz_mu_mB_m \qquad (8.4) \\ &\text{(cycle (8.4) generates no } as) \Rightarrow^s_G cw_1cw_2. \end{aligned}$$

Next, modify the derivation on the left, the derivation in components generating cw_1, so that the a-generating cycle (8.3) is repeated $(l+1)$-times. Similarly, modify the derivation on the right, the derivation in the other components, so that the no-a-generating cycle (8.4) is repeated $(k+1)$-times. Thus, the modified left-hand side derivation is of length

$$p + k(l+1) + r + l + s = p + k + r + l(k+1) + s,$$

which is the length of the modified right-hand side derivation. Moreover, the modified left-hand side derivation generates more as in w_1 than the right-hand side derivation in w_2 — a contradiction.

Induction hypothesis. Suppose that the theorem holds for all $k \leq n$, for some $n \geq 1$.

Induction step. Consider $n + 1$, and let

$$\{(cw)^{n+1} \mid w \in \{a, b\}^*\} \in {}_l\mathbf{L_{FSFA}},$$

for some $l \geq 1$. As ${}_l\mathbf{L_{FSFA}}$ is closed under the right quotient with a regular language, and language $\{cw \mid w \in \{a, b\}^*\}$ is regular, we obtain

$$\{(cw)^n \mid w \in \{a, b\}^*\} \in {}_l\mathbf{L_{FSFA}} \subseteq \mathbf{L_{FSFA}},$$

which is a contradiction. □

Self-regulating pushdown automata

Definition 8.7. A *self-regulating pushdown automaton* (SPDA for short) M is a 9-tuple

$$M = (Q, \Delta, \Gamma, \delta, q_0, q_t, Z_0, F, R),$$

where $(Q, \Delta, \Gamma, \delta, q_0, Z_0, F)$ is a pushdown automaton entering a final state and emptying its pushdown, $q_t \in Q$ is a *turn state*, and $R \subseteq \Psi \times \Psi$ is a finite relation, where Ψ is an alphabet of rule labels. ■

Definition 8.8. Let $n \geq 0$ and

$$M = (Q, \Delta, \Gamma, \delta, q_0, q_t, Z_0, F, R)$$

be a self-regulating pushdown automaton. M is said to be an *n-turn first-move self-regulating pushdown automaton, n-first-SPDA* if every $w \in L(M)$ is accepted by M in the following way:

$$Z_0 q_0 w \vdash_M^* f\ [\mu]$$

such that

$$\mu = r_1^0 \cdots r_k^0 r_1^1 \cdots r_k^1 \cdots r_1^n \cdots r_k^n,$$

where $k \geq 1$, r_k^0 is the first rule of the form $Zqx \rightarrow \gamma q_t$, for some $Z \in \Gamma$, $q \in Q$, $x \in \Delta^*$, $\gamma \in \Gamma^*$, and

$$(r_1^j, r_1^{j+1}) \in R$$

for all $0 \leq j < n$. ■

The family of languages accepted by n-first-SPDAs is denoted by ${}_n\mathbf{L}_{\mathbf{FSPDA}}$.

Definition 8.9. Let $n \geq 0$ and

$$M = \big(Q,\Delta,\Gamma,\delta, q_0, q_t, Z_0, F, R\big)$$

be a self-regulating pushdown automaton. M is said to be an *n-turn all-move self-regulating pushdown automaton* (n-all-SPDA for short) if every $w \in L(M)$ is accepted by M in the following way:

$$Z_0 q_0 w \vdash_M^* f\,[\mu]$$

such that

$$\mu = r_1^0 \cdots r_k^0 r_1^1 \cdots r_k^1 \cdots r_1^n \cdots r_k^n,$$

where $k \geq 1$, r_k^0 is the first rule of the form $Zqx \to \gamma q_t$, for some $Z \in \Gamma$, $q \in Q$, $x \in \Delta^*$, $\gamma \in \Gamma^*$, and

$$(r_i^j, r_i^{j+1}) \in R,$$

for all $1 \leq i \leq k$, $0 \leq j < n$. ■

The family of languages accepted by n-all-SPDAs is denoted by ${}_n\mathbf{L}_{\mathbf{ASPDA}}$.

Accepting power

As every n-all-SPDA without any turn state represents, in effect, an ordinary pushdown automaton, we obtain the following theorem.

Theorem 8.8. ${}_0\mathbf{L}_{\mathbf{ASPDA}} = \mathbf{L}_{\mathbf{CF}}$ ■

However, if we consider 1-all-SPDAs, their power is that of phrase-structure grammars.

Theorem 8.9. ${}_1\mathbf{L}_{\mathbf{ASPDA}} = \mathbf{L}_{\mathbf{RE}}$.

Proof. For any $L \in RE$, $L \subseteq \Delta^*$, there are context-free languages $L(G)$ and $L(H)$ and a homomorphism $h : \Delta^* \to \Delta^*$ such that

$$L = h\big(L(G) \cap L(H)\big)$$

(see Harrison, 1978, Theorem 10.3.1). Suppose that $G = (V_G, T, P_G, S_G)$ and $H = (V_H, T, P_H, S_H)$ are in the Greibach normal form (see Definition 6.17); that is, all rules are of the form $A \to a\alpha$, where A is a nonterminal, a is a terminal, and α is a (possibly empty) string of nonterminals. Let us construct an 1-all-SPDA:

$$M = (\{q_0, q, q_t, p, f\}, T, V_G \cup V_H \cup \{Z\}, \delta, q_0, Z, \{f\}, R)$$

where $Z \notin V_G \cup V_H$, with R constructed by performing (1)–(4), stated as follows:

(1) Add $(Zq_0 \to ZS_Gq, Zq_t \to ZS_Hp)$ to R.
(2) Add $(Aq \to B_n \cdots B_1aq, Cp \to D_m \cdots D_1ap)$ to R if $A \to aB_1 \cdots B_n \in P_G$ and $C \to aD_1 \cdots D_m \in P_H$.
(3) Add $(aqh(a) \to q, ap \to p)$ to R.
(4) Add $(Zq \to Zq_t, Zp \to f)$ to R.

Moreover, δ contains only the rules from the definition of R.

Next, we prove that $w \in h(L(G) \cap L(H))$ if and only if $w \in L(M)$.

Only If. Let $w \in h(L(G) \cap L(H))$. There are $a_1, a_2, \ldots, a_n \in T$ such that

$$a_1a_2 \cdots a_n \in L(G) \cap L(H)$$

and $w = h(a_1a_2 \cdots a_n)$, for some $n \geq 0$. There are leftmost derivations

$$S_G \Rightarrow {}^n_G a_1a_2 \cdots a_n$$

and

$$S_H \Rightarrow {}^n_H a_1a_2 \cdots a_n$$

of length n in G and H, respectively, because in every derivation step, exactly one terminal element is derived. Thus, M accepts

$h(a_1)h(a_2)\cdots h(a_n)$ as

$$\begin{aligned}
& Zq_0h(a_1)h(a_2)\cdots h(a_n) \\
\vdash_M\ & ZS_Gqh(a_1)h(a_2)\cdots h(a_n) \\
& \vdots \\
\vdash_M\ & Za_nqh(a_n) \\
\vdash_M\ & Zq \\
\vdash_M\ & Zq_t \\
\vdash_M\ & ZS_Hp \\
& \vdots \\
\vdash_M\ & Za_np \\
\vdash_M\ & Zp \\
\vdash_M\ & f.
\end{aligned}$$

In state q, by using its pushdown, M simulates a derivation of $a_1\cdots a_n$ in G but reads $h(a_1)\cdots\ h(a_n)$ as the input. In p, M simulates a derivation of $a_1a_2\cdots a_n$ in H but reads no input. As $a_1a_2\cdots a_n$ can be derived in both G and H by making the same number of steps, the automaton can successfully complete the acceptance of w.

If. Note that in one step, M can read only $h(a) \in T^*$, for some $a \in T$. Let $w \in L(M)$, then $w = h(a_1)h(a_2)\cdots h(a_n)$, for some $a_1, a_2, \ldots, a_n \in T$. Consider the following acceptance of w in M:

$$\begin{aligned}
& Zq_0h(a_1)h(a_2)\cdots h(a_n) \\
\vdash_M\ & ZS_Gqh(a_1)h(a_2)\cdots h(a_n) \\
& \vdots \\
\vdash_M\ & Za_nqh(a_n) \\
\vdash_M\ & Zq \\
\vdash_M\ & Zq_t \\
\vdash_M\ & ZS_Hp \\
& \vdots \\
\vdash_M\ & Za_np \\
\vdash_M\ & Zp \\
\vdash_M\ & f.
\end{aligned}$$

As stated above, in q, M simulates a derivation of $a_1a_2 \cdots a_n$ in G, and then in p, M simulates a derivation of $a_1a_2 \cdots a_n$ in H. It successfully completes the acceptance of w only if $a_1a_2 \cdots a_n$ can be derived in both G and H. Hence, the if part holds too. □

Open problems

Although the fundamental results about self-regulating automata have been achieved in this chapter, there still remain several open problems concerning them.

Open problem 1. What is the language family accepted by n-turn first-move self-regulating pushdown automata when $n \geq 1$ (see Definition 8.8)? ■

Open problem 2. By analogy with the standard deterministic finite and pushdown automata (see Meduna, 2000, pp. 145, 437), introduce the deterministic versions of self-regulating automata. What is their power? ■

Open problem 3. Discuss the closure properties of other language operations, such as the reversal. ■

8.2 Automata Regulated by Control Languages

This section discusses automata in which the application of rules is regulated by control languages. First, it studies this topic in terms of finite automata, after which it investigates pushdown automata regulated in this way. More precisely, the current section starts its discussion by studying finite automata working under two kinds of regulation: *state-controlled regulation* and *transition-controlled regulation.* It establishes conditions under which any state-controlled finite automaton can be turned to an equivalent transition-controlled finite automaton, and vice versa. Then, it proves that under either of the two regulations, finite automata controlled by regular languages characterize the family of regular languages, and an analogical result is then reformulated in terms of context-free languages. Concerning regulated pushdown automata, this section first shows that pushdown automata regulated by regular languages are as powerful as ordinary pushdown automata. Then, however, it proves that pushdown automata regulated by linear languages characterize the family of recursively enumerable languages.

Finite automata regulated by control languages

This section studies finite automata regulated by control languages. In fact, it studies two kinds of this regulation: *state-controlled regulation* and *transition-controlled regulation*. To give an insight into these two types of regulation, consider a finite automaton M controlled by a language C and a sequence $\tau \in C$ that result in the acceptance of an input word w. Working under the former regulation, M has C defined over the set of states, and it accepts w by going through all the states in τ and ending up in a final state. Working under the latter regulation, M has C defined over the set of transitions, and it accepts w by using all the transitions in τ and ending up in a final state.

First, this section formally defines these two types of controlled finite automata. After that, it establishes conditions under which it is possible to convert any state-controlled finite automaton to an equivalent transition-controlled finite automaton, and vice versa (Theorem 8.10). Then, it proves that under both regulations, finite automata controlled by regular languages characterize the family of regular languages (Theorem 8.11 and Corollary 8.4). Finally, this section shows that finite automata controlled by context-free languages define the family of context-free languages (Theorem 8.12 and Corollary 8.5). In a briefer way, this section also discusses pushdown automata regulated by linear languages. It points out that even some restricted versions of these automata are computationally complete (see Theorem 8.15).

Definitions

We begin by formally defining state-controlled and transition-controlled finite automata.

Definition 8.10. Let $M = (Q, \Delta, R, s, F)$ be a finite automaton. Based on $\vdash_M$, we define a relation $\triangleright_M$ over $Q\Delta^* \times Q^*$ as follows:

If $\alpha \in Q^*$ and $pax \vdash_M qx$, where $p, q \in Q$, $x \in \Delta^*$, and $a \in \Delta \cup \{\varepsilon\}$, then

$$(pax, \alpha) \triangleright_M (qx, \alpha p).$$

Let $\triangleright_M^n$, $\triangleright_M^*$, and $\triangleright_M^+$ denote the nth power of $\triangleright_M$, for some $n \geq 0$, the reflexive–transitive closure of $\triangleright_M$, and the transitive closure of $\triangleright_M$, respectively.

Let $C \subseteq Q^*$ be a *control language*. The *state-controlled language of M with respect to C* is denoted by ${}_{\triangleright}L(M, C)$ and defined as

$${}_{\triangleright}L(M, C) = \big\{w \in \Delta^* \mid (sw, \varepsilon) \triangleright_M^* (f, \alpha), f \in F, \alpha \in C\big\}.$$

The pair (M, C) is called a *state-controlled finite automaton.* ∎

Before defining transition-controlled finite automata, recall the rule labels from Definition 3.1.

Definition 8.11. Let $M = (Q, \Delta, \Psi, R, s, F)$ be a finite automaton. Based on $\vdash_M$, we define a relation $\blacktriangleright_M$ over $Q\Delta^* \times \Psi^*$ as follows: If $\beta \in \Psi^*$ and $pax \vdash_M qx\ [r]$, where $r\colon pa \to q \in R$ and $x \in \Delta^*$, then

$$(pax, \beta) \blacktriangleright_M (qx, \beta r).$$

Let $\blacktriangleright_M^n$, $\blacktriangleright_M^*$, and $\blacktriangleright_M^+$ denote the nth power of $\blacktriangleright_M$, for some $n \geq 0$, the reflexive–transitive closure of $\blacktriangleright_M$, and the transitive closure of $\blacktriangleright_M$, respectively.

Let $C \subseteq \Psi^*$ be a *control language*. The *transition-controlled language of M with respect to C* is denoted by ${}_{\blacktriangleright}L(M, C)$ and defined as

$${}_{\blacktriangleright}L(M, C) = \{w \in \Delta^* \mid (sw, \varepsilon) \blacktriangleright_M^* (f, \beta), f \in F, \beta \in C\}.$$

The pair (M, C) is called a *transition-controlled finite automaton.* ∎

For any family of languages $\mathbf{L}$, $\mathbf{L}_{\mathbf{SCFA}}(\mathbf{L})$ and $\mathbf{L}_{\mathbf{TCFA}}(\mathbf{L})$ denote the language families defined by state-controlled finite automata controlled by languages from $\mathbf{L}$ and transition-controlled finite automata controlled by languages from $\mathbf{L}$, respectively.

Conversions

First, we show that under certain circumstances, it is possible to convert any state-controlled finite automaton to an equivalent transition-controlled finite automaton, and vice versa. These conversions will be helpful because, to prove that $\mathbf{L}_{\mathbf{SCFA}}(\mathbf{L}) = \mathbf{L}_{\mathbf{TCFA}}(\mathbf{L}) = \mathbf{L}$, where $\mathbf{L}$ satisfies the required conditions, we only have to prove that either $\mathbf{L}_{\mathbf{SCFA}}(\mathbf{L}) = \mathbf{L}$ or $\mathbf{L}_{\mathbf{TCFA}}(\mathbf{L}) = \mathbf{L}$.

Lemma 8.8. *Let* $\mathbf{L}$ *be a language family that is closed under finite* ε*-free substitution. Then,* $\mathbf{L}_{\mathbf{SCFA}}(\mathbf{L}) \subseteq \mathbf{L}_{\mathbf{TCFA}}(\mathbf{L})$.

Proof. Let $\mathbf{L}$ be a language family that is closed under finite ε-free substitution, $M = (Q, \Delta, R, s, F)$ be a finite automaton, and $C \in \mathbf{L}$ be a control language. Without any loss of generality, assume that $C \subseteq Q^*$. We next construct a finite automaton M' and a language $C' \in \mathbf{L}$ such that ${}_{\triangleright}L(M, C) = {}_{\blacktriangleright}L(M', C')$. Define

$$M' = \bigl(Q, \Delta, \Psi, R', s, F\bigr),$$

where

$$\begin{aligned} \Psi &= \{\langle p, a, q\rangle \mid pa \to q \in R\}, \\ R' &= \{\langle p, a, q\rangle : pa \to q \mid pa \to q \in R\}. \end{aligned}$$

Define the finite ε-free substitution π from Q^* to Ψ^* as

$$\pi(p) = \bigl\{\langle p, a, q\rangle \mid pa \to q \in R\bigr\}.$$

Let $C' = \pi(C)$. Since $\mathbf{L}$ is closed under finite ε-free substitution, $C' \in \mathbf{L}$. Observe that $(sw, \varepsilon) \triangleright^n_M (f, \alpha)$, where $w \in \Delta^*$, $f \in F$, $\alpha \in C$, and $n \geq 0$ if and only if $(sw, \varepsilon) \blacktriangleright^n_{M'} (f, \beta)$, where $\beta \in \pi(\alpha)$. Hence, ${}_{\triangleright}L(M, C) = {}_{\blacktriangleright}L(M', C')$, so the lemma holds. □

Lemma 8.9. *Let* $\mathbf{L}$ *be a language family that contains all finite languages and is closed under concatenation. Then,* $\mathbf{L}_{\mathbf{TCFA}}(\mathbf{L}) \subseteq \mathbf{L}_{\mathbf{SCFA}}(\mathbf{L})$.

Proof. Let $\mathbf{L}$ be a language family that contains all finite languages and is closed under concatenation, $M = (Q, \Delta, \Psi, R, s, F)$ be a finite automaton, and $C \in \mathbf{L}$ be a control language. Without any loss of generality, assume that $C \subseteq \Psi^*$. We next construct a finite automaton M' and a language $C' \in \mathbf{L}$ such that ${}_\blacktriangleright L(M, C) = {}_\triangleright L(M', C')$. Define

$$M' = (Q', \Delta, R', s', F'),$$

where

$$\begin{aligned}
Q' &= \Psi \cup \{s', \ell\} \quad (s', \ell \notin \Psi), \\
R' &= \{s' \to r \mid r\colon sa \to q \in R\} \ \cup \\
&\quad\ \{ra \to t \mid r\colon pa \to q, t\colon qb \to m \in R\} \ \cup \\
&\quad\ \{ra \to \ell \mid r\colon pa \to q \in R, q \in F\}, \\
F' &= \{r \mid r\colon pa \to q \in R, q \in F\} \cup \{\ell\}.
\end{aligned}$$

Finally, if $s \in F$, then add s' to F'. Set $C' = \{s',\varepsilon\}C$. Since $\mathbf{L}$ is closed under concatenation and contains all finite languages, $C' \in \mathbf{L}$. Next, we argue that ${}_\blacktriangleright L(M, C) = {}_\triangleright L(M', C')$. First, note that $s \in F$ if and only if $s' \in F$. Hence, by the definition of C', it is sufficient to consider non-empty sequences of moves of both M and M'. Indeed, $(s, \varepsilon) \blacktriangleright_M^0 (s, \varepsilon)$ with $s \in F$ and $\varepsilon \in C$ if and only if $(s', \varepsilon) \triangleright_{M'}^0 (s', \varepsilon)$ with $s' \in F$ and $\varepsilon \in C'$. Observe that

$$(sw, \varepsilon) \blacktriangleright_M (p_1 w_1, r_1) \blacktriangleright_M (p_2 w_2, r_1 r_2) \blacktriangleright_M \cdots \blacktriangleright_M (p_n w_n, r_1 r_2 \cdots r_n)$$

by

$$\begin{aligned}
r_1\colon &\quad p_0 a_1 \to p_1, \\
r_2\colon &\quad p_1 a_2 \to p_2, \\
&\quad\ \vdots \\
r_n\colon &\, p_{n-1} a_n \to p_n,
\end{aligned}$$

where $w \in \Delta^*$, $p_i \in Q$ for $i = 1, 2, \ldots, n$, $p_n \in F$, $w_i \in \Delta^*$ for $i = 1, 2, \ldots, n$, $a_i \in \Delta \cup \{\varepsilon\}$ for $i = 1, 2, \ldots n$, and $n \geq 1$ if and only if

$$(s'w, \varepsilon) \triangleright_{M'} (r_1 w, s') \triangleright_{M'} (r_2 w_1, s' r_1) \triangleright_{M'} \cdots \triangleright_{M'} (r_{n+1} w_n, s' r_1 r_2 \cdots r_n)$$

by

$$\begin{aligned} s' &\to r_1, \\ r_1 a_1 &\to r_2, \\ r_2 a_2 &\to r_3, \\ &\vdots \\ r_n a_n &\to r_{n+1}, \end{aligned}$$

with $r_{n+1} \in F'$ (recall that $p_n \in F$). Hence, ${}_{\blacktriangleright}L(M, C) = {}_{\triangleright}L(M', C')$, and this lemma holds. □

Theorem 8.10. *Let* $\mathbf{L}$ *be a language family that is closed under finite* ε*-free substitution, contains all finite languages, and is closed under concatenation. Then,* $\mathbf{L}_{\mathbf{SCFA}}(\mathbf{L}) = \mathbf{L}_{\mathbf{TCFA}}(\mathbf{L})$.

Proof. This theorem follows directly from Lemmas 8.8 and 8.9. □

Regular-controlled finite automata

Initially, we consider finite automata controlled by regular control languages.

Lemma 8.10. $\mathbf{L}_{\mathbf{SCFA}}(\mathbf{L}_{\mathbf{REG}}) \subseteq \mathbf{L}_{\mathbf{REG}}$.

Proof. Let $M = (Q, \Delta, R, s, F)$ be a finite automaton and $C \subseteq Q^*$ be a regular control language. Since C is regular, there is a complete finite automaton $H = (\hat{Q}, Q, \hat{R}, \hat{s}, \hat{F})$ such that $L(H) = C$. We next construct a finite automaton M' such that $L(M') = {}_{\triangleright}L(M', L(H))$. Define

$$M' = \big(Q', \Delta, R', s', F'\big),$$

where

$$\begin{aligned} Q' &= \{\langle p, q\rangle \mid p \in Q, q \in \hat{Q}\}, \\ R' &= \{\langle p, r\rangle a \to \langle q, t\rangle \mid pa \to q \in R, rp \to t \in \hat{R}\}, \\ s' &= \langle s, \hat{s}\rangle, \\ F' &= \{\langle p, q\rangle \mid p \in F, q \in \hat{F}\}. \end{aligned}$$

Observe that a move in M' by $\langle p, r\rangle a \to \langle q, t\rangle \in R'$ simultaneously simulates a move in M by $pa \to q \in R$ and a move in H by

$rp \to t \in \hat{R}$. Based on this observation, it is rather easy to see that M' accepts an input string $w \in \Delta^*$ if and only if M reads w and enters a final state after going through a sequence of states from $L(H)$. Therefore, $L(M') = {}_{\triangleright}L(M, L(H))$. A rigorous proof of the identity $L(M') = {}_{\triangleright}L(M, L(H))$ is left to the reader. □

The following theorem shows that finite automata controlled by regular languages are of little or no interest because they are as powerful as ordinary finite automata.

Theorem 8.11. $\mathbf{L}_{\mathbf{SCFA}}(\mathbf{L}_{\mathbf{REG}}) = \mathbf{L}_{\mathbf{REG}}$.

Proof. The inclusion $\mathbf{L}_{\mathbf{REG}} \subseteq \mathbf{L}_{\mathbf{SCFA}}(\mathbf{L}_{\mathbf{REG}})$ is obvious. The converse inclusion follows from Lemma 8.10. □

Combining Theorems 8.10 and 8.11, we obtain the following corollary (recall that $\mathbf{L}_{\mathbf{REG}}$ satisfies all the conditions from Theorem 8.10).

Corollary 8.4. $\mathbf{L}_{\mathbf{TCFA}}(\mathbf{L}_{\mathbf{REG}}) = \mathbf{L}_{\mathbf{REG}}$. ■

Context-free-controlled finite automata

Next, we consider finite automata controlled by context-free control languages.

Lemma 8.11. $\mathbf{L}_{\mathbf{SCFA}}(\mathbf{L}_{\mathbf{CF}}) \subseteq \mathbf{L}_{\mathbf{CF}}$.

Proof. Let $M = (Q, \Delta, R, s, F)$ be a finite automaton and $C \subseteq Q^*$ be a context-free control language. Since C is context-free, there is a pushdown automaton $H = (\hat{Q}, Q, \Gamma, \hat{R}, \hat{s}, \hat{Z}, \hat{F})$ such that $L(H) = C$. Without any loss of generality, we assume that $bpa \to wq \in \hat{R}$ implies $a \neq \varepsilon$ (see Wood, 1987, Lemma 5.2.1). We next construct a pushdown automaton M' such that $L(M') = {}_{\triangleright}L(M, L(H))$. Define

$$M' = (Q', \Delta, \Gamma, R', s', Z, F'),$$

where

$$\begin{aligned} Q' &= \{\langle p, q\rangle \mid p \in Q, q \in \hat{Q}\}, \\ R' &= \{b\langle p, r\rangle a \to w\langle q, t\rangle \mid pa \to q \in R, bpr \to wt \in \hat{R}\}, \\ s' &= \langle s, \hat{s}\rangle, \\ F' &= \{\langle p, q\rangle \mid p \in F, q \in \hat{F}\}. \end{aligned}$$

By a similar reasoning as in Lemma 8.10, we can prove that $L(M') = {}_{\triangleright}L(M, L(H))$. A rigorous proof of the identity $L(M') = {}_{\triangleright}L(M, L(H))$ is left to the reader. □

The following theorem says that even though finite automata controlled by context-free languages are more powerful than finite automata, they cannot accept any non-context-free language.

Theorem 8.12. $\mathbf{L}_{\mathbf{SCFA}}(\mathbf{L}_{\mathbf{CF}}) = \mathbf{L}_{\mathbf{CF}}$.

Proof. The inclusion $\mathbf{L}_{\mathbf{CF}} \subseteq \mathbf{L}_{\mathbf{SCFA}}(\mathbf{L}_{\mathbf{CF}})$ is obvious. The converse inclusion follows from Lemma 8.11. □

Combining Theorems 8.10 and 8.12, we obtain the following corollary (recall that $\mathbf{L}_{\mathbf{CF}}$ satisfies all the conditions from Theorem 8.10).

Corollary 8.5. $\mathbf{L}_{\mathbf{TCFA}}(\mathbf{L}_{\mathbf{CF}}) = \mathbf{L}_{\mathbf{CF}}$. ■

Pushdown automata regulated by control languages

First, we define pushdown automata that regulate the application of their rules by control languages. Then, we demonstrate that this regulation has no effect on the power of pushdown automata if the control languages are regular. Considering this result, we point out that pushdown automata regulated by regular languages are of little interest because their power coincides with the power of ordinary pushdown automata. After that, however, we prove that pushdown automata increase their power remarkably if they are regulated by linear languages; indeed, they characterize the family of recursively enumerable languages. Finally, we continue with the discussion of regulated pushdown automata, but we narrow our attention to their special cases, such as one-turn pushdown automata.

Definitions

Without further ado, we next define pushdown automata regulated by control languages. Recall the rule labels from Definition 4.1 because this formalization is often used throughout this section.

Definition 8.12. Let $M = (Q, \Delta, \Gamma, R, s, S, F)$ be a pushdown automaton, and let Ψ be an alphabet of its rule labels. Let Ξ be a *control language* over Ψ, i.e., $\Xi \subseteq \Psi^*$. With Ξ, M defines the following three types of accepted languages:

- $L(M, \Xi, 1)$ — the *language accepted by final state*,
- $L(M, \Xi, 2)$ — the *language accepted by empty pushdown*,
- $L(M, \Xi, 3)$ — the *language accepted by final state and empty pushdown*,

defined as follows. Let $\chi \in \Gamma^* Q \Delta^*$. If $\chi \in \Gamma^* F$, $\chi \in Q$, $\chi \in F$, then χ is a 1-*final configuration*, 2-*final configuration*, 3-*final configuration*, respectively. For $i = 1, 2, 3$, we define $L(M, \Xi, i)$ as

$$L(M, \Xi, i) = \{w \mid w \in \Delta^* \text{ and } Ssw \vdash_M^* \chi\ [\sigma] \text{ for an } i\text{-final configuration } \chi \text{ and } \sigma \in \Xi\}.$$

The pair (M, Ξ) is called a *controlled pushdown automaton*. ■

For any family of languages $\mathbf{L}$ and $i \in \{1, 2, 3\}$, define

$$\mathbf{L_{RPDA}}(\mathbf{L}, i) = \{L \mid L = L(M, \Xi, i), \text{ where } M \text{ is a pushdown automaton and } \Xi \in \mathbf{L}\}.$$

We demonstrate that

$$\mathbf{L_{CF}} = \mathbf{L_{RPDA}}(\mathbf{L_{REG}}, 1) = \mathbf{L_{RPDA}}(\mathbf{L_{REG}}, 2) = \mathbf{L_{RPDA}}(\mathbf{L_{REG}}, 3)$$

and

$$\mathbf{L_{RE}} = \mathbf{L_{RPDA}}(\mathbf{L_{LIN}}, 1) = \mathbf{L_{RPDA}}(\mathbf{L_{LIN}}, 2) = \mathbf{L_{RPDA}}(\mathbf{L_{LIN}}, 3).$$

Some of the following proofs involve several grammars and automata. To avoid any confusion, these proofs sometimes specify a regular grammar G as $G = (V_G, T_G, P_G, S_G)$ because this specification clearly expresses that V_G, T_G, P_G, and S_G represent the components of G. Other grammars and automata are specified analogously whenever any confusion may exist.

Regular-Controlled Pushdown Automata

Next, this section proves that if the control languages are regular, then the regulation of pushdown automata has no effect on their power. The proof of the following lemma presents a transformation that converts any regular grammar G and any pushdown automaton K to an ordinary pushdown automaton M such that $L(M) = L(K, L(G), 1)$.

Lemma 8.12. *For every regular grammar G and every pushdown automaton K, there exists a pushdown automaton M such that $L(M) = L(K, L(G), 1)$.*

Proof. Let $G = (V_G, T_G, P_G, S_G)$ be any regular grammar, and let $K = (Q_K, \Delta_K, \Gamma_K, R_K, s_K, S_K, F_K)$ be any pushdown automaton. Next, we construct a pushdown automaton M that simultaneously simulates G and K so that $L(M) = L(K, L(G), 1)$.

Let f be a new symbol. Define M as

$$M = (Q_M, \Delta_M, \Gamma_M, R_M, s_M, S_M, F_M),$$

where

$$\begin{aligned}
Q_M &= \{\langle qB\rangle \mid q \in Q_K, B \in (V_G - T_G) \cup \{f\}\},\\
\Delta_M &= \Delta_K,\\
\Gamma_M &= \Gamma_K,\\
s_M &= \langle s_K S_G\rangle,\\
S_M &= S_K,\\
F_M &= \{\langle qf\rangle \mid q \in F_K\},\\
R_M &= \{C\langle qA\rangle b \to x\langle pB\rangle \mid a\colon Cqb \to xp \in R_K, A \to aB \in P_G\},\\
&\quad \cup \{C\langle qA\rangle b \to x\langle pf\rangle \mid a\colon Cqb \to xp \in R_K, A \to a \in P_G\}.
\end{aligned}$$

Observe that a move in M according to $C\langle qA\rangle b \to x\langle pB\rangle \in R_M$ simulates a move in K according to $a\colon Cqb \to xp \in R_K$, where a is generated in G by using $A \to aB \in P_G$. Based on this observation, it is rather easy to see that M accepts an input string w if and only if K reads w and enters a final state after using a complete string of $L(G)$; therefore, $L(M) = L(K, L(G), 1)$. A rigorous proof that $L(M) = L(K, L(G), 1)$ is left to the reader. $\square$

Theorem 8.13. *For $i \in \{1, 2, 3\}$, $\mathbf{L_{CF}} = \mathbf{L_{RPDA}}(\mathbf{L_{REG}}, i)$.*

Proof. To prove that $\mathbf{L_{CF}} = \mathbf{L_{RPDA}}(\mathbf{L_{REG}}, 1)$, note that $\mathbf{L_{RPDA}}(\mathbf{L_{REG}}, 1) \subseteq \mathbf{L_{CF}}$ follows from Lemma 8.12. Clearly,

$\mathbf{L_{CF}} \subseteq \mathbf{L_{RPDA}}(\mathbf{L_{REG}}, 1)$, so $\mathbf{L_{RPDA}}(\mathbf{L_{REG}}, 1) = \mathbf{L_{CF}}$. By analogy with the demonstration of $\mathbf{L_{RPDA}}(\mathbf{L_{REG}}, 1) = \mathbf{L_{CF}}$, we can prove that $\mathbf{L_{CF}} = \mathbf{L_{RPDA}}(\mathbf{L_{REG}}, 2)$ and $\mathbf{L_{CF}} = \mathbf{L_{RPDA}}(\mathbf{L_{REG}}, 3)$. □

As follows from Theorem 8.13, pushdown automata regulated by regular languages are of little or no interest because they are as powerful as ordinary pushdown automata. However, this power increases under the linear-language regulation, as pointed out next.

Linear-Controlled Pushdown Automata

In its conclusion, this section points out that pushdown automata regulated by linear control languages are more powerful than ordinary pushdown automata. It only gives the fundamental results while omitting their proofs, which are to be found in Chapter 16 in the work of Meduna and Zemek (2014).

Pushdown automata regulated by linear control languages characterize RE as follows from the following theorem.

Theorem 8.14. $\mathbf{L_{RE}} = \mathbf{L_{RPDA}}(\mathbf{L_{LIN}}, 1) = \mathbf{L_{RPDA}}(\mathbf{L_{LIN}}, 2) = \mathbf{L_{RPDA}}(\mathbf{L_{LIN}}, 3)$. ■

In fact, this characterization holds true even if it is based on restricted versions of one-turn pushdown automata. To give an insight into them, consider two consecutive moves made by an ordinary pushdown automaton M. If during the first move M does not shorten its pushdown and during the second move it does, then M makes a *turn* during the second move. A pushdown automaton is *one-turn* if it makes no more than one turn with its pushdown during any computation starting from a start configuration. One-turn pushdown automata characterize the family of linear languages (see Harrison, 1978) while their unrestricted versions characterize the family of context-free languages (see Meduna, 2000). As a result, one-turn pushdown automata are less powerful than the pushdown automata.

As the most surprising result, we demonstrate that linear-regulated versions of one-turn pushdown automata characterize the family of recursively enumerable languages. Thus, as opposed to the ordinary one-turn pushdown automata, one-turn linear-regulated pushdown automata are as powerful as linear-regulated pushdown automata that can make any number of turns.

In fact, this characterization holds even for some restricted versions of one-turn regulated pushdown automata, including their atomic and reduced versions, which are sketched in the following:

(I) During a move, an *atomic* one-turn regulated pushdown automaton changes a state and, in addition, performs exactly one of the following three actions:
  (a) it pushes a symbol onto the pushdown;
  (b) it pops a symbol from the pushdown;
  (c) it reads an input symbol.

(II) A *reduced* one-turn regulated pushdown automaton has a limited number of some components, such as the number of states, pushdown symbols, or transition rules.

We give the above-mentioned characterization in terms of acceptance by final state and empty pushdown, acceptance by final state, and acceptance by empty pushdown.

Definition 8.13. An *atomic pushdown automaton* is a septuple

$$M = (Q, \Delta, \Gamma, R, s, \$, F),$$

where Q, Δ, Γ, $s \in Q$, and $F \subseteq Q$ are defined as in a pushdown automaton. $\$$ is the *pushdown-bottom marker*, $\$ \notin Q \cup \Delta \cup \Gamma$, and R is a finite set of *rules* of the form $Apa \to wq$, where $p, q \in Q$, $A, w \in \Gamma \cup \{\varepsilon\}$, $a \in \Delta \cup \{\varepsilon\}$, such that $|Aaw| = 1$. That is, R is a finite set of rules such that each of them has one of the following forms:

(a) $Ap \to q$ (*popping rule*),
(b) $p \to wq$ (*pushing rule*),
(c) $pa \to q$ (*reading rule*).

Let Ψ be an alphabet of *rule labels* such that $|\Psi| = |R|$ and ψ be a bijection from R to Ψ. For simplicity, to express that ψ maps a rule, $Apa \to wq \in R$, to r, where $r \in \Psi$, we write $r\colon Apa \to wq \in R$; in other words, $r\colon Apa \to wq$ means $\psi(Apa \to wq) = r$. A *configuration* of M, χ, is any string from $\{\$\}\Gamma^* Q \Delta^*$; χ is a *start* *configuration* if $\chi = \$sw$, where $w \in \Delta^*$. For every $x \in \Gamma^*$, $y \in \Delta^*$, and $r\colon Apa \to wq \in R$, M makes a *move* from configuration $\$xApay$ to configuration $\$xwqy$ according to r, written as

$$\$xApay \vdash_M \$xwqy \; [r]$$

or, simply, $\$xApay \vdash_M \$xwqy$.

Let χ be any configuration of M. M makes *zero moves* from χ to χ according to ε, symbolically written as

$$\chi \vdash_M^0 \chi \; [\varepsilon].$$

Let there exist a sequence of configurations $\chi_0, \chi_1, \ldots, \chi_n$, for some $n \geq 1$ such that $\chi_{i-1} \vdash_M \chi_i \; [r_i]$, where $r_i \in \Psi$, for $i = 1, \ldots, n$, then M makes n *moves* from χ_0 to χ_n according to $r_1 \cdots r_n$, symbolically written as

$$\chi_0 \vdash_M^n \chi_n \; [r_1 \cdots r_n]$$

or, simply, $\chi_0 \vdash_M^n \chi_n$. Define $\vdash_M^*$ and $\vdash_M^+$ in the standard manner.

Let $x, x', x'' \in \Gamma^*$, $y, y', y'' \in \Delta^*$, $q, q', q'' \in Q$, and $\$xqy \vdash_M \$x'q'y' \vdash_M \$x''q''y''$. If $|x| \leq |x'|$ and $|x'| > |x''|$, then $\$x'q'y' \vdash_M \$x''q''y''$ is a *turn*. If M makes no more than one turn during any sequence of moves starting from a start configuration, then M is said to be *one-turn*.

Let Ξ be a *control language* over Ψ, i.e., $\Xi \subseteq \Psi^*$. With Ξ, M defines the following three types of accepted languages:

- $L(M, \Xi, 1)$ — the *language accepted by final state*,
- $L(M, \Xi, 2)$ — the *language accepted by empty pushdown*,
- $L(M, \Xi, 3)$ — the *language accepted by final state and empty pushdown*,

defined as follows. Let $\chi \in \{\$\}\Gamma^*Q\Delta^*$. If $\chi \in \{\$\}\Gamma^*F$, $\chi \in \{\$\}Q$, $\chi \in \{\$\}F$, then χ is a 1-*final configuration*, 2-*final configuration*, 3-*final configuration*, respectively. For $i = 1, 2, 3$, define $L(M, \Xi, i)$ as

$$L(M, \Xi, i) = \big\{w \mid w \in \Delta^*, \text{ and } sw \vdash_M^* \chi \; [\sigma] \text{ for an } i\text{-final configuration } \chi \text{ and } \sigma \in \Xi\big\}.$$

The pair (M, Ξ) is called a *controlled pushdown automaton*. ■

For any family of language $\mathbf{L}$ and $i \in \{1, 2, 3\}$, define

$$\mathbf{L_{OA\text{-}RPDA}}(\mathbf{L}, i) = \big\{L \mid L = L(M, \Xi, i), \text{ where } M \text{ is a one-turn atomic pushdown automaton and } \Xi \in \mathbf{L}\big\}.$$

One-turn atomic pushdown automata regulated by linear languages characterize **RE** even if the size of their components is strongly limited, as stated in the following.

Theorem 8.15. *For every* $L \in \mathbf{L_{RE}}$*, there is a linear language* Ξ *and a one-turn atomic pushdown automaton* $M = (Q, \Delta, \Gamma, R, s, \$, F)$ *such that* $|Q| \leq 1$, $|\Gamma| \leq 2$, $|R| \leq |\Delta| + 4$, *and* $L(M, \Xi, 1) = L, i = 1, \ldots, 3$.

From the previous theorem, we obtain the following corollary.

Corollary 8.6. *For* $i \in \{1, 2, 3\}$, $\mathbf{L_{RE}} = \mathbf{L_{OA\text{-}RPDA}}(\mathbf{L_{LIN}}, i)$. ■

We close this section by suggesting some open problem areas concerning regulated automata.

Open problem 4. For $i = 1, \ldots, 3$, consider $\mathbf{L_{RPDA}}(\mathbf{L}, i)$, where $\mathbf{L}$ is a language family satisfying $\mathbf{L_{REG}} \subset \mathbf{L} \subset \mathbf{L_{LIN}}$. For instance, consider $\mathbf{L}$ as the family of *minimal linear languages* (see Salomaa, 1973, p. 76). Compare $\mathbf{L_{RE}}$ with $\mathbf{L_{RPDA}}(\mathbf{L}, i)$. ■

Open problem 5. Investigate special cases of regulated pushdown automata, such as their deterministic versions. ■

Open problem 6. By analogy with regulated pushdown automata, introduce and study some other types of regulated automata. ■

Chapter 9

Jumping Automata

This chapter introduces and studies jumping finite automata. In essence, these automata work just like classical finite automata, except that they do not read their input strings in a symbol-by-symbol, left-to-right way. Instead, after reading a symbol, they can jump in either direction within their input tapes and continue making jumps from there. Once an occurrence of a symbol is read, it cannot be reread later on. Otherwise, their concept coincides with that of standard finite automata.

Organized into four sections, this chapter gives a systematic body of knowledge concerning jumping finite automata. First, it formalizes them (see Section 9.1). Then, it demonstrates their fundamental properties and generative power (see Section 9.2). Naturally, this chapter also establishes several results concerning jumping finite automata with respect to commonly studied areas of automata theory, such as closure properties and decidability (see Section 9.3). In addition, it establishes an infinite hierarchy of language families resulting from these automata. Finally, it studies some special topics and features, such as one-directional jumps and various start configurations (see Section 9.4). Throughout its discussion, this chapter points out several open questions regarding these automata, which may represent a new area of investigation in automata theory in the future.

9.1 Definitions and Examples

To see the difference between the classical notion of a finite automaton and its jumping version, recall that the former works by making a sequence of moves on the input word (see Chapter 3). Each move is made according to a computational rule that describes how the current state is changed and whether the current input symbol is read. If the symbol is read, the read head is shifted precisely one symbol to the right on the word. If the automaton can read the input word by making a sequence of moves from the start state to a final state, it accepts the input word. Compared to this classical concept, the jumping version is conceptualized such that it does not read the input string in a strictly symbol-by-symbol left-to-right way. That is, after reading a symbol, it can jump over a portion of the input word in either direction and continue making jumps from there. Otherwise, both concepts coincide.

Definition 9.1. A *general jumping finite automaton* (GJFA) is a quintuple

$$M = \big(Q, \Delta, R, s, F\big),$$

where Q is a finite set of *states*, Δ is the *input alphabet*, $\Delta \cap Q = \emptyset$, $R \subseteq Q \times \Delta^* \times Q$ is a finite relation, $s \in Q$ is the *start state*, and F is a set of *final states*. The members of R are referred to as *rules* of M, and instead of $(p, y, q) \in R$, we write $py \to q \in R$. A *configuration* of M is any string in $\Delta^* Q \Delta^*$. The binary *jumping relation*, symbolically denoted by $\curvearrowright_M$, over $\Delta^* Q \Delta^*$, is defined as follows. Let x, z, x', $z' \in \Delta^*$ such that $xz = x'z'$ and $py \to q \in R$; then, M makes a *jump* from $xpyz$ to $x'qz'$, written as

$$xpyz \curvearrowright_M x'qz'.$$

When there is no danger of confusion, we simply write $\curvearrowright$ instead of $\curvearrowright_M$. In the standard manner, we extend $\curvearrowright$ to $\curvearrowright^m$, where $m \geq 0$. Let $\curvearrowright^+$ and $\curvearrowright^*$ denote the transitive closure of $\curvearrowright$ and the reflexive-transitive closure of $\curvearrowright$, respectively. The *language* accepted by M, denoted by $L(M)$, is defined as

$$L(M) = \{uv \mid u, v \in \Delta^*, usv \curvearrowright^* f, f \in F\}.$$

Let $w \in \Delta^*$. We say that M *accepts* w if and only if $w \in L(M)$. M *rejects* w if and only if $w \in \Delta^* - L(M)$. Two GJFAs M and M' are said to be *equivalent* if and only if $L(M) = L(M')$. ■

Definition 9.2. Let $M = (Q, \Delta, R, s, F)$ be a GJFA. M is an *ε-free GJFA* if $py \to q \in R$ implies that $|y| \geq 1$. M is of *degree* n, where $n \geq 0$, if $py \to q \in R$ implies that $|y| \leq n$. M is a *jumping finite automaton* (JFA for short) if its degree is 1. ■

Definition 9.3. Let $M = (Q, \Delta, R, s, F)$ be a JFA. Analogously to a GJFA, M is an *ε-free JFA* if $py \to q \in R$ implies that $|y| = 1$. M is a *deterministic JFA* (a *DJFA* for short) if: (1) it is an ε-free JFA; and (2) for each $p \in Q$ and each $a \in \Delta$, there is no more than one $q \in Q$ such that $pa \to q \in R$. M is a *complete JFA* (CJFA for short) if: (1) it is a DJFA; and (2) for each $p \in Q$ and each $a \in \Delta$, there is precisely one $q \in Q$ such that $pa \to q \in R$. ■

Definition 9.4. Let $M = (Q, \Delta, R, s, F)$ be a GJFA. The *transition graph* of M, denoted by $\Delta(M)$, is a multigraph, where nodes are states from Q, and there is an edge from p to q labeled with y if and only if $py \to q \in R$. A state $q \in Q$ is *reachable* if there is a walk from s to q in $\Delta(M)$; q is *terminating* if there is a walk from q to some $f \in F$. If there is a walk from p to q, $p = q_1, q_2, \ldots, q_n = q$, for some $n \geq 2$, where $q_i y_i \to q_{i+1} \in R$ for all $i = 1, \ldots, n-1$, then we write

$$py_1y_2 \cdots y_n \rightsquigarrow q.$$ ■

Next, we illustrate the previous definitions with two examples.

Example 9.1. Consider the DJFA

$$M = (\{s, r, t\}, \Delta, R, s, \{s\})$$

where $\Delta = \{a, b, c\}$ and

$$R = \{sa \to r, rb \to t, tc \to s\}.$$

Starting from s, M has to read some a, some b, and some c, entering again the start (and also the final) state s. All these occurrences of a, b, and c can appear anywhere in the input string. Therefore, the accepted language is clearly

$$L(M) = \{w \in \Delta^* \mid occur(w, a) = occur(w, b) = occur(w, c)\}. \quad \blacksquare$$

Recall that $L(M)$ in Example 9.1 is a well-known non-context-free, context-sensitive language.

Example 9.2. Consider the GJFA

$$M = (\{s, t, f\}, \{a, b\}, R, s, \{f\}),$$

where

$$R = \{sba \to f, fa \to f, fb \to f\}.$$

Starting from s, M has to read the string ba, which can appear anywhere in the input string. Then, it can read an arbitrary number of symbols a and b, including no symbols. Therefore, the accepted language is $L(M) = \{a, b\}^*\{ba\}\{a, b\}^*$. $\blacksquare$

Denotation of language families

Throughout the rest of this chapter, $\mathbf{L_{GJFA}}$, $\mathbf{L^{-\varepsilon}_{GJFA}}$, $\mathbf{L_{JFA}}$, $\mathbf{L^{-\varepsilon}_{JFA}}$, and $\mathbf{L_{DJFA}}$ denote the families of languages accepted by GJFAs, ε-free GJFAs, JFAs, ε-free JFAs, and DJFAs, respectively.

9.2 Accepting Power

In this section, we discuss the generative power of GJFAs and JFAs and some other basic properties of these automata.

Theorem 9.1. *For every DJFA M, there is a CJFA M' such that $L(M) = L(M')$.*

Proof. Let $M = (Q, \Delta, R, s, F)$ be a DJFA. We next construct a CJFA M' such that $L(M) = L(M')$. Without any loss of generality, we assume that $\bot \notin Q$. Initially, set

$$M' = (Q \cup \{\bot\}, \Delta, R', s, F),$$

where $R' = R$. Next, for each $a \in \Delta$ and each $p \in Q$ such that $pa \to q \notin R$ for all $q \in Q$, add $pa \to \bot$ to R'. For each $a \in \Delta$, add $\bot a \to \bot$ to R'. Clearly, M' is a CJFA, and $L(M) = L(M')$. $\square$

Lemma 9.1. *For every GJFA M of degree $n \geq 0$, there is an ε-free GJFA M' of degree n such that $L(M') = L(M)$.*

Proof. This lemma can be demonstrated by using the standard conversion of finite automata to ε-free finite automata (see Meduna, 2000, Algorithm 3.2.2.3). □

Theorem 9.2. $\mathbf{L_{GJFA}} = \mathbf{L^{-\varepsilon}_{GJFA}}$.

Proof. $\mathbf{L^{-\varepsilon}_{GJFA}} \subseteq \mathbf{L_{GJFA}}$ follows from the definition of a GJFA. $\mathbf{L_{GJFA}} \subseteq \mathbf{L^{-\varepsilon}_{GJFA}}$ follows from Lemma 9.1. □

Theorem 9.3. $\mathbf{L_{JFA}} = \mathbf{L^{-\varepsilon}_{JFA}} = \mathbf{L_{DJFA}}$.

Proof. $\mathbf{L_{JFA}} = \mathbf{L^{-\varepsilon}_{JFA}}$ can be proved by analogy with the proof of Theorem 9.2, so we only prove that $\mathbf{L^{-\varepsilon}_{JFA}} = \mathbf{L_{DJFA}}$. $\mathbf{L_{DJFA}} \subseteq \mathbf{L^{-\varepsilon}_{JFA}}$ follows from the definition of a DJFA. The converse inclusion can be proved by using the standard technique of converting ε-free finite automata to deterministic finite automata (see Meduna, 2000, Algorithm 3.2.3.1). □

The following theorem shows a property of languages accepted by GJFAs with unary input alphabets.

Theorem 9.4. *Let $M = (Q, \Delta, R, s, F)$ be a GJFA such that $|\Delta| = 1$. Then, $L(M)$ is regular.*

Proof. Let $M = (Q, \Delta, R, s, F)$ be a GJFA such that $|\Delta| = 1$. Since $|\Delta| = 1$, without any loss of generality, we can assume that the acceptance process for $w \in \Delta^*$ starts from the configuration sw, and M does not jump over any symbols. Therefore, we can treat M as an equivalent finite automaton (see Definition 3.1). As finite automata accept only regular languages (see Theorem 3.8), $L(M)$ is regular. □

As a consequence of Theorem 9.4, we obtain the following corollary (recall that K in the following is not regular).

Corollary 9.1. *The language $K = \{a^p \mid p \text{ is a prime number}\}$ cannot be accepted by any GJFA.* ■

The following theorem gives a necessary condition for a language to be in $\mathbf{L_{JFA}}$.

Theorem 9.5. *Let K be an arbitrary language. Then, $K \in \mathbf{L_{JFA}}$ only if $K = \text{perm}(K)$.*

Proof. Let $M = (Q, \Delta, R, s, F)$ be a JFA. Without any loss of generality, we assume that M is a DJFA (recall that $\mathbf{L_{JFA}} = \mathbf{L_{DJFA}}$ by Theorem 9.3). Let $w \in L(M)$. We next prove that $\text{perm}(w) \subseteq L(M)$. If $w = \varepsilon$, then $\text{perm}(\varepsilon) = \varepsilon \in L(M)$, so we assume that $w \neq \varepsilon$. Then, $w = a_1 a_2 \cdots a_n$, where $a_i \in \Delta$ for all $i = 1, \ldots, n$, for some $n \geq 1$. Since $w \in L(M)$, R contains

$$\begin{aligned} s a_{i_1} &\to s_{i_1} \\ s_{i_1} a_{i_2} &\to s_{i_2} \\ &\vdots \\ s_{i_{n-1}} a_{i_n} &\to s_{i_n}, \end{aligned}$$

where $s_j \in Q$ for all $j \in \{i_1, i_2, \ldots, i_n\}$, $(i_1, i_2, \ldots, i_n)$ is a permutation of $(1, 2, \ldots, n)$ and $s_{i_n} \in F$. However, this implies that $a_{k_1} a_{k_2} \cdots a_{k_n} \in L(M)$, where $(k_1, k_2, \ldots, k_n)$ is a permutation of $(1, 2, \ldots, n)$, so $\text{perm}(w) \subseteq L(M)$. □

From Theorem 9.5, we obtain the following two corollaries, which are used in subsequent proofs.

Corollary 9.2. *There is no JFA that accepts $\{ab\}^*$.* ■

Corollary 9.3. *There is no JFA that accepts $\{a, b\}^* \{ba\} \{a, b\}^*$.* ■

Consider the language of primes K from Corollary 9.1. Since $K = \text{perm}(K)$, the condition from Theorem 9.5 is not sufficient for a language to be in $\mathbf{L_{JFA}}$. This is stated in the following corollary.

Corollary 9.4. *There is a language K satisfying $K = \text{perm}(K)$ that cannot be accepted by any JFA.* ■

The following theorem gives both a necessary and sufficient condition for a language to be accepted by a JFA.

Theorem 9.6. *Let L be an arbitrary language. $L \in \mathbf{L_{JFA}}$ if and only if $L = \text{perm}(K)$, where K is a regular language.*

Proof. The proof is divided into the only-if part and the if part.

Only If. Let M be a JFA. Consider M as a finite automaton M'. Set $K = L(M')$. K is regular, and $L(M) = \text{perm}(K)$. Hence, the only-if part holds.

If. Take $\text{perm}(K)$, where K is any regular language. Let $K = L(M)$, where M is a finite automaton. Consider M as a JFA M'. Observe that $L(M') = \text{perm}(K)$, which proves the if part of the proof. □

Finally, we show that GJFAs are stronger than JFAs.

Theorem 9.7. $\mathbf{L_{JFA}} \subset \mathbf{L_{GJFA}}$

Proof. $\mathbf{L_{JFA}} \subseteq \mathbf{L_{GJFA}}$ follows from the definition of a JFA. From Corollary 9.3, $\mathbf{L_{GJFA}} - \mathbf{L_{JFA}} \neq \emptyset$ because $\{a, b\}^*\{ba\}\{a, b\}^*$ is accepted by the GJFA from Example 9.2. □

Open Problem 1. Is there a necessary and sufficient condition for a language to be in $\mathbf{L_{GJFA}}$? ■

Now, we establish relations between $\mathbf{L_{GJFA}}$, $\mathbf{L_{JFA}}$, and some well-known language families, including $\mathbf{L_{FIN}}$, $\mathbf{L_{REG}}$, $\mathbf{L_{CF}}$, and $\mathbf{L_{CS}}$.

Theorem 9.8. $\mathbf{L_{FIN}} \subset \mathbf{L_{GJFA}}$.

Proof. Let $K \in \mathbf{L_{FIN}}$. Since K is a finite, there exists $n \geq 0$ such that $|K| = n$. Therefore, we can express K as $K = \{w_1, w_2, \ldots, w_n\}$. Define the GJFA as

$$M = (\{s, f\}, \Delta, R, s, \{f\}),$$

where $\Delta = \text{alph}(K)$ and $R = \{sw_1 \to f, sw_2 \to f, \ldots, sw_n \to f\}$. Clearly, $L(M) = K$. Therefore, $\mathbf{L_{FIN}} \subseteq \mathbf{L_{GJFA}}$. From Example 9.1, $\mathbf{L_{GJFA}} - \mathbf{L_{FIN}} \neq \emptyset$, which proves the theorem. □

Lemma 9.2. *There is no GJFA that accepts* $\{a\}^*\{b\}^*$.

Proof. By contradiction. Let $K = \{a\}^*\{b\}^*$. Assume that there is a GJFA, $M = (Q, \Delta, R, s, F)$, such that $L(M) = K$. Let $w = a^n b$, where n is the degree of M. Since $w \in K$, during an acceptance of w, a rule, $pa^i b \to q \in R$, where $p, q \in Q$ and $0 \leq i < n$, has to be used. However, then M also accepts from the configuration $a^i bsa^{n-i}$. Indeed, as $a^i b$ is read in a single step and all the other symbols in w

are just as, a^iba^{n-i} may be accepted by using the same rules as during an acceptance of w. This implies that $a^iba^{n-i} \in K$ — a contradiction with the assumption that $L(M) = K$. Therefore, there is no GJFA that accepts $\{a\}^*\{b\}^*$. □

Theorem 9.9. *$\mathbf{L_{REG}}$ and $\mathbf{L_{GJFA}}$ are incomparable.*

Proof. $\mathbf{L_{GJFA}} \not\subseteq \mathbf{L_{REG}}$ follows from Example 9.1. $\mathbf{L_{REG}} \not\subseteq \mathbf{L_{GJFA}}$ follows from Lemma 9.2. □

Theorem 9.10. *$\mathbf{L_{CF}}$ and $\mathbf{L_{GJFA}}$ are incomparable.*

Proof. $\mathbf{L_{GJFA}} \not\subseteq \mathbf{L_{CF}}$ follows from Example 9.1, and $\mathbf{L_{CF}} \not\subseteq \mathbf{L_{GJFA}}$ follows from Lemma 9.2. □

Theorem 9.11. $\mathbf{L_{GJFA}} \subset \mathbf{L_{CS}}$.

Proof. Clearly, jumps of GJFAs can be simulated by context-sensitive grammars, so $\mathbf{L_{GJFA}} \subseteq \mathbf{L_{CS}}$. From Lemma 9.2, it follows that $\mathbf{L_{CS}} - \mathbf{L_{GJFA}} \neq \emptyset$. □

Theorem 9.12. *$\mathbf{L_{FIN}}$ and $\mathbf{L_{JFA}}$ are incomparable.*

Proof. $\mathbf{L_{JFA}} \not\subseteq \mathbf{L_{FIN}}$ follows from Example 9.1. Consider the finite language $K = \{ab\}$. By Theorem 9.5, $K \notin \mathbf{L_{JFA}}$, so $\mathbf{L_{FIN}} \not\subseteq \mathbf{L_{JFA}}$. □

9.3 Properties

In this section, we show the closure properties of the families $\mathbf{L_{GJFA}}$ and $\mathbf{L_{JFA}}$ under various operations.

Theorem 9.13. *Both $\mathbf{L_{GJFA}}$ and $\mathbf{L_{JFA}}$ are not closed under end-marking.*

Proof. Consider the language $K = \{a\}^*$. Clearly, $K \in \mathbf{L_{JFA}}$. A proof that no GJFA accepts $K\{\#\}$, where # is a symbol such that $\# \neq a$, can be made by analogy with the proof of Lemma 9.2. □

Theorem 9.13 implies that both families are not closed under concatenation. Indeed, observe that the JFA

$$M = (\{s, f\}, \{\#\}, \{s\# \to f\}, s, \{f\})$$

accepts $\{\#\}$.

Corollary 9.5. *Both* $\mathbf{L_{GJFA}}$ *and* $\mathbf{L_{JFA}}$ *are not closed under concatenation.* ■

Theorem 9.14. $\mathbf{L_{JFA}}$ *is closed under shuffle.*

Proof. Let $M_1 = (Q_1, \Delta_1, R_1, s_1, F_1)$ and $M_2 = (Q_2, \Delta_2, R_2, s_2, F_2)$ be two JFAs. Without any loss of generality, we assume that $Q_1 \cap Q_2 = \emptyset$. Define the JFA as

$$H = \big(Q_1 \cup Q_2, \Delta_1 \cup \Delta_2, R_1 \cup R_2 \cup \{f \to s_2 \mid f \in F_1\}, s_1, F_2\big).$$

To see that $L(H) = \text{shuffle}(L(M_1), L(M_2))$, observe how H works. On an input string, $w \in (\Delta_1 \cup \Delta_2)^*$, H first runs M_1 on w, and if it ends in a final state, then it runs M_2 on the rest of the input. If M_2 ends in a final state, H accepts w. Otherwise, it rejects w. By Theorem 9.5, $L(M_i) = \text{perm}(L(M_i))$ for all $i \in \{1, 2\}$. Based on these observations, since H can jump anywhere after a symbol is read, we see that $L(H) = \text{shuffle}(L(M_1), L(M_2))$. □

Note that the construction used in the previous proof coincides with the standard construction of a concatenation of two finite automata (see Meduna, 2000).

Theorem 9.15. *Both* $\mathbf{L_{GJFA}}$ *and* $\mathbf{L_{JFA}}$ *are closed under union.*

Proof. Let $M_1 = (Q_1, \Delta_1, R_1, s_1, F_1)$ and $M_2 = (Q_2, \Delta_2, R_2, s_2, F_2)$ be two GJFAs. Without any loss of generality, we assume that $Q_1 \cap Q_2 = \emptyset$ and $s \notin (Q_1 \cup Q_2)$. Define the GJFA as

$$H = \big(Q_1 \cup Q_2 \cup \{s\}, \Delta_1 \cup \Delta_2, R_1 \cup R_2 \cup \{s \to s_1, s \to s_2\}, s, F_1 \cup F_2\big).$$

Clearly, $L(H) = L(M_1) \cup L(M_2)$, and if both M_1 and M_2 are JFAs, then H is also a JFA. □

Theorem 9.16. $\mathbf{L}_{\mathbf{GJFA}}$ *is not closed under complement.*

Proof. Consider the GJFA M from Example 9.2. Observe that the complement of $L(M)$ (with respect to $\{a,b\}^*$) is $\{a\}^*\{b\}^*$, which cannot be accepted by any GJFA (see Lemma 9.2). □

Theorem 9.17. $\mathbf{L}_{\mathbf{JFA}}$ *is closed under complement.*

Proof. Let $M = (Q, \Delta, R, s, F)$ be a JFA. Without any loss of generality, we assume that M is a CJFA ($\mathbf{L}_{\mathbf{JFA}} = \mathbf{L}_{\mathbf{DJFA}}$ by Theorem 9.3, and every DJFA can be converted to an equivalent CJFA by Theorem 9.1). Then, the JFA

$$M' = \big(Q, \Delta, R, s, Q - F\big)$$

accepts $\overline{L(M)}$. □

By using De Morgan's laws, we obtain the following two corollaries of Theorems 9.15, 9.16, and 9.17.

Corollary 9.6. $\mathbf{L}_{\mathbf{GJFA}}$ *is not closed under intersection.* ■

Corollary 9.7. $\mathbf{L}_{\mathbf{JFA}}$ *is closed under intersection.* ■

Theorem 9.18. *Both* $\mathbf{L}_{\mathbf{GJFA}}$ *and* $\mathbf{L}_{\mathbf{JFA}}$ *are not closed under intersection with regular languages.*

Proof. Consider the language $J = \{a,b\}^*$, which can be accepted by both GJFAs and JFAs. Consider the regular language $K = \{a\}^*\{b\}^*$. Since $J \cap K = K$, this theorem follows from Lemma 9.2. □

Theorem 9.19. $\mathbf{L}_{\mathbf{JFA}}$ *is closed under reversal.*

Proof. Let $K \in \mathbf{L}_{\mathbf{JFA}}$. Since $\text{perm}(w) \subseteq K$ by Theorem 9.5 for all $w \in K$, also $reversal(()w) \in K$ for all $w \in K$, so the theorem holds. □

Theorem 9.20. $\mathbf{L}_{\mathbf{JFA}}$ *is not closed under Kleene star or under Kleene plus.*

Proof. Consider the language $K = \{ab, ba\}$, which is accepted by the JFA:

$$M = \big(\{s, r, f\}, \{a, b\}, \{sa \to r, rb \to f\}, s, \{f\}\big).$$

However, by Theorem 9.5, there is no JFA that accepts K^* or K^+ (note that, for example, $abab \in K^+$, but $aabb \notin K^+$). □

Lemma 9.3. *There is no GJFA that accepts* $\{a\}^*\{b\}^* \cup \{b\}^*\{a\}^*$.

Proof. This lemma can be proved by analogy with the proof of Lemma 9.2. □

Theorem 9.21. *Both* $\mathbf{L_{GJFA}}$ *and* $\mathbf{L_{JFA}}$ *are not closed under substitution.*

Proof. Consider the language $K = \{ab, ba\}$, which is accepted by the JFA M from the proof of Theorem 9.20. Define the substitution σ from $\{a,b\}^*$ to $2^{\{a,b\}^*}$ as $\sigma(a) = \{a\}^*$ and $\sigma(b) = \{b\}^*$. Clearly, both $\sigma(a)$ and $\sigma(b)$ can be accepted by JFAs. However, $\sigma(K)$ cannot be accepted by any GJFA (see Lemma 9.3). □

Since the substitution σ in the proof of Theorem 9.21 is regular, we obtain the following corollary.

Corollary 9.8. *Both* $\mathbf{L_{GJFA}}$ *and* $\mathbf{L_{JFA}}$ *are not closed under regular substitution.* ■

Theorem 9.22. $\mathbf{L_{GJFA}}$ *is closed under finite substitution.*

Proof. Let $M = (Q, \Delta, R, s, F)$ be a GJFA, Γ be an alphabet, and σ be a finite substitution from Δ^* to 2^{Γ^*}. The language $\sigma(L(M))$ is accepted by the GJFA

$$M' = (Q, \Gamma, R', s, F),$$

where $R' = \{py' \to q \mid y' \in \sigma(y), py \to q \in R\}$. □

Since homomorphism is a special case of finite substitution, we obtain the following corollary of Theorem 9.22.

Corollary 9.9. $\mathbf{L_{GJFA}}$ *is closed under homomorphism.* ■

Theorem 9.23. $\mathbf{L_{JFA}}$ *is not closed under* ε*-free homomorphism.*

Proof. Define the ε-free homomorphism φ from $\{a\}$ to $\{a,b\}^+$ as $\varphi(a) = ab$, and consider the language $\{a\}^*$, which is accepted by the JFA

$$M = (\{s\}, \{a\}, \{sa \to s\}, \{s\}).$$

Note that $\varphi(L(M)) = \{ab\}^*$, which cannot be accepted by any JFA (see Corollary 9.2). □

Since ε-free homomorphism is a special case of homomorphism and since homomorphism is a special case of finite substitution, we obtain the following corollary of Theorem 9.23.

Corollary 9.10. $\mathbf{L_{JFA}}$ *is not closed under homomorphism or under finite substitution.* ■

Theorem 9.24. *Both* $\mathbf{L_{GJFA}}$ *and* $\mathbf{L_{JFA}}$ *are closed under inverse homomorphism.*

Proof. Let $M = (Q, \Gamma, R, s, F)$ be a GJFA, Δ be an alphabet, and φ be a homomorphism from Δ^* to Γ^*. We next construct a JFA M' such that $L(M') = \varphi^{-1}(L(M))$. Define

$$M' = \left(Q, \Delta, R', s, F\right),$$

where

$$R' = \left\{pa \to q \mid a \in \Delta, p\varphi(a) \rightsquigarrow q \text{ in } \Delta(M)\right\}.$$

Observe that $w_1 s w_2 \curvearrowright^* q$ in M if and only if $w_1' s w_2' \curvearrowright^* q$ in M', where $w_1 w_2 = \varphi(w_1' w_2')$ and $q \in Q$, so $L(M') = \varphi^{-1}(L(M))$. A fully rigorous proof is left to the reader. □

A summary of the closure properties of the families $\mathbf{L_{GJFA}}$ and $\mathbf{L_{JFA}}$ is given in Table 9.1, where + marks closure, − marks non-closure, and ? means that the closure property represents an open problem. It is worth noting that $\mathbf{L_{REG}}$, characterized by finite automata, is closed under all of these operations.

Open Problem 2. Is $\mathbf{L_{GJFA}}$ closed under shuffle, Kleene star, Kleene plus, and reversal? ■

Decidability

In this section, we prove the decidability of some decision problems with regard to $\mathbf{L_{GJFA}}$ and $\mathbf{L_{JFA}}$.

Lemma 9.4. *Let* $M = (Q, \Delta, R, s, F)$ *be a GJFA. Then,* $L(M)$ *is infinite if and only if* $py \rightsquigarrow p$ *in* $\Delta(M)$*, for some* $y \in \Delta^+$ *and* $p \in Q$ *such that* p *is both reachable and terminating in* $\Delta(M)$*.*

Table 9.1. Summary of closure properties.

	$\mathbf{L}_{\mathbf{GJFA}}$	$\mathbf{L}_{\mathbf{JFA}}$
Endmarking	−	−
Concatenation	−	−
Shuffle	?	+
Union	+	+
Complement	−	+
Intersection	−	+
Int. with regular languages	−	−
Kleene star	?	−
Kleene plus	?	−
Reversal	?	+
Substitution	−	−
Regular substitution	−	−
Finite substitution	+	−
Homomorphism	+	−
ε-free homomorphism	+	−
Inverse homomorphism	+	+

Proof. *If.* Let $M = (Q, \Delta, R, s, F)$ be a GJFA such that $py \rightsquigarrow p$ in $\Delta(M)$, for some $y \in \Delta^+$ and $p \in Q$, such that p is both reachable and terminating in $\Delta(M)$. Then,

$$w_1 s w_2 \curvearrowright^* upv \curvearrowright^+ xpz \curvearrowright^* f,$$

where $w_1 w_2 \in L(M)$, $u, v, x, z \in \Delta^*$, $p \in Q$, and $f \in F$. Consequently,

$$w_1 s w_2 \curvearrowright^* upvy' \curvearrowright^+ xpz \curvearrowright^* f,$$

where $y' = y^n$ for all $n \geq 0$. Therefore, $L(M)$ is infinite, so the if part holds.

Only If. Let $M = (Q, \Delta, R, s, F)$ be a GJFA such that $L(M)$ is infinite. Without any loss of generality, we assume that M is ε-free (see Lemma 9.1). Then,

$$w_1 s w_2 \curvearrowright^* upv \curvearrowright^+ xpz \curvearrowright^* f,$$

for some $w_1 w_2 \in L(M)$, $u, v, x, z \in \Delta^*$, $p \in Q$, and $f \in F$. This implies that p is both terminating and reachable in $\Delta(M)$. Let $y \in \Delta^+$ be a string read by M during $upv \curvearrowright^+ xpz$. Then, $py \rightsquigarrow p$ in $\Delta(M)$, so the only-if part holds. □

Theorem 9.25. *Both finiteness and infiniteness are decidable for* $\mathbf{L_{GJFA}}$.

Proof. Let $M = (Q, \Delta, R, s, F)$ be a GJFA. By Lemma 9.4, $L(M)$ is infinite if and only if $py \rightsquigarrow p$ in $\Delta(M)$, for some $y \in \Delta^+$ and $p \in Q$, such that p is both reachable and terminating in $\Delta(M)$. This condition can be checked by any graph searching algorithm, such as breadth-first search (see Russell and Norvig, 2002, p. 73). Therefore, the theorem holds. □

Corollary 9.11. *Both finiteness and infiniteness are decidable for* $\mathbf{L_{JFA}}$. ■

Observe that since there is no deterministic version of a GJFA, the following proof of Theorem 9.26 is not as straightforward as in terms of regular languages and classical deterministic finite automata.

Theorem 9.26. *The membership problem is decidable for* $\mathbf{L_{GJFA}}$.

Proof. Let $M = (Q, \Delta, R, s, F)$ be a GJFA, and let $x \in \Delta^*$. Without any loss of generality, we assume that M is ε-free (see Theorem 9.2). If $x = \varepsilon$, then $x \in L(M)$ if and only if $s \in F$, so assume that $x \neq \varepsilon$. Set

$$\Gamma = \{(x_1, x_2, \ldots, x_n) \mid x_i \in \Delta^+, 1 \leq i \leq n, x_1 x_2 \cdots x_n = x, n \geq 1\}$$

and

$$\Gamma_p = \{(y_1, y_2, \ldots, y_n) \mid (x_1, x_2, \ldots, x_n) \in \Gamma, n \geq 1, (y_1, y_2, \ldots, y_n) \text{ is a permutation of } (x_1, x_2, \ldots, x_n)\}.$$

If there exist $(y_1, y_2, \ldots, y_n) \in \Gamma_p$ and $q_1, q_2, \ldots, q_{n+1} \in Q$, for some n, $1 \leq n \leq |x|$, such that $s = q_1$, $q_{n+1} \in F$, and $q_i y_i \rightarrow q_{i+1} \in R$, for all $i = 1, 2, \ldots, n$, then $x \in L(M)$; otherwise, $x \notin L(M)$. Since both Q and Γ_p are finite, this check can be performed in finite time. □

Corollary 9.12. *The membership problem is decidable for* $\mathbf{L_{JFA}}$. ■

Theorem 9.27. *The emptiness problem is decidable for* $\mathbf{L_{GJFA}}$.

Proof. Let $M = (Q, \Delta, R, s, F)$ be a GJFA. Then, $L(M)$ is empty if and only if no $f \in F$ is reachable in $\Delta(M)$. This check can be done by any graph searching algorithm, such as breadth-first search (see Russell and Norvig, 2002, p. 73). □

Table 9.2. Summary of decidability properties.

	$\mathbf{L_{GJFA}}$	$\mathbf{L_{JFA}}$
Membership	+	+
Emptiness	+	+
Finiteness	+	+
Infiniteness	+	+

Corollary 9.13. *The emptiness problem is decidable for* $\mathbf{L_{JFA}}$. ■

A summary of the decidability properties of the families $\mathbf{L_{GJFA}}$ and $\mathbf{L_{JFA}}$ is given in Table 9.2, where + marks decidability.

An Infinite Hierarchy of Language Families

Next, we establish an infinite hierarchy of language families resulting from GJFAs of degree n, where $n \geq 0$. Let ${}_n\mathbf{L_{GJFA}}$ and ${}_n\mathbf{L}^{-\varepsilon}_{\mathbf{GJFA}}$ denote the families of languages accepted by GJFAs of degree n and by ε-free GJFAs of degree n, respectively. Observe that ${}_n\mathbf{L_{GJFA}} = {}_n\mathbf{L}^{-\varepsilon}_{\mathbf{GJFA}}$ by the definition of a GJFA and by Lemma 9.1, for all $n \geq 0$.

Lemma 9.5. *Let Δ be an alphabet such that $|\Delta| \geq 2$. Then, for any $n \geq 1$, there is a GJFA of degree n, $M_n = (Q, \Delta, R, s, F)$, such that $L(M_n)$ cannot be accepted by any GJFA of degree $n-1$.*

Proof. Let Δ be an alphabet such that $|\Delta| \geq 2$, and let $a, b \in \Delta$ such that $a \neq b$. The case of $n = 1$ follows immediately from the definition of a JFA, so we assume that $n \geq 2$. Define the GJFA of degree n as

$$M_n = (\{s, f\}, \Delta, \{sw \to r\}, s, \{r\}),$$

where $w = ab(a)^{n-2}$. Clearly, $L(M_n) = \{w\}$. We next prove that $L(M_n)$ cannot be accepted by any GJFA of degree $n-1$.

Suppose, for the sake of contradiction, that there is a GJFA of degree $n-1$, $H = (Q, \Delta, R, s', F)$, such that $L(H) = L(M_n)$. Without any loss of generality, we assume that H is ε-free (see Lemma 9.1).

Since $L(H) = L(M_n) = \{w\}$ and $|w| > n - 1$, there has to be

$$us'xv \curvearrowright^m f$$

in H, where $w = uxv$, $u, v \in \Delta^*$, $x \in \Delta^+$, $f \in F$, and $m \geq 2$. Thus,

$$s'xuv \curvearrowright^m f$$

and

$$uvs'x \curvearrowright^m f$$

in H, which contradicts the assumption that $L(H) = \{w\}$. Therefore, $L(M_n)$ cannot be accepted by any GJFA of degree $n - 1$. □

Theorem 9.28. ${}_n\mathbf{L}_{\mathbf{GJFA}} \subset {}_n\mathbf{L}_{\mathbf{GJFA}} + 1$ *for all* $n \geq 0$.

Proof. ${}_n\mathbf{L}_{\mathbf{GJFA}} \subseteq {}_{n+1}\mathbf{L}_{\mathbf{GJFA}}$ follows from the definition of a GJFA of degree n, for all $n \geq 0$. From Lemma 9.5, ${}_{n+1}\mathbf{L}_{\mathbf{GJFA}} - {}_n\mathbf{L}_{\mathbf{GJFA}} \neq \emptyset$, which proves the theorem. □

Taking Lemma 9.1 into account, we obtain the following corollary of Theorem 9.28.

Corollary 9.14. ${}_n\mathbf{L}^{-\varepsilon}_{\mathbf{GJFA}} \subset {}_{n+1}\mathbf{L}^{-\varepsilon}_{\mathbf{GJFA}}$ *for all* $n \geq 0$. ■

Left and Right Jumps

We define two special cases of the jumping relation.

Definition 9.5. Let $M = (Q, \Delta, R, s, F)$ be a GJFA. Let $w, x, y, z \in \Delta^*$ and $py \to q \in R$. Then: (1) M makes a *left jump* from $wxpyz$ to $wqxz$, symbolically written as

$$wxpyz \ {}_l\!\curvearrowright wqxz;$$

and (2) M makes a *right jump* from $wpyxz$ to $wxqz$, written as

$$wpyxz \ {}_r\!\curvearrowright wxqz.$$

Let $u, v \in \Delta^* Q \Delta^*$; then, $u \curvearrowright v$ if and only if $u \ {}_l\!\curvearrowright v$ or $u \ {}_r\!\curvearrowright v$. Extend ${}_l\!\curvearrowright$ and ${}_r\!\curvearrowright$ to ${}_l\!\curvearrowright^m$, ${}_l\!\curvearrowright^*$, ${}_l\!\curvearrowright^+$, ${}_r\!\curvearrowright^m$, ${}_r\!\curvearrowright^*$, and ${}_r\!\curvearrowright^+$, where $m \geq 0$, by analogy with extending $\curvearrowright$. Set

$${}_lL(M) = \{uv \mid u, v \in \Delta^*, usv \ {}_l\!\curvearrowright^* f \text{ with } f \in F\}$$

and

$${}_rL(M) = \{uv \mid u, v \in \Delta^*, usv \ {}_r\!\curvearrowright^* f \text{ with } f \in F\}.$$ ■

Let ${}_{l}\mathbf{L_{GJFA}}$, ${}_{l}\mathbf{L_{JFA}}$, ${}_{r}\mathbf{L_{GJFA}}$, and ${}_{r}\mathbf{L_{JFA}}$ denote the families of languages accepted by GJFAs using only left jumps, JFAs using only left jumps, GJFAs using only right jumps, and JFAs using only right jumps, respectively.

Theorem 9.29. ${}_{r}\mathbf{L_{GJFA}} = {}_{r}\mathbf{L_{JFA}} = \mathbf{L_{REG}}$.

Proof. We first prove that ${}_{r}\mathbf{L_{JFA}} = \mathbf{L_{REG}}$. Consider any JFA, $M = (Q, \Delta, R, s, F)$. Observe that if M occurs in a configuration of the form xpy, where $x \in \Delta^*$, $p \in Q$, and $y \in \Delta^*$, then it cannot read the symbols in x anymore because M can make only right jumps. Also, observe that this covers the situation when M starts to accept $w \in \Delta^*$ from a different configuration than sw. Therefore, to read the whole input, M has to start in configuration sw, and it cannot jump to skip some symbols. Consequently, M behaves like an ordinary finite automaton, reading the input from the left to the right, so $L(M)$ is regular and, therefore, ${}_{r}\mathbf{L_{JFA}} \subseteq \mathbf{L_{REG}}$. Conversely, any finite automaton can be viewed as a JFA that starts from configuration sw and does not jump to skip some symbols. Therefore, $\mathbf{L_{REG}} \subseteq {}_{r}\mathbf{L_{JFA}}$, which proves that ${}_{r}\mathbf{L_{JFA}} = \mathbf{L_{REG}}$. A proof that ${}_{r}\mathbf{L_{GJFA}} = \mathbf{L_{REG}}$ is left as an exercise. □

Next, we show that JFAs using only left jumps accept some non-regular languages.

Theorem 9.30. ${}_{l}\mathbf{L_{JFA}} - \mathbf{L_{REG}} \neq \emptyset$.

Proof. Consider the JFA

$$M = \big(\{s, p, q\}, \{a, b\}, R, s, \{s\}\big),$$

where

$$R = \big\{sa \to p, pb \to s, sb \to q, qa \to s\big\}.$$

We argue that

$${}_{l}L(M) = \big\{w \mid occur(w, a) = occur(w, b)\big\}.$$

With $w \in \{a, b\}^*$ on its input, M starts over the last symbol. M reads this symbol by using $sa \to p$ or $sb \to q$ and jumps to the left in front of the rightmost occurrence of b or a, respectively. Then, it consumes it by using $pb \to s$ or $qa \to s$, respectively. If this read symbol was the rightmost one, it jumps one letter to the left and repeats the process. Otherwise, it makes no jumps at all. Observe that, in this way, every configuration is of the form urv, where $r \in \{s, p, q\}$, $u \in \{a, b\}^*$, and either $v \in \{a, \varepsilon\}\{b\}^*$ or $v \in \{b, \varepsilon\}\{a\}^*$.

Based on the previous observations, we see that

$$_lL(M) = \{w \mid occur(w, a) = occur(w, b)\}.$$

Since $L(M)$ is not regular, $_l\mathbf{L_{JFA}} - \mathbf{L_{REG}} \neq \emptyset$, so the theorem holds. □

Open Problem 3. Study the effect of left jumps on the acceptance power of JFAs and GJFAs. ■

9.4 A Variety of Start Configurations

In general, a GJFA can start its computation anywhere in the input string (see Definition 9.1). In this section, we consider the impact of various start configurations on the acceptance power of GJFAs and JFAs.

Definition 9.6. Let $M = (Q, \Delta, R, s, F)$ be a GJFA. Set

$$^bL(M) = \{w \in \Delta^* \mid sw \curvearrowright^* f \text{ with } f \in F\},$$
$$^aL(M) = \{uv \mid u, v \in \Delta^*, usv \curvearrowright^* f \text{ with } f \in F\},$$
$$^eL(M) = \{w \in \Delta^* \mid ws \curvearrowright^* f \text{ with } f \in F\}.$$

■

Intuitively, b, a, and e stand for *beginning*, *anywhere*, and *end*, respectively; in this way, we express where the acceptance process starts. Observe that we simplify $^aL(M)$ to $L(M)$ because we pay principal attention to the languages accepted in this way in this chapter. Let $^b\mathbf{L_{GJFA}}$, $^a\mathbf{L_{GJFA}}$, $^e\mathbf{L_{GJFA}}$, $^b\mathbf{L_{JFA}}$, $^a\mathbf{L_{JFA}}$, and $^e\mathbf{L_{JFA}}$ denote the families of languages accepted by GJFAs starting at the beginning, GJFAs starting anywhere, GJFAs starting at the end, JFAs starting at the beginning, JFAs starting anywhere, and JFAs starting at the end, respectively.

We show that:

(1) starting at the beginning increases the acceptance power of GJFAs and JFAs, and
(2) starting at the end does not increase the acceptance power of GJFAs and JFAs.

Theorem 9.31. $^{a}\mathbf{L_{JFA}} \subset {}^{b}\mathbf{L_{JFA}}$.

Proof. Let $M = (Q, \Delta, R, s, F)$ be a JFA. The JFA

$$M' = \big(Q, \Delta, R \cup \{s \to s\}, s, F\big)$$

clearly satisfies $^{a}L(M) = {}^{b}L(M')$, so $^{a}\mathbf{L_{JFA}} \subseteq {}^{b}\mathbf{L_{JFA}}$. We prove that this inclusion is, in fact, proper. Consider the language $K = \{a\}\{b\}^*$. The JFA

$$H = \big(\{s, f\}, \{a, b\}, \{sa \to f, fb \to f\}, s, \{f\}\big)$$

satisfies $^{b}L(H) = K$. However, observe that $^{a}L(H) = \{b\}^*\{a\}\{b\}^*$, which differs from K. By Theorem 9.5, for every JFA N, it holds that $^{a}L(N) \neq K$. Hence, $^{a}\mathbf{L_{JFA}} \subset {}^{b}\mathbf{L_{JFA}}$. □

Theorem 9.32. $^{a}\mathbf{L_{GJFA}} \subset {}^{b}\mathbf{L_{GJFA}}$.

Proof. This theorem can be proved by analogy with the proof of Theorem 9.31. □

Lemma 9.6. *Let M be a GJFA of degree $n \geq 0$. Then, there is a GJFA M' of degree n such that $^{a}L(M) = {}^{e}L(M')$.*

Proof. Let $M = (Q, \Delta, R, s, F)$ be a GJFA of degree n. Then, the GJFA

$$M' = \big(Q, \Delta, R \cup \{s \to s\}, s, F\big)$$

is of degree n and satisfies $^{a}L(M) = {}^{e}L(M')$. □

Lemma 9.7. *Let M be a GJFA of degree $n \geq 0$. Then, there is a GJFA $\hat{M}$ of degree n such that $^{e}L(M) = {}^{a}L(\hat{M})$.*

Proof. Let $M = (Q, \Delta, R, s, F)$ be a GJFA of degree n. If ${}^{e}L(M) = \emptyset$, then the GJFA

$$M' = (\{s\}, \Delta, \emptyset, s, \emptyset)$$

is of degree n and satisfies ${}^{a}L(M') = \emptyset$. If ${}^{e}L(M) = \{\varepsilon\}$, then the GJFA

$$M'' = (\{s\}, \Delta, \emptyset, s, \{s\})$$

is of degree n and satisfies ${}^{a}L(M'') = \{\varepsilon\}$. Therefore, assume that $w \in {}^{e}L(M)$, where $w \in \Delta^{+}$. Then, $s \to p \in R$, for some $p \in Q$. Indeed, observe that either ${}^{e}L(M) = \emptyset$ or ${}^{e}L(M) = \{\varepsilon\}$, which follows from the observation that if M starts at the end of an input string, then it first has to jump to the left to be able to read some symbols.

Define the GJFA $\hat{M} = (Q, \Delta, \hat{R}, s, F)$, where

$$\hat{R} = R - \big\{su \to q \mid u \in \Delta^{+}, q \in Q, \text{ and there is no } x \in \Delta^{+} \text{ such that } sx \rightsquigarrow s \text{ in } \Delta(M)\big\}.$$

The reason for excluding such $su \to q$ from $\hat{R}$ is that M first has to use a rule of the form $s \to p$, where $p \in Q$ (see the argumentation above). However, since $\hat{M}$ starts anywhere in the input string, we need to force it to use $s \to p$ as the first rule, thus changing the state from s to p, just like M does.

Clearly, $\hat{M}$ is of degree n and satisfies ${}^{e}L(M) = {}^{a}L(\hat{M})$, so the lemma holds. □

Theorem 9.33. ${}^{e}\mathbf{L}_{\mathbf{GJFA}} = {}^{a}\mathbf{L}_{\mathbf{GJFA}}$ *and* ${}^{e}\mathbf{L}_{\mathbf{JFA}} = {}^{a}\mathbf{L}_{\mathbf{JFA}}$.

Proof. This theorem follows from Lemmas 9.6 and 9.7. □

We also consider combinations of left jumps, right jumps, and various start configurations. For this purpose, by analogy with the previous denotations, we define ${}^{b}_{l}\mathbf{L}_{\mathbf{GJFA}}$, ${}^{a}_{l}\mathbf{L}_{\mathbf{GJFA}}$, ${}^{e}_{l}\mathbf{L}_{\mathbf{GJFA}}$, ${}^{b}_{r}\mathbf{L}_{\mathbf{GJFA}}$, ${}^{a}_{r}\mathbf{L}_{\mathbf{GJFA}}$, ${}^{e}_{r}\mathbf{L}_{\mathbf{GJFA}}$, ${}^{b}_{l}\mathbf{L}_{\mathbf{JFA}}$, ${}^{a}_{l}\mathbf{L}_{\mathbf{JFA}}$, ${}^{e}_{l}\mathbf{L}_{\mathbf{JFA}}$, ${}^{b}_{r}\mathbf{L}_{\mathbf{JFA}}$, ${}^{a}_{r}\mathbf{L}_{\mathbf{JFA}}$, and ${}^{e}_{r}\mathbf{L}_{\mathbf{JFA}}$. For example, ${}^{b}_{r}\mathbf{L}_{\mathbf{GJFA}}$ denotes the family of languages accepted by GJFAs that perform only right jumps and starts at the beginning.

Theorem 9.34. ${}^{a}_{r}\mathbf{L}_{\mathbf{GJFA}} = {}^{a}_{r}\mathbf{L}_{\mathbf{JFA}} = {}^{b}_{r}\mathbf{L}_{\mathbf{GJFA}} = {}^{b}_{r}\mathbf{L}_{\mathbf{JFA}} = {}^{b}_{l}\mathbf{L}_{\mathbf{GJFA}} = {}^{b}_{l}\mathbf{L}_{\mathbf{JFA}} = \mathbf{L}_{\mathbf{REG}}$.

Proof. Theorem 9.29, in fact, states that ${}^{a}_{r}\mathbf{L}_{\mathbf{GJFA}} = {}^{a}_{r}\mathbf{L}_{\mathbf{JFA}} = \mathbf{L}_{\mathbf{REG}}$. Furthermore, ${}^{b}_{r}\mathbf{L}_{\mathbf{GJFA}} = {}^{b}_{r}\mathbf{L}_{\mathbf{JFA}} = \mathbf{L}_{\mathbf{REG}}$ follows from the proof of Theorem 9.29 because M has to start the acceptance process of a string w from the configuration sw, i.e., it starts at the beginning of w. ${}^{b}_{l}\mathbf{L}_{\mathbf{GJFA}} = {}^{b}_{l}\mathbf{L}_{\mathbf{JFA}} = \mathbf{L}_{\mathbf{REG}}$ can be proved analogously. □

Theorem 9.35. ${}^{e}_{r}\mathbf{L}_{\mathbf{GJFA}} = {}^{e}_{r}\mathbf{L}_{\mathbf{JFA}} = \{\emptyset, \{\varepsilon\}\}$.

Proof. Consider JFAs $M = (\{s\}, \{a\}, \emptyset, s, \emptyset)$ and $M' = (\{s\}, \{a\}, \emptyset, s, \{s\})$ to see that $\{\emptyset, \{\varepsilon\}\} \subseteq {}^{e}_{r}\mathbf{L}_{\mathbf{GJFA}}$ and $\{\emptyset, \{\varepsilon\}\} \subseteq {}^{e}_{r}\mathbf{L}_{\mathbf{JFA}}$. The converse inclusion also holds. Indeed, any GJFA that starts the acceptance process of a string w from ws and can make only right jumps accepts either $\emptyset$ or $\{\varepsilon\}$. □

Open Problem 4. What are the properties of ${}^{e}_{l}\mathbf{L}_{\mathbf{GJFA}}$ and ${}^{e}_{l}\mathbf{L}_{\mathbf{JFA}}$? ■

Note that Open Problem 3, in fact, suggests an investigation of the properties of ${}^{a}_{l}\mathbf{L}_{\mathbf{GJFA}}$ and ${}^{a}_{l}\mathbf{L}_{\mathbf{JFA}}$.

A summary of open problems

We have already pointed out several specific open problems concerning them. We close this chapter by pointing out some crucially important open problem areas as suggested topics for future investigations:

(I) Concerning closure properties, study the closure of $\mathbf{L}_{\mathbf{GJFA}}$ under shuffle, Kleene star, Kleene plus, and under reversal.

(II) Regarding decision problems, investigate other decision properties of $\mathbf{L}_{\mathbf{GJFA}}$ and $\mathbf{L}_{\mathbf{JFA}}$, such as equivalence, universality, inclusion, or regularity. Furthermore, study their computational complexity. Do there exist undecidable problems for $\mathbf{L}_{\mathbf{GJFA}}$ or $\mathbf{L}_{\mathbf{JFA}}$?

(III) Section 9.3 has demonstrated that GJFAs and JFAs using only right jumps define the family of regular languages. How precisely do left jumps affect the acceptance power of JFAs and GJFAs?

(IV) Broaden the results of Section 9.4 concerning various start configurations by investigating the properties of ${}^{e}_{l}\mathbf{L}_{\mathbf{GJFA}}$ and ${}^{e}_{l}\mathbf{L}_{\mathbf{JFA}}$.

(V) Determinism represents a crucially important area of investigation in terms of all types of automata. In essence, the nondeterministic versions of automata can make several different moves from the same configuration, while their deterministic counterparts cannot; that is, they make no more than one move from any configuration. More specifically, the deterministic versions of classical finite automata require that for any state q and any input symbol a, there exists no more than one rule with qa on its left-hand side; in this way, they make no more than one move from any configuration. As a result, with any input string w, they make a unique sequence of moves. As should be obvious, in terms of jumping finite automata, this requirement does not guarantee their determinism in the above sense. Modify the requirement so that it guarantees determinism.

Chapter 10

Deep Pushdown Automata

This chapter deals with deep pushdown automata, which represent a natural modification of ordinary pushdown automata. While the ordinary versions can expand only the very topmost pushdown symbol, deep pushdown automata can make expansions deeper in the pushdown; otherwise, they both work identically.

This chapter proves that the power of deep pushdown automata is stronger than that of ordinary pushdown automata but less than that of linear-bounded automata (LBA). More precisely, they give rise to an infinite hierarchy of language families coinciding with the hierarchy resulting from n-limited state grammars (see Section 6.3). For every positive integer n, however, there always exist some languages that are not accepted by any deep pushdown automata, although they are accepted by LBA.

This chapter is divided into two sections. Section 10.1 defines and illustrates deep pushdown automata. Section 10.2 establishes their power.

10.1 Definitions and Examples

Without further ado, we define the notion of a deep pushdown automata, after which we illustrate it with an example.

Definition 10.1. A *deep pushdown automaton* is a septuple

$$M = \big(Q, \Delta, \Gamma, R, s, S, F\big),$$

where:

- Q is a finite set of *states*;
- Δ is an *input alphabet*;
- Γ is a *pushdown alphabet*, $\mathbb{N}$, Q, and Γ are pairwise disjoint, $\Delta \subseteq \Gamma$, and $\Gamma - \Delta$ contains a special *bottom symbol*, denoted by $\#$;
- $$R \subseteq (\mathbb{N} \times Q \times (\Gamma - (\Delta \cup \{\#\})) \times Q \times (\Gamma - \{\#\})^+) \cup (\mathbb{N} \times Q \times \{\#\} \times Q \times (\Gamma - \{\#\})^*\{\#\})$$ is a finite relation;
- $s \in Q$ is the *start state*;
- $S \in \Gamma$ is the *start pushdown symbol*;
- $F \subseteq Q$ is the set of *final states*.

Instead of $(m, q, A, p, v) \in R$, we write $mqA \to pv \in R$ and call $mqA \to pv$ a *rule*; accordingly, R is referred to as the *set of rules* of M. A *configuration* of M is a triple in $Q \times T^* \times (\Gamma - \{\#\})^*\{\#\}$. Let χ denote the set of all configurations of M. Let $x, y \in \chi$ be two configurations. M *pops* its pushdown from x to y, symbolically written as

$$x \;{}_p{\vdash}\; y$$

if $x = (q, au, az)$, $y = (q, u, z)$, where $a \in \Delta$, $u \in \Delta^*$, $z \in \Gamma^*$. M *expands* its pushdown from x to y, symbolically written as

$$x \;{}_e{\vdash}\; y$$

if $x = (q, w, uAz)$, $y = (p, w, uvz)$, $mqA \to pv \in R$, where $q, p \in Q$, $w \in \Delta^*$, $A \in \Gamma$, $u, v, z \in \Gamma^*$, and $occur(u, \Gamma - \Delta) = m - 1$. To express that M makes $x \;{}_e{\vdash}\; y$ according to $mqA \to pv$, we write

$$x \;{}_e{\vdash}\; y \; [mqA \to pv].$$

We say that $mqA \to pv$ is a *rule of depth* m; accordingly, $x \;{}_e{\vdash}\; y \; [mqA \to pv]$ is an *expansion of depth* m. M makes a *move* from x to y, symbolically written as

$$x \vdash y$$

if M makes either $x \;{}_e{\vdash}\; y$ or $x \;{}_p{\vdash}\; y$. If $n \in \mathbb{N}$ is the minimal positive integer such that each rule of M is of depth n or less, we say that M *is of depth* n, symbolically written as ${}_nM$. In the standard manner, we extend ${}_p{\vdash}$, ${}_e{\vdash}$, and $\vdash$ to ${}_p{\vdash}^m$, ${}_e{\vdash}^m$, and $\vdash^m$, respectively, for $m \geq 0$;

then, based on ${}_p\vdash^m$, ${}_e\vdash^m$, and $\vdash^m$, we define ${}_p\vdash^+$, ${}_p\vdash^*$, ${}_e\vdash^+$, ${}_e\vdash^*$, $\vdash^+$, and $\vdash^*$.

Let M be of depth n, for some $n \in \mathbb{N}$. We define the *language accepted by* ${}_nM$, $L({}_nM)$, as

$$L({}_nM) = \{w \in \Delta^* \mid (s, w, S\#) \vdash^* (f, \varepsilon, \#) \text{ in } {}_nM \text{ with } f \in F\}.$$

In addition, we define the *language that* ${}_nM$ *accepts by empty pushdown*, $E({}_nM)$, as

$$E({}_nM) = \{w \in \Delta^* \mid (s, w, S\#) \vdash^* (q, \varepsilon, \#) \text{ in } {}_nM \text{ with } q \in Q\}. \blacksquare$$

For every $k \geq 1$, ${}_{deep}\mathbf{L_{PDA}}_k$ denotes the family of languages defined by deep pushdown automata of depth i, where $1 \leq i \leq k$. Analogously, ${}^{empty}_{deep}\mathbf{L_{PDA}}_k$ denotes the family of languages defined by deep pushdown automata of depth i by empty pushdown, where $1 \leq i \leq k$.

The following example gives a deep pushdown automaton accepting a language from

$$\left({}_{deep}\mathbf{L_{PDA2}} \cap {}^{empty}_{deep}\mathbf{L_{PDA2}} \cap \mathbf{L_{CS}}\right) - \mathbf{L_{CF}}.$$

Example 10.1. Consider the deep pushdown automaton

$${}_2M = (\{s, q, p\}, \{a, b, c\}, \{A, S, \#\}, R, s, S, \{f\}),$$

with R containing the following five rules:

$$\begin{array}{lll} 1sS \to qAA, & 1qA \to fab, & 1fA \to fc. \\ 1qA \to paAb, & 2pA \to qAc, & \end{array}$$

On *aabbcc*, M makes

$$\begin{array}{rll} (s, aabbcc, S\#) & {}_e\vdash (q, aabbcc, AA\#) & [1sS \to qAA] \\ & {}_e\vdash (p, aabbcc, aAbA\#) & [1qA \to paAb] \\ & {}_p\vdash (p, abbcc, AbA\#) & \\ & {}_e\vdash (q, abbcc, AbAc\#) & [2pA \to qAc] \\ & {}_e\vdash (q, abbcc, abbAc\#) & [1qA \to fab] \\ & {}_p\vdash (f, bcc, bAc\#) & \\ & {}_p\vdash (f, cc, Ac\#) & \\ & {}_e\vdash (f, cc, Ac\#) & [1fA \to fc] \\ & {}_p\vdash (f, cc, cc\#) & \\ & {}_p\vdash (f, c, c\#) & \\ & {}_p\vdash (f, \varepsilon, \#). & \end{array}$$

In brief, $(s, aabbcc, S\#) \vdash^* (f, \varepsilon, \#)$. Observe that $L({}_2M) = E({}_2M) = \{a^n b^n c^n \mid n \geq 1\}$, which belongs to $\mathbf{L_{CS}} - \mathbf{L_{CF}}$. ■

10.2 Accepting Power

In this section, we establish the main results of this chapter. That is, we demonstrate that deep pushdown automata that make expansions of depth m or less, where $m \geq 1$, are equivalent to m-limited state grammars, so these automata accept a proper subfamily of the language family accepted by deep pushdown automata that make expansions of depth $m+1$ or less. Then, we point out that the resulting infinite hierarchy of language families obtained in this way occurs between the families of context-free and context-sensitive languages. However, we also show that there always exist some context-sensitive languages that cannot be accepted by any deep pushdown automata that make expansions of depth n or less, for every positive integer n.

To rephrase these results briefly and formally, we prove that

$$_{deep}\mathbf{L_{PDA1}} = {}_{deep}^{empty}\mathbf{L_{PDA1}} = \mathbf{L_{CF}},$$

and for every $n \geq 1$,

$$_{deep}^{empty}\mathbf{L_{PDA}}n = {}_{deep}\mathbf{L_{PDA}}n \subset {}_{deep}^{empty}\mathbf{L_{PDA}}n{+}1 = {}_{deep}\mathbf{L_{PDA}}n{+}1 \subset \mathbf{L_{CS}}.$$

After proving all these results, we formulate several open problem areas, including some suggestions concerning new deterministic and generalized versions of deep pushdown automata.

Lemma 10.1. *For every state grammar G and for every $n \geq 1$, there exists a deep pushdown automaton of depth n, ${}_nM$, such that $L(G, n) = L({}_nM)$.*

Proof. Let $G = (V, W, T, P, S)$ be a state grammar and let $n \geq 1$. Set $N = V - T$. Define the homomorphism f over $(\{\#\} \cup V)^*$ as $f(A) = A$, for every $A \in \{\#\} \cup N$, and $f(a) = \varepsilon$, for every $a \in T$. Introduce the deep pushdown automaton of depth n

$${}_nM = \big(Q, T, \{\#\} \cup V, R, s, S, \{\$\}\big),$$

where

$$Q = \{S, \$\} \cup \{\langle p, u\rangle \mid p \in W, u \in N^*\{\#\}^n, |u| \leq n\}$$

and R is constructed by performing the following four steps:

(1) for each $(p, S) \to (q, x) \in P$, $p, q \in W$, $x \in V^+$, add
$1sS \to \langle p, S\rangle S$ to R;

(2) if $(p, A) \to (q, x) \in P$, $\langle p, uAv\rangle \in Q$, $p, q \in W$, $A \in N$, $x \in V^+$, $u \in N^*$, $v \in N^*\{\#\}^*$, $|uAv| = n$, $p \notin \text{states}_G(u)$, add
$|uA|\langle p, uAv\rangle A \to \langle q, \text{prefix}(uf(x)v,\, n)\rangle x$ to R;

(3) if $A \in N$, $p \in W$, $u \in N^*$, $v \in \{\#\}^*$, $|uv| \leq n-1$, $p \notin \text{states}_G(u)$, add
$|uA|\langle p, uv\rangle A \to \langle p, uAv\rangle A$ and
$|uA|\langle p, uv\rangle\# \to \langle p, uv\#\rangle\#$ to R;

(4) for each $q \in W$, add
$1\langle q, \#^n\rangle\# \to \$\#$ to R.

${}_nM$ simulates n-limited derivations of G, so it always records the first n nonterminals occurring in the current sentential form in its state (if there appear fewer than n nonterminals in the sentential form, it completes them to n in the state by $\#$s from behind). ${}_nM$ simulates a derivation step in the pushdown and, simultaneously, records the newly generated nonterminals in the state. When G successfully completes the generation of a terminal string, ${}_nM$ completes reading the string, empties its pushdown, and enters the final state \$.

To establish $L(G, n) = L({}_nM)$, we first prove two claims.

Claim A. Let $(p, S)\ {}_n\!\Rightarrow^m\ (q, dy)$ in G, where $d \in T^*, y \in (NT^*)^*, p, q \in W, m \geq 0$. Then, $(\langle p, S\rangle, d, S\#) \vdash^* (\langle q, \text{prefix}(f(y\#^n), n)\rangle, \varepsilon, y\#)$ in ${}_nM$.

Proof of Claim A. This claim is proved by induction on $m \geq 0$.

Basis. Let $i = 0$, so $(p, S)\ {}_n\!\Rightarrow^0\ (p, S)$ in $G, d = \varepsilon$ and $y = S$. By using rules introduced in steps (1) and (4),

$$(\langle p, S\rangle, \varepsilon, S\#) \vdash^* (\langle p, \text{prefix}(f(S\#^n), n)\rangle, \varepsilon, S\#) \text{ in } {}_nM,$$

so the basis holds.

Induction hypothesis. Assume that the claim holds for all m, $0 \leq m \leq k$, where k is a non-negative integer.

Induction step. Let $(p, S)\ {}_n\!\Rightarrow^{k+1}\ (q, dy)$ in G, where $d \in T^*$, $y \in (NT^*)^*$, $p, q \in W$. Since $k + 1 \geq 1$, express $(p, S)\ {}_n\!\Rightarrow^{k+1}\ (q, dy)$ as

$$(p, S)\ {}_n\!\Rightarrow^{k}\ (h, buAo)\ {}_n\!\Rightarrow\ (q, buxo)\ [(h, A) \rightarrow (q, x)],$$

where $b \in T^*$, $u \in (NT^*)^*$, $A \in N$, $h, q \in W$, $(h, A) \rightarrow (q, x) \in P$, max-suffix($buxo$, $(NT^*)^*$) $= y$, and max-prefix($buxo$, T^*) $= d$. By the induction hypothesis,

$$(\langle p, S\rangle, w, S\#) \vdash^* (\langle h, \text{prefix}(f(uAo\#^n), n)\rangle, \varepsilon, uAo\#) \text{ in } M,$$

where $w =$ max-prefix($buAo$, T^*). As $(p, A) \rightarrow (q, x) \in P$, step (2) of the construction introduces the rule

$$|uA|\langle h, \text{prefix}(f(uAo\#^n), n)\rangle A \rightarrow \langle q, \text{prefix}(f(uxo\#^n), n)\rangle x \text{ to } R.$$

By using this rule, ${}_nM$ simulates $(buAo, h)\ {}_n\!\Rightarrow\ (buxo, q)$ by making

$$(\langle h, \text{prefix}(f(uAo\#^n), n)\rangle, \varepsilon, uAo\#) \vdash (\langle q, z\rangle, \varepsilon, uxo\#),$$

where $z =$ prefix($f(uxo\#^n)$, n) if $x \in V^+ - T^+$ and $z =$ prefix($f(uxo\#^n)$, $n - 1$) $=$ prefix($f(uo\#^n)$, $n - 1$) if $x \in T^+$. In the latter case, ($z =$ prefix($f(uo\#^n)$, $n - 1$), so $|z| = n - 1$), ${}_nM$ makes

$$(\langle q, \text{prefix}(f(uo\#^n), n-1)\rangle, \varepsilon, uxo\#) \vdash (\langle q, \text{prefix}(f(uo\#^n), n)\rangle, \varepsilon, uxo\#)$$

by a rule introduced in (3). If $uxo \in (NT^*)^*$, $uxo = y$, and the induction step is completed. Therefore, assume that $uxo \neq y$, so $uxo = ty$ and $d = wt$, for some $t \in T^+$. Observe that prefix($f(uxo\#^n)$, n) $=$ prefix($f(y\#^n)$, n) at this point. Then, ${}_nM$ removes t by making $|t|$ popping moves so that

$$(\langle q, \text{prefix}(f(uxo\#^n), n)\rangle, t, ty\#)\ {}_p\!\vdash^{|t|}\ (\langle q, \text{prefix}(f(y\#^n), n)\rangle, \varepsilon, y\#^n).$$

Thus, putting the previous sequences of moves together, we obtain

$$\begin{aligned}(p, wt, S\#^n) \vdash^* &\ (\langle q, \text{prefix}(f(uxo\#^n), n)\rangle, t, ty\#)\ [1sS \rightarrow qAA] \\ {}_p\!\vdash^{|t|} &\ (\langle q, \text{prefix}(f(y\#^n), n)\rangle, \varepsilon, y\#),\end{aligned}$$

which completes the induction step.

By the previous claim for $y = \varepsilon$, if $(p, S)\ {}_n{\Rightarrow}^* (q, d)$ in G, where $d \in T^*$, $p, q \in W$, then

$$(\langle p, S\rangle, d, S\#) \vdash^* (\langle q, \text{prefix}(f(\#^n), n)\rangle, \varepsilon, \#) \text{ in } {}_nM.$$

As $\text{prefix}(f(\#^n), n) = \#$ and R contains rules introduced in (1) and (4), we also have

$$\begin{aligned}(s, d, S\#) &\vdash (\langle p, S\rangle, d, S\#)\\ &\vdash^* (\langle q, \#^n, n)\rangle, \varepsilon, \#)\\ &\vdash^* (\$, \varepsilon, \#) \text{ in } {}_nM.\end{aligned}$$

Thus, $d \in L(G)$ implies that $d \in L({}_nM)$, so $L(G, n) \subseteq L({}_nM)$.

Claim B. Let $(\langle p, S\#^{n-1}\rangle, c, S\#) \vdash^m (\langle q, \text{prefix}(f(y\#^n), n)\rangle, \varepsilon, by\#)$ in ${}_nM$, with $c, b \in T^*$, $y \in (NT^*)^*$, $p, q \in W$, and $m \geq 0$. Then, $(p, S)\ {}_n{\Rightarrow}^* (q, cby)$ in G.

Proof of Claim B. This claim is proved by induction on $m \geq 0$.

Basis. Let $m = 0$. Then, $c = b = \varepsilon$, $y = S$, and

$$(\langle p, S\#^{n-1}\rangle, \varepsilon, S\#) \vdash^0 (\langle q, \text{prefix}(f(S\#^n), n)\rangle, \varepsilon, S\#) \text{ in } {}_nM.$$

As $(p, S)\ {}_n{\Rightarrow}^0 (p, S)$ in G, the basis holds.

Induction hypothesis. Assume that the claim holds for all m, $0 \leq m \leq k$, where k is a non-negative integer.

Induction step. Let

$$(\langle p, S\#^{n-1}\rangle, c, S\#) \vdash^{k+1} (\langle q, \text{prefix}(f(y\#^n), n)\rangle, \varepsilon, by\#) \text{ in } {}_nM,$$

where $c, b \in T^*, y \in (NT^*)^*, p, q \in W$ in ${}_nM$. Since $k + 1 \geq 1$, we can express

$$(\langle p, S\#^{n-1}\rangle, c, S\#) \vdash^{k+1} (\langle q, \text{prefix}(f(y\#^n), n)\rangle, \varepsilon, by\#)$$

as

$$\begin{aligned}(\langle p, S\#^{n-1}\rangle, c, S\#) &\vdash^k \alpha\\ &\vdash (\langle q, \text{prefix}(f(y\#^n), n)\rangle, \varepsilon, by\#) \text{ in } {}_nM,\end{aligned}$$

where α is a configuration of ${}_nM$ whose form depends on whether the last move is: (i) a popping move or (ii) an expansion, as described in the following:

(i) Assume that $\alpha \ {}_p\vdash (\langle q, \text{prefix}(f(y\#^n),\ n)\rangle, \varepsilon, by\#)$ in ${}_nM$. In a greater detail, let $\alpha = (\langle q, \text{prefix}(f(y\#^n),\ n)\rangle, a, aby\#)$, with $a \in T$ such that $c = \text{prefix}(c,\ |c| - 1)a$. Thus,

$$(\langle p, S\#^{n-1}\rangle, c, S\#) \vdash^k (\langle q, \text{prefix}(f(y\#^n), n)\rangle, a, aby\#)$$
$$\qquad {}_p\vdash (\langle q, \text{prefix}(f(y\#^n), n)\rangle, \varepsilon, by\#).$$

Since $(\langle p, S\#^{n-1}\rangle, c, S\#) \vdash^k (\langle q, \text{prefix}(f(y\#^n),\ n)\rangle, a, aby\#)$, we have

$$(\langle p, S\#^{n-1}\rangle, \text{prefix}(c, |c| - 1), S\#) \vdash^k (\langle q, \text{prefix}(f(y\#^n), n)\rangle, \varepsilon, aby\#).$$

By the induction hypothesis, $(p, S)\ {}_n\Rightarrow^* (q, \text{prefix}(c,\ |c| - 1)aby)$ in G. As $c = \text{prefix}(c,\ |c| - 1)a$, $(p, S)\ {}_n\Rightarrow^* (q, cby)$ in G.

(ii) Assume that $\alpha\ {}_e\vdash (\langle q, \text{prefix}(f(y\#^n),\ n)\rangle, \varepsilon, by\#)$ in ${}_nM$. Observe that this expansion cannot be made by rules introduced in steps (1) or (4). If this expansion is made by a rule introduced in (3), which does not change the pushdown contents at all, the induction step follows from the induction hypothesis. Finally, suppose that this expansion is made by a rule introduced in step (2). In a greater detail, suppose that

$$\alpha = (\langle o, \text{prefix}(f(uAv\#^n), n)\rangle, \varepsilon, uAv\#)$$

and ${}_nM$ makes

$$(\langle o, \text{prefix}(f(uAv\#^n), n)\rangle, \varepsilon, uAv\#)\ {}_e\vdash (\langle q, \text{prefix}(f(uxv\#^n), n)\rangle,$$
$$\varepsilon, uxv\#)$$

by using

$$|f(uA)|\langle o, \text{prefix}(f(uAv\#^n), n)\rangle A \rightarrow \langle q, \text{prefix}(f(uxv\#^n), n)\rangle x \in R$$

introduced in step (2) of the construction, where $A \in N$, $u \in (NT^*)^*$, $v \in (N \cup T)^*$, $o \in W$, $|f(uA)| \leq n$, $by\# = uxv\#$. By the induction hypothesis,

$$(\langle p, S\#^{n-1}\rangle, c, S\#) \vdash^k (\langle o, \text{prefix}(f(uAv\#^n), n)\rangle, \varepsilon, uAv\#) \text{ in } {}_nM$$

implies that $(p, S)\ {}_n\Rightarrow^* (o, cuAv)$ in G. From

$$|f(uA)|\langle o, \text{prefix}(f(uAv\#^n), n)\rangle A \rightarrow \langle q, \text{prefix}(f(uxv\#^n), n)\rangle x \in R,$$

it follows that $(o, A) \rightarrow (q, x) \in P$ and $A \notin \text{states}_G(f(u))$. Thus,

$$\begin{aligned}(p, S) \;{}_n\!\Rightarrow^* &\; (o, cuAv)\\ {}_n\!\Rightarrow &\; (q, cuxv) \text{ in } G.\end{aligned}$$

Therefore, $(p, S) \;{}_n\!\Rightarrow^* (q, cby)$ in G because $by\# = uxv\#$.

Consider the previous claim for $b = y = \varepsilon$ to see that

$$(\langle p, S\#^{n-1}\rangle, c, S\#) \vdash^* (\langle q, \text{prefix}(f(\#), n)\rangle, \varepsilon, \#^n) \text{ in } {}_nM$$

implies $(p, S) \;{}_n\!\Rightarrow^* (q, c)$ in G. Let $c \in L({}_nM)$. Then,

$$(s, c, S\#) \vdash^* (\$, \varepsilon, \#) \text{ in } {}_nM.$$

Examine the construction of ${}_nM$ to see that $(s, c, S) \vdash^* (\$, \varepsilon, \#)$ starts by using a rule introduced in (1), so $(s, c, S) \vdash^* (\langle p, S\#^{n-1}\rangle, c, S\#)$. Furthermore, note that this sequence of moves ends with $(s, c, S) \vdash^* (\$, \varepsilon, \varepsilon)$ by using a rule introduced in step (4). Thus, we can express

$$(s, c, \#) \vdash^* (\$, \varepsilon, \#)$$

as

$$\begin{aligned}(s, c, \#) \vdash^* &\; (\langle p, S\#^{n-1}\rangle, c, S\#)\\ \vdash^* &\; (\langle q, \text{prefix}(f(\#^n), n)\rangle, \varepsilon, \#)\\ \vdash &\; (\$, \varepsilon, \#) \text{ in } {}_nM.\end{aligned}$$

Therefore, $c \in L({}_nM)$ implies that $c \in L(G, n)$, so $L({}_nM) \subseteq L(G, n)$.

As $L({}_nM) \subseteq L(G, n)$ and $L(G, n) \subseteq L({}_nM)$, $L(G, n) = L({}_nM)$. Thus, Lemma 10.1 holds. □

Lemma 10.2. *For every $n \geq 1$ and every deep pushdown automaton ${}_nM$, there exists a state grammar G such that $L(G, n) = L({}_nM)$.*

Proof. Let $n \geq 1$ and ${}_nM = (Q, T, V, R, s, S, F)$ be a deep pushdown automaton. Let Z and $\$$ be two new symbols that occur in no component of ${}_nM$. Set $N = V - T$. Introduce the sets

$$C = \{\langle q, i, \triangleright\rangle \mid q \in Q, 1 \leq i \leq n-1\}$$

and

$$D = \{\langle q, i, \triangleleft\rangle \mid q \in Q, 0 \leq i \leq n-1\}.$$

Moreover, introduce an alphabet W such that $|V| = |W|$, and for all i, $1 \leq i \leq n$, an alphabet U_i such that $|U_i| = |N|$. Without any loss of generality, assume that V, Q, and all these newly introduced sets and alphabets are pairwise disjoint. Set $U = \bigcup_{i=1}^{n} U_i$. For each i, $1 \leq i \leq n-1$, set $C_i = \{\langle q, i, \triangleright\rangle \mid q \in Q\}$, and for each i, $0 \leq i \leq n-1$, set $D_i = \{\langle q, i, \triangleleft\rangle \mid q \in Q\}$. Introduce a bijection h from V to W. For each i, $1 \leq i \leq n$, introduce a bijection ${}_ig$ from N to U_i. Define the state grammar

$$G = \big(V \cup W \cup U \cup \{Z\}, Q \cup C \cup D \cup \{\$\}, T, P, Z\big),$$

where P is constructed by performing the following steps:

(1) add $(s, Z) \to (\langle s, 1, \triangleright\rangle, h(S))$ to P;
(2) for each $q \in Q$, $A \in N$, $1 \leq i \leq n-1$, $x \in V^+$, add

$$(\langle q, i, \triangleright\rangle, A) \to (\langle q, i+1, \triangleright\rangle, {}_ig(A)) \text{ and}$$
$$(\langle q, i, \triangleleft\rangle, {}_ig(A)) \to (\langle p, i-1, \triangleleft\rangle, A) \text{ to } P;$$

(3) if $ipA \to qxY \in R$, for some $p, q \in Q$, $A \in N$, $x \in V^*$, $Y \in V$, $i = 1, \ldots, n$, add

$$(\langle p, i, \triangleright\rangle, A) \to (\langle q, i-1, \triangleleft\rangle, xY) \text{ and}$$
$$(\langle p, i, \triangleright\rangle, h(A)) \to (\langle q, i-1, \triangleleft\rangle, xh(Y)) \text{ to } P;$$

(4) for each $q \in Q$, $A \in N$, add

$$(\langle q, 0, \triangleleft\rangle, A) \to (\langle q, 1, \triangleright\rangle, A) \text{ and}$$
$$(\langle q, 0, \triangleleft\rangle, h(Y)) \to (\langle q, 1, \triangleright\rangle, h(Y)) \text{ to } P;$$

(5) for each $q \in F$, $a \in T$, add

$$(\langle q, 0, \triangleleft\rangle, h(a)) \to (\$, a) \text{ to } P.$$

G simulates the application of $ipA \to qy \in R$, so it makes a left-to-right scan of the sentential form, counting the occurrences of nonterminals until it reaches the ith occurrence of a nonterminal. If this occurrence equals A, it replaces this A with y and returns to the beginning of the sentential form in order to analogously simulate a move from q. Throughout the simulation of moves of ${}_nM$ by G, the rightmost symbol of every sentential form is from W. G completes the simulation of an acceptance of a string x by ${}_nM$, so it uses a rule introduced in step (5) of the construction of P to change the rightmost symbol of x, $h(a)$, to a and, thereby, to generate x.

We next establish $L(G,n) = L({}_nM)$. To keep the rest of the proof as readable as possible, we omit some details in what follows. The reader can easily fill them in.

Claim A. $L(G,n) \subseteq L({}_nM)$.

Proof of Claim A. Consider any $w \in L(G,n)$. Observe that G generates w as

$$\begin{aligned}(p,Z) {}_n\Rightarrow\ & (\langle s,1,\triangleright\rangle, h(S))\ [(s,Z) \rightarrow (\langle s,1,\triangleright\rangle, h(S))]\\ {}_n\Rightarrow^*\ & (f, yh(a))\\ {}_n\Rightarrow\ & (\$, w),\end{aligned}$$

where $f \in F$, $a \in T$, $y \in T^*$, $ya = w$, $(s,Z) \rightarrow (\langle s,1,\triangleright\rangle, h(S))$ in step (1) of the construction of P, $(\langle q,0,\triangleleft\rangle, h(a)) \rightarrow (\$, a)$ in (5), every

$$u \in \text{strings}\big((\langle s,1,\triangleright\rangle, h(S))\ {}_n\vdash^* (q, yh(a))\big)$$

satisfies $u \in (V \cup U)^*W$, and every step in

$$(\langle s,1,\triangleright\rangle, h(S))\ {}_n\vdash^* (f, yh(S))$$

is made by a rule introduced in (2)–(4). Indeed, the rule constructed in (1) is always used in the first step and a rule constructed in (5) is always used in the very last step of any successful generation in G; during any other step, neither of them can be applied. Note that the rule in (1) generates $h(S)$. Furthermore, examine the rules in (2)–(4) to see that by their use, G always produces a string that has exactly one occurrence of a symbol from W in any string, and this occurrence appears as the rightmost symbol of the string; formally,

$$u \in \text{strings}\big((\langle s,1,\triangleright\rangle, h(S))\ {}_n\Rightarrow^* (f, yh(a))\big)$$

implies that $u \in (V \cup U)^*W$. In a greater detail,

$$(\langle s,1,\triangleright\rangle, h(S))\ {}_n\Rightarrow^* (f, yh(a))$$

can be expressed as

$$\begin{array}{lllll} (q_0, z_0) & {}_n\!\Rightarrow^* (c_0, y_0) & {}_n\!\Rightarrow (d_0, u_0) & {}_n\!\Rightarrow^* (p_0, v_0) & {}_n\!\Rightarrow \\ (q_1, z_1) & {}_n\!\Rightarrow^* (c_1, y_1) & {}_n\!\Rightarrow (d_1, u_1) & {}_n\!\Rightarrow^* (p_1, v_1) & {}_n\!\Rightarrow \\ & \vdots & \vdots & \vdots & \vdots \\ (q_m, z_m) & {}_n\!\Rightarrow^* (c_m, y_m) & {}_n\!\Rightarrow (d_m, u_m) & {}_n\!\Rightarrow^* (p_m, v_m) & {}_n\!\Rightarrow \\ (q_{m+1}, z_{m+1}), & & & & \end{array}$$

for some $m \geq 1$, where $z_0 = h(S)$, $z_{m+1} = yh(a)$, and $f = q_{m+1}$, and for each j, $0 \leq j \leq m$, $q_j \in C_1$, $p_j \in D_0$, and $z_j \in V^*W$, and there exists $i_j \in \{1, \ldots, n\}$ so that $c_j \in C_{i_j}$, $y_j \in T^*C_1T^*C_2 \cdots T^*C_{i_j-1}V^*W$, $d_j \in D_{i_j-1}$, $u_j \in T^*C_1T^*C_2 \cdots T^*D_{i_j-1}V^*W$, and

$$(q_j, z_j) \;{}_n\!\Rightarrow^* (c_j, y_j) \;{}_n\!\Rightarrow (d_j, u_j) \;{}_n\!\Rightarrow^* (p_j, v_j) \;{}_n\!\Rightarrow (q_{j+1}, z_{j+1})$$

satisfies (i)–(iv), which is given in the following.

For brevity, we first introduce the following notation. Let w be any string. For $i = 1, \ldots, |w|$, $\lfloor w, i, N \rfloor$ denotes the ith occurrence of a nonterminal from N in w, and if such a nonterminal does not exist, $\lfloor w, i, W \rfloor = 0$; for instance, $\lfloor ABABC, 2, \{A, C\} \rfloor$ denotes the underlined $\underline{A}$ in $AB\underline{A}BC$.

(i) $(q_j, z_j) \;{}_n\!\Rightarrow^* (c_i, y_i)$ consists of $i_j - 1$ steps during which G changes $\lfloor z_j, 1, N \rfloor, \ldots, \lfloor z_j, i_j - 1, N \rfloor$ to ${}_1g(\langle \lfloor z_j, 1, N \rfloor, 2 \rangle), \ldots, {}_{i_j}g(\langle \lfloor z_j, i_j - 1, N \rfloor, i_j - 1 \rangle)$, respectively, by using the rules in (2.1) in the construction of P;

(ii) if $i_j \leq occur(z_j, N)$, then $(c_j, y_j) \;{}_n\!\Rightarrow (d_j, u_j)$ have to consist of a step according to $(\langle q, i, \triangleright \rangle, A_j) \rightarrow (\langle q, i-1, \triangleleft \rangle, x_jX_j)$, where $\lfloor z_j, i_j, N \rfloor$ is an occurrence of $A_j, x_j \in V^*, X_j \in V$, and if $i_j = occur(z_j, N \cup W)$, then $(c_j, y_j) \;{}_n\!\Rightarrow (d_j, u_j)$ consists of a step according to $(\langle p, i, \triangleright \rangle, h(A_j)) \rightarrow (\langle q, i-1, \triangleleft \rangle, x_jh(X_j))$ constructed in (3), where $\lfloor z_j, i_j, N \cup W \rfloor$ is an occurrence of $h(A_j), x_j \in V^*, X_j \in V$;

(iii) $(d_j, u_j) \;{}_n\!\Rightarrow^* (p_j, v_j)$ consists of $i_j - 1$ steps, during which G changes from ${}_{i_j}g(\langle \lfloor z_j, i_j - 1, N \rfloor, i_j - 1 \rangle), \ldots, {}_1g(\langle \lfloor z_j, 1, N \rfloor, 1 \rangle)$ back to $\lfloor z_j, i_j - 1, N \rfloor, \ldots, \lfloor z_j, 1, N \rfloor$, respectively, in a right-to-left way by using rules constructed in (2.2);

(iv) $(p_j, v_j) \;{}_n\!\Rightarrow (q_{j+1}, z_{j+1})$ is made by a rule constructed in (4).

For every

$$\begin{aligned}(q_j, z_j) \ {}_n\!\Rightarrow^* &\ (c_j, y_j)\\ {}_n\!\Rightarrow &\ (d_j, u_j)\\ {}_n\!\Rightarrow^* &\ (p_j, v_j)\\ {}_n\!\Rightarrow &\ (q_{j+1}, z_{j+1}) \text{ in } G,\end{aligned}$$

where $0 \leq j \leq m$, ${}_nM$ makes

$$(q_j, o_j, \text{suffix}(z_j, t_j)) \vdash^* (q_{j+1}, o_{j+1}, \text{suffix}(z_{j+1}, t_{j+1}))$$

with $o_0 = w$, $z_0 = S\#$, $t_{j+1} = |\,\text{max-prefix}(z_{j+1}, T^*)|$, $o_{j+1} = \text{suffix}(o_j, |o_j| + t_{j+1})$, where $o_0 = w$, $z_0 = S\#$, and $t_0 = |z_0|$. In this sequence of moves, the first move is an expansion made according to $i_j q_j A_j \to q_{j+1} x_j X_j$ (see steps (2) and (3) of the construction), followed by t_{j+1} popping moves (note that $i_j \geq 2$ implies that $t_{j+1} = 0$). As $f \in F$ and $ya = w$, $w \in L({}_nM)$. Therefore, $L(G, n) \subseteq L({}_nM)$.

Claim B. $L({}_nM) \subseteq L(G, n)$.

Proof of Claim B. This proof is simple and left to the reader.

As $L({}_nM) \subseteq L(G, n)$ and $L(G, n) \subseteq L({}_nM)$, we have $L(G, n) = L({}_nM)$, so this lemma holds true. □

Theorem 10.1. *For every $n \geq 1$ and for every language $L, L = L(G, n)$, for a state grammar G, if and only if $L = L({}_nM)$ for a deep pushdown automaton ${}_nM$.*

Proof. This theorem follows from Lemmas 10.1 and 10.2. □

By analogy with the demonstration of Theorem 10.1, we can establish the following theorem.

Theorem 10.2. *For every $n \geq 1$ and for every language $L, L = L(G, n)$, for a state grammar G, if and only if $L = E({}_nM)$ for a deep pushdown automaton ${}_nM$.* ■

Following is the main result of this chapter.

Corollary 10.1. *For every $n \geq 1$,*

$${}^{empty}_{deep}\mathbf{L}_{\mathbf{PDA}n} = {}_{deep}\mathbf{L}_{\mathbf{PDA}n} \subset {}_{deep}\mathbf{L}_{\mathbf{PDA}n+1} = {}^{empty}_{deep}\mathbf{L}_{\mathbf{PDA}n+1}.$$

Proof. This corollary follows from Theorems 10.1 and 10.2 above and from Theorem 6.27, which says that the m-limited state grammars generate a proper subfamily of the family generated by $(m+1)$-limited state grammars, for every $m \geq 1$. □

Finally, we state two results concerning $\mathbf{L_{CF}}$ and $\mathbf{L_{CS}}$.

Corollary 10.2. $_{deep}\mathbf{L_{PDA1}} = {}^{empty}_{deep}\mathbf{L_{PDA1}} = \mathbf{L_{CF}}$.

Proof. This corollary follows from Lemmas 10.1 and 10.2 for $n = 1$, and from Theorem 6.27, which says that one-limited state grammars characterize $\mathbf{L_{CF}}$. □

Corollary 10.3. *For every* $n \geq 1$, $_{deep}\mathbf{L_{PDA}}_n = {}^{empty}_{deep}\mathbf{L_{PDA}}_n \subset \mathbf{L_{CS}}$.

Proof. This corollary follows from Lemmas 10.1 and 10.2, Theorems 10.1 and 10.2, and from Theorem 6.27, which says that $_m\mathbf{L_{ST}}$, for every $m \geq 1$, is properly included in $\mathbf{L_{CS}}$. □

Finally, we suggest two open problem areas concerning deep pushdown automata.

Determinism

This chapter has discussed a general version of deep pushdown automata, which work non-deterministically. Undoubtedly, future investigations of these automata should pay special attention to their deterministic versions, which fulfill a crucial role in practice. In fact, we can introduce a variety of deterministic versions, including the following two types. First, we consider the fundamentally strict form of determinism.

Definition 10.2. Let $M = (Q, \Delta, \Gamma, R, s, S, F)$ be a deep pushdown automaton. We say that M is *deterministic* if for every $mqA \rightarrow pv \in R$,

$$\left|\{mqA \rightarrow ow \mid mqA \rightarrow ow \in R, o \in Q, w \in \Gamma^+\} - \{mqA \rightarrow pv\}\right| = 0.$$

■

As a weaker form of determinism, we obtain the following definition.

Definition 10.3. Let $M = (Q, \Delta, \Gamma, R, s, S, F)$ be a deep pushdown automaton. We say that M is *deterministic with respect to the depth of its expansions* if for every $q \in Q$,

$$\left|\{m \mid mqA \to pv \in R, A \in \Gamma, p \in Q, v \in \Gamma^+\}\right| \leq 1$$

because at this point, from the same state, all expansions that M can make are of the same depth. ■

To illustrate, consider, for instance, the deep pushdown automaton ${}_2M$ from Example 10.1. This automaton is deterministic with respect to the depth of its expansions; however, it does not satisfy the strict determinism. Note that ${}_nM$ constructed in the proof of Lemma 10.1 is deterministic with respect to the depth of its expansions, so we obtain this corollary.

Corollary 10.4. *For every state grammar G and for every $n \geq 1$, there exists a deep pushdown automaton ${}_nM$ such that $L(G, n) = L({}_nM)$ and ${}_nM$ is deterministic with respect to the depth of its expansions.* ■

Open Problem 1. Can an analogical statement to Corollary 10.4 be established in terms of the strict determinism? ■

Generalization

Let us note that throughout this chapter, we have considered only true pushdown expansions in the sense that the pushdown symbol is replaced with a non-empty string rather than with an empty string; at this point, no pushdown expansion can result in shortening the pushdown length. Nevertheless, the moves that allow deep pushdown automata to replace a pushdown symbol with ε and, thereby, shorten its pushdown represent a natural generalization of deep pushdown automata discussed in this chapter.

Open Problem 2. What is the language family defined by deep pushdown automata generalized in this way? ■

Part 4

Applications

This two-chapter part of the book describes some applications of the formal models covered in Part 2. First, thoroughly and in detail, Chapter explains the use of finite and pushdown automata as programming language analyzers. Then, in a briefer way, Chapter 11 sketches grammatical applications in natural language processing and musicology. Compared to Parts 2 and 3, the current application-oriented part differs in style; indeed, all of its concepts are presented in a descriptive way rather than in the form of the strictly rigorous mathematical notions used in the previous two parts. On the other hand, this part primarily concentrates its attention on the use of the formal models in practice.

Chapter 11

Applications of Automata

As demonstrated in Chapter 7, acting ultimately as general models of computation, Turing machines underlie the theory of computation in its entirety. On the other hand, the other two major types of automata covered in Part 2, namely finite and pushdown automata discussed in Chapters 3 and 4, respectively, fulfill a central role in practice. Out of a large variety of their applications, the current two-section chapter explains how they underlie computer-science engineering techniques used in programming language analysis performed by common software units, such as compilers. More specifically, by using finite automata, supported by regular expressions, Section 11.1 builds up lexical analyzers, which recognize lexical units and verify that they are properly formed. Then, based on pushdown automata, Section 11.2 creates syntax analyzers, which recognize syntactic structures in computer programs and verify that they are correctly written according to context-free grammatical rules that specify the syntax.

11.1 Finite Automata and Lexical Analysis

Finite automata underlie many computer-science units, ranging from hardware components, such as switching circuits, up to software tools, such as text editors. Out of this large variety, this section narrows its attention to explaining how they are applied in the lexical analysis of high-level programming languages. First, it describes the implementation of CSDFAs. Then, making use of this implementation, it explains how to build up lexical analyzers.

Implementation of finite automata

Next, we explain how to implement CSDFAs because grasping the construction of a lexical analyzer presupposes familiarity with this implementation. As a matter of fact, we give two alternative algorithms that implement these automata. The first algorithm actually implements them based on their state tables. The other algorithm is based on a nested **case** statement. In both of them, we make use of the following convention.

Convention 11.1. Algorithms 11.1 and 11.2 assume that a given input string w is followed by $\lhd$, which thus acts as an *input end marker.* As their output, the algorithms announces **ACCEPT** if $w \in L(M)$ and **REJECT** if $w \notin L(M)$. ■

As is obvious, in practice, the end of a given input string is always somehow specified, so $\lhd$ only symbolically expresses this end. For instance, if the input string is written as a file, then the end of the file fulfills the role of $\lhd$.

A table-based implementation

First, we give the algorithm that implements any completely specified DFA, M, based on its state table. In this algorithm, we assume that its states are numbered 1–n, for some $n \in \mathbb{N}$, where 1 is the start state.

Algorithm 11.1. FA implementation — A tabular method.

Input: A CSDFA $M = (Q, \Delta, R, s, F)$ with $Q = \{1, \ldots, n\}$,
$\Delta = \{a_1, \ldots, a_m\}, s = 1$, and $w\lhd$ with $w \in \Delta^*$.
Output: **ACCEPT** if $w \in L(M)$, and **REJECT** if $w \notin L(M)$.
Method:

type
states $= 1..n$
symbols $= a_1..a_m$

```
    rules = array [states, symbols] of states
    stateset = set of states
var
    state: states
    symbol: symbols
    rule: rules
    finalstates: stateset
begin
    for all i = 1, ..., n and all j = 1, ..., m
      if ia_j → k ∈ R then
        set rule[i, a_j] = k

    set finalstates to F
    set state = 1

    read (symbol)

    while symbol ≠ ◁ do
      state = rule[state, symbol]
      read (symbol)

    if state ∈ finalstates then
      ACCEPT {w ∈ L(M)}
    else
      REJECT {w ∉ L(M)}
end
```

In Algorithm 11.1, a rule, $ia_j \to k \in R_M$, is represented by $rule[i, a_j] = k$. If $state = i, symbol = a_j$, and $rule[i, a_j] = k$, the **while** loop of this algorithm sets *state* to k to simulate the application of $ia_j \to k$. When this loop reads $\triangleleft$, it exits, and the **if** statement tests whether *state* represents a final state. If so, $w \in L(M)$; otherwise, $w \notin L(M)$.

Example 11.1. Let us consider the CSDFA represented by the state table given in Table 11.1. Its start state is 1, and the final states are 2 and 3. As obvious, it accepts the language $\{a\}^*\{b\}^*$. ■

Table 11.1. State table.

State	a	b
1	2	3
2	2	3
3	4	3
4	4	3

With $abb\#$ as its input, the **while** loop in Algorithm 11.1 makes these three iterations. The first iteration begins with $state = 1$ and $symbol = a$. Thus, it sets $state$ to 2 because $rule[1, a] = 2$. The second iteration has $state = 2$ and $symbol = b$, so it sets $state$ to 3 because $rule[2, b] = 3$. The third iteration starts with $state = 3$ and $symbol = b$; therefore, it sets $state = 3$ because $rule[3, b] = 3$. The next symbol is $\lhd$, so the **while** loop exits, and the **if** statement determines that $state$ belongs to $finalstates$ because $state = 3$ and $finalstates = [2, 3]$. Therefore, this statement writes **ACCEPT**.

A case-statement implementation

The previous algorithm represents the state table of its input automaton as a two-dimensional array. The following algorithm is based on a nested **case** statement, which frees this implementation from using any array; indeed, the state table is hardwired into the program structure by using case statements. Later on, this section illustrates this algorithm with several examples that put it firmly into the implementation of a lexical analyzer.

Algorithm 11.2. FA implementation — A **case**-statement-based method.

Input: A CSDFA $M = (Q, \Delta, R, s, F)$ with $Q = \{1, \ldots, n\}$,
$\Delta = \{a_1, \ldots, a_m\}$, $s = 1$, and $w\lhd$ with $w \in \Delta^*$.
Output: **ACCEPT** if $w \in L(M)$, and **REJECT** if $w \notin L(M)$.
Method:

type
  $states = 1..n$
  $symbols = a_1..a_m$
  $stateset =$ **set of** $states$

```
var
  state: states
  symbol: symbols
  rule: rules
  finalstates: stateset

begin
  set finalstates to F
  set state = 1

  read (symbol)

  while symbol ≠ ◁ do
    case state of
      1 : ...
      ⋮
      i : case symbol of
        a^1 : ...
        ⋮
        a^j : state = k if ia_j → k ∈ R
        ⋮
        a^m : ...
    read(symbol)

  if state ∈ finalstates then
    ACCEPT {w ∈ L(M)}
  else
    REJECT {w ∉ L(M)}
end
```

Algorithm 11.2 has an important advantage over Algorithm 11.1. Indeed, while Algorithm 11.1 strictly requires CSDFA as its input automaton, Algorithm 11.2 does not; that is, the latter can also implement incompletely specified DFAs without any problems.

Introduction to lexical analysis

A lexical analyzer or, more briefly and customarily, a scanner breaks up a programming-language code into logically cohesive lexical entities, referred to as lexical units. That is, it reads the string of characters that make up the program in order to:

(1) recognize the instances of lexical units the program consists of;
(2) verify their correct construction according to regular expressions that define them;
(3) categorize them according to their types, such as identifiers or integers.

This section demonstrates how to define lexical units using regular expressions and design a scanner based on finite automata.

Lexical units and regular expressions

The lexical units of a typical high-level programming language are usually specified by regular expressions. Many of them contain several identical sub-expressions. Therefore, to simplify their specification, we often name some elementary expressions, such as l and d in Convention 11.2, and use these names in the definitions of more complex regular expressions.

Convention 11.2. Throughout this chapter, **letter** and **digit** stand for any letter and digit, respectively; furthermore, l and d are regular expressions standing for $(\mathrm{A}|\cdots|\mathrm{Z}|\mathrm{a}|\cdots|\mathrm{z})$ and $(0|\cdots|9)$, so **letter** and **digit** are denoted by l and d, respectively. ∎

In Table 11.2, we define identifiers, positive integers, and positive real numbers by regular expressions, which make use of l and d from Convention 11.2. First, identifiers are defined as arbitrarily long alphanumerical strings that begin with a letter. Second, integers are

Table 11.2. Definition of lexical units by regular expressions.

Lexical Unit	Regular Expression	Example
Identifier	$l(d\|l)^*$	$a21$
Integer number	d^+	10
Real number	$d^+.d^*$	0.01

defined as arbitrarily long non-empty numerical strings. Finally, real numbers correspond to the positive real numbers in mathematics. That is, they have the form $x.y$, where x is a non-empty numeric string and y is a numeric string. The case when $y = \varepsilon$ means $y = 0$; for instance, 2. has the same meaning as 2.0.

Scanners and finite automata

A scanner recognizes every instance of a lexical unit occurring in the program, so it scans the sequence of characters that make up the program in order to recognize the next instance of a lexical unit, verify its correct form, and categorize it according to its type. It is usually based on completely specified DFAs (see Definition 3.6), described by their state diagrams in what follows.

Convention 11.3. In Figures 11.1 and 11.2, by labeling an edge leaving a state with **others**, we mean any symbol that denotes no other edge leaving or entering the state. ■

Figure 11.1 shows a DFA that acts as a scanner for identifiers. In practice, a scanner is usually designed for several types of lexical units. To illustrate a multi-lexical-unit scanner like this, Figure 11.2 presents a DFA that acts as a scanner for both integer numbers and

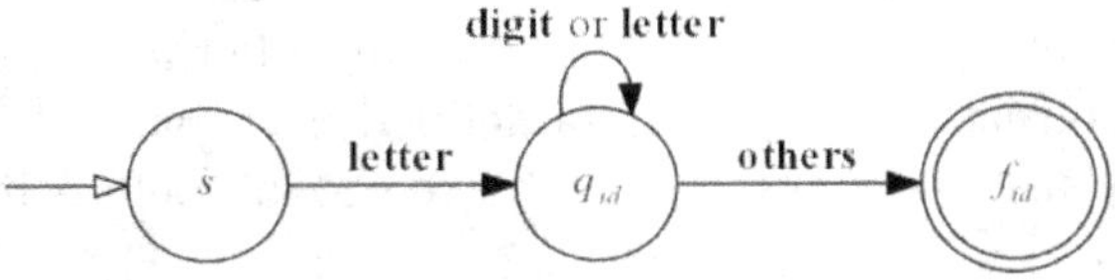

Figure 11.1. A scanner for identifiers.

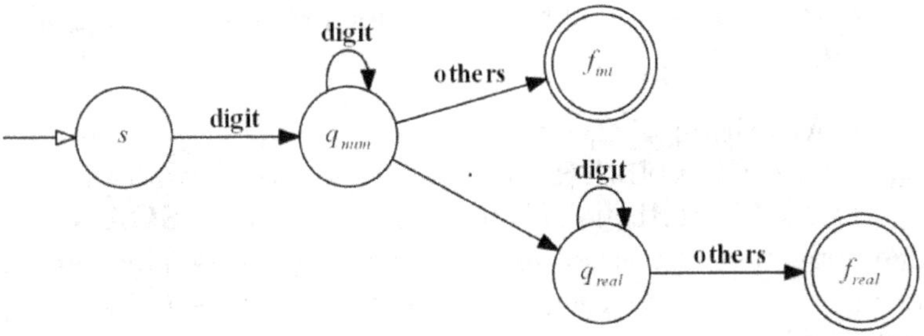

Figure 11.2. A scanner for integers and real numbers.

real numbers. It distinguishes these lexical units by their two final states: f_{int} and f_{real}. Indeed, ending in f_{int} means that an integer has been recognized while ending in $freal$ implies the recognition of a real number.

A scanner recognizes a lexical unit as the longest possible string denoted by its regular expression. Indeed, as is obvious, ab represents a single identifier, not two identifiers, a and b. However, recognizing the longest string of this kind necessities reading an extra symbol that actually follows the string. As a result, the lexical analyzer has to return this extra character to the scanned file and start the next scan in front of it. To illustrate, note that scanners in Figures 11.1 and 11.2 complete the recognition of a lexical unit by performing a move on **others** and, thereby, reading an extra character, which thus has to be returned.

On the other hand, the file that contains a high-level program code usually contains sequences of characters irrelevant as far as the lexical analysis is concerned. These irrelevant text passages include comments as well as sequences of characters that result in empty space when printed, such as white-space characters, including blanks, carriage returns, and line feeds. The scanner removes them all.

Implementation of a scanner

Next, we sketch an implementation of a scanner by using a pseudocode. For simplicity, we suppose that any file that contains a program code consists of lexical units defined in Table 11.2 and blanks are the only useless characters to be removed. Before implementation, some conventions are needed.

Convention 11.4. Throughout the current chapter, ch is a variable that stores a single *ch*aracter, and a blank is denoted by a □. Furthermore, lu is a string of characters, which represents a *l*exical *u*nit. ■

Based on Algorithm 11.2, we sketch the scanner implementation as the procedure **SCANNER**, which makes use of two other procedures: **IDENTIFIER** and **NUMBER**. Typically, **SCANNER** is applied repeatedly in order to recognize a whole sequence of identifiers and numbers. Consequently, whenever ch contains an extra character, which follows the recognized lexical unit, it pushes it back onto the input, so the next application of **SCANNER** starts by reading it. For instance, $1A$ consists of the integer 1, followed by the identifier A. SCANNER actually recognizes 1 when it reads A, which is thus returned onto the input; without this return, **SCANNER** would improperly omit A during its next application. We suppose that the input sequence of identifiers and numbers is ended by $\lhd$ (see Convention 11.1).

As is obvious, in practice, the end of a given input string is always somehow specified, so $\lhd$ only symbolically expresses this end. For instance, if the input string is written as a file, then the end of the file fulfills the role of $\lhd$.

Strictly, Algorithm 11.2 requires that its input DFA be completely specified. In the following implementation, this complete specification is arranged by the **otherwise** branch in the case statement, which is entered in all cases corresponding to a lexical error.

```
procedure SCANNER
begin
  repeat
    read (ch)
  until ch ≠ □
  if ch ≠ ◁ then
    set lu to ε
    case ch of
      letter: IDENTIFIER
      digit: NUMBER
      otherwise: write('lexicalerror')
end
```

IDENTIFIER, given next, implements a scanner that recognizes identifiers. **SCANNER** calls **IDENTIFIER**, with ch containing a letter and lu set to ε.

```
procedure IDENTIFIER
begin
  repeat
    extend lu by concatenating it with ch;
    read (ch)
  until ch ∉ letter ∪ digit
  push back ch
  write('identifier ', lu, ' is recognized')
end
```

Based on Figure 11.2, procedure **NUMBER** recognizes and categorizes positive integer and real numbers.

```
procedure NUMBER
begin
  repeat
    extend lu by concatenating it with ch;
    read (ch)
  until ch ∉ digit
  if ch ≠ '.'
    push back ch onto the standard input
    write('identifier ', lu, ' is recognized')
  else
    repeat
      extend lu by concatenating it with ch;
      read (ch)
    until ch ∉ digit
    push back ch
    write('identifier ', lu, ' is recognized')
end
```

11.2 Pushdown Automata and Syntax Analysis

In this section, we demonstrate the use of PDAs in practice. More specifically, we apply them to build up the syntax analyzers of programming languages. First, we explain how to specify the programming language syntax by using CFGs. Then, we construct the syntax analyzers based on PDAs in accordance with this grammatical specification.

The syntax of a programming language L is almost always specified by a CFG, $G = (V, T, P, S)$, satisfying $L = L(G)$. In essence, a syntax analyzer for G is a PDA M that decides whether a string $w \in L(G)$. M makes this decision by accepting w exactly when G generates w; consequently, $L(M) = L(G)$. In greater detail, to demonstrate that $w \in L(G), M$ simulates the construction of $S \Rightarrow^* w$ $[\rho]$, where ρ represents a *parse* of w, i.e., a sequence of rules from P

by which G derives w in G. If M successfully completes this construction, it usually produces the parse ρ as its output, hence it is customarily referred to as a G-based parser. Typically, M is designed in such a way that it constructs $S \Rightarrow^* w$ either in a leftmost way or in a rightmost way. Accordingly, there exist two different approaches to parsing, customarily referred to as *top-down parsing* and *bottom-up parsing*, which produce left parses and right parses, respectively. Next, (I) and (II) describe both approaches together with the two notions of parses:

(I) M simulates $S\ {}_{lm}\!\Rightarrow^*\ w\ [\rho]$, so it starts from S and proceeds toward w by simulating the leftmost derivation steps according to rules from P. If and when it completes $S\ {}_{lm}\!\Rightarrow^*\ w\ [\rho]$, it usually also produces ρ as the left parse of w corresponding to $S, {}_{lm}\!\Rightarrow^*\ w\ [\rho]$ — the sequence of rules according to which G makes this leftmost derivation. In terms of the corresponding derivation tree $(S\ {}_{lm}\!\Rightarrow^*\ w)$, this approach can be naturally rephrased as the construction of $(S\ {}_{lm}\!\Rightarrow^*\ w)$ so that it starts from the root and proceeds down toward the frontier; hence, a parser that works in this top-down way is called a *G-based top-down parser*.

(II) If M simulates $S\ {}_{rm}\!\Rightarrow^*\ w\ [\rho]$, it makes this simulation in reverse. That is, it starts from w and proceeds toward S by making the rightmost derivation steps, each of which is performed in reverse by reducing the right-hand side of a rule to its left-hand side. If and when it completes this reverse construction of $S\ {}_{rm}\!\Rightarrow^*\ w\ [\rho]$, it produces reversal($\rho$) as the right parse of w corresponding to $S\ {}_{rm}\!\Rightarrow^*\ w\ [\rho]$ — the reverse sequence of rules according to which G makes this rightmost derivation. To express this construction in terms of $(S\ {}_{rm}\!\Rightarrow^* w)$, a parser like this constructs this tree so that it starts from its frontier w and proceeds up toward the root; hence, a parser that works in this way is referred to as a *G-based bottom-up parser*.

Whichever way a parser is designed, it is always based on a PDA. The following Convention 11.5 simplifies the upcoming discussion about parsers by considering only one-state PDAs, which are equivalent to ordinary PDAs, as follows from Algorithm 6.11, Theorem 6.17, and Theorem 6.19. In addition, some pragmatically motivated conventions concerning configurations are introduced too.

Convention 11.5. Throughout this chapter, we assume that every PDA $M = (Q, \Delta, \Gamma, R, s, S, F)$ has a single state denoted by $\Diamond$, so $Q = F = \{\Diamond\}$. Instead of a configuration of the form $x\Diamond y$ from X (see Definition 4.1), we write $\triangleright x\Diamond y\triangleleft$, where $\triangleright$ and $\triangleleft$ are two special symbols such that $\{\triangleright,\triangleleft\} \cap (Q \cup \Delta \cup \Gamma) = \emptyset$. By *pd*, we refer to the pushdown $\triangleright x$, whose rightmost symbol represents the *pd* top symbol and $\triangleright$ is called the *pd* bottom. We consider *pd* empty if $x = \varepsilon$; therefore, $\triangleright$ occurs on the pushdown top. By ins, we refer to the input symbol defined as the leftmost symbol of $y\triangleleft$. When $ins = \triangleleft$, referred to as the input end, all the input string has been read. As a result, M always accepts in a configuration of the form $\triangleright\Diamond\triangleleft$. ■

11.3 Syntax Specified by Context-Free Grammars

Of course, parsers can verify the syntax of a programming language only if it is precisely specified. Today, CFGs are almost exclusively used for this purpose. The following example illustrates how to specify the syntax of common syntactical structures, such as logical expressions, by using CFGs.

Example 11.2. We want to describe a logical expression of the form

$$\iota_0 o_1 \iota_1 o_2 \iota_2 \cdots \iota_{n-1} o_n \iota_n,$$

where $o_j \in \{\vee, \wedge\}$ and ι_k is a logical variable, symbolically denoted by i (intuitively, i stands for an identifier) or another logical expression enclosed in parentheses, for all $1 \leq j \leq n$ and $0 \leq k \leq n$, for some $n \in {}_0\mathbb{N}$. For instance, $(i \vee i) \wedge i$ is an expression like this. A logical variable can be simply derived by the rule

$$S \to i.$$

We also introduce

$$S \to (S)$$

to derive a parenthesized expression, i.e., any valid expression enclosed by parentheses. To derive logical operators $\vee$ and $\wedge$, we add the two rules

$$S \to S \vee S \quad \text{and} \quad S \to S \wedge S.$$

As a result, we define the expressions by the four-rule CFG defined as

$$1 : S \to S \vee S, \quad 2 : S \to S \wedge S, \quad 3 : S \to (S), \quad 4 : S \to i.$$

This CFG is obviously ambiguous (see Definition 6.6); for instance, $S \ {}_{lm}\!\Rightarrow^* i \vee i \wedge i$ [14244] as well as $S \ {}_{lm}\!\Rightarrow^* i \vee i \wedge i$ [21444]. Observe, however, that the same language that the four-rule CFG generates is also generated by the unambiguous six-rule CFG defined as

$$1 : S \to S \vee A, \quad 2 : S \to A, \quad 3 : A \to A \wedge B, \quad 4 : A \to B,$$
$$5 : B \to (S), \quad 6 : B \to i.$$

Both previous CFGs are left-recursive, however (see Definition 6.15). Some important methods of syntax analysis work only with non-left-recursive CFGs; in fact, all the methods described later in this section are of this kind. Therefore, we give one more equivalent non-left-recursive unambiguous CFG, obtained from the previous CFG by Algorithm 6.9:

$$1 : S \to AC, \quad 2 : C \to \vee AC, \quad 3 : C \to \varepsilon, \quad 4 : A \to BD,$$
$$5 : D \to \wedge BD, \quad 6 : D \to \varepsilon, \quad 7 : B \to (S), \quad 8 : B \to i.$$

Compare the three equivalent CFGs introduced in this example. Intuitively, we obviously see that in the syntax analysis, any grammatical ambiguity may represent an undesirable phenomenon, and if it does, we prefer the second CFG to the first CFG. On the other hand, the definition of the former is more succinct than that of the latter because the former contains a single nonterminal and four rules, while the latter has three nonterminals and six rules. As already noted, some methods of syntax analysis necessitates using non-left-recursive CFGs, in which case we obviously use the third CFG, which has more nonterminals and rules than the other two CFGs in this example. Simply put, all three CFGs have their pros and cons. ■

Consequently, from a broader perspective, the previous example illustrates a typical process of designing an appropriate grammatical specification for a programming language in practice. Indeed, we often design several equivalent CFGs that generate the language, carefully consider their advantages and disadvantages, and based on this consideration, we choose the CFG that is optimal under given circumstances.

In what follows, we make use of the CFGs from the previous example so often that we introduce the following convention for the sake of brevity.

Convention 11.6. Consider the CFGs defined in Example 11.2. Throughout the rest of this chapter, E, H, and J denotes its first, second, and third CFG, respectively. That is,

(A) E denotes $1 : S \to S \vee S, 2 : S \to S \wedge S, 3 : S \to (S), 4 : S \to i$;
(B) H denotes $1 : S \to S \vee A, 2 : S \to A, 3 : A \to A \wedge B, 4 : A \to B, 5 : B \to (S), 6 : B \to i$;
(C) J denotes $1 : S \to AC, 2 : C \to \vee AC, 3 : C \to \varepsilon, 4 : A \to BD, 5 : D \to \wedge BD, 6 : D \to \varepsilon, 7 : B \to (S), 8 : B \to i$. ■

Top-Down Parsing

Let $I \in$ **CFG**. Given $w \in T_I^*$, an I-based top-down parser works on w so that it simulates a leftmost derivation of w in I. If there is no leftmost derivation of w in I, the parser rejects w because $w \notin L(I)$. On the other hand, if the parser successfully completes the simulation of $S\ {}_{lm}\Rightarrow^* w\ [\rho]$ in I, which means that $w \in L(I)$, the parser accepts w to express that I generates w; in addition, it often produces the left parse ρ as output too.

Algorithm 11.3, as follows, transforms any CFG I into an equivalent PDA O that acts as an I-based top-down parser. In many respects, this transformation resembles Algorithm 6.11, reformulated in terms of Convention 11.5. Note that O performs only the following two pushdown operations, popping and expanding:

(A) If a terminal a occurs as the pushdown top symbol, O pops a off the pushdown by a rule of the form $a\Diamond a \to \Diamond$, by which O removes a from the pushdown top and, simultaneously, reads the input symbol a, so it actually verifies their identity.
(B) If a nonterminal A occurs as the pushdown top symbol, O simulates a leftmost derivation step made by a rule $r : A \to X_1X_2 \ldots X_n \in P_I$, where each $X_i \in V_I, 1 \leq i \leq n$, for

some $n \in {}_0\mathbb{N}$ ($n = 0$ means $X_1X_2X_n = \varepsilon$). O performs this simulation so that it expands its pushdown by using $A\Diamond \to reversal(X_1X_2X_n)\Diamond$ and it replaces the pushdown top A with X_nX_1. To describe an expansion like this more formally, consider $S_I\ {}_{lm}\Rightarrow^* w\ [\rho]$ in I, where $w \in T_I^*$, expressed as

$$\begin{aligned} S_I\ {}_{lm}\Rightarrow^*&\ vAy \\ {}_{lm}\Rightarrow&\ vX_1X_2X_ny \\ {}_{lm}\Rightarrow^*&\ vu, \end{aligned}$$

where $w = vu$, so $v, u \in T_I^*$. Suppose that I has just simulated the first portion of this derivation, $S\ {}_{lm}\Rightarrow^* vAy$; at this point, the pushdown of O contains Ay in reverse, and u is the remaining input to be read. In symbols, the PDA O occurs in the configuration $\triangleright reversal(y)A\Diamond u\triangleleft$, from which it simulates $vAy\ {}_{lm} \Rightarrow vX_1X_2X_ny\ [r]$ by performing the *expansion*

$$\triangleright reversal(y)A\Diamond u\triangleleft \quad \Rightarrow \quad reversal(y)X_nX_1\Diamond u\triangleleft$$

according to the rule $A\Diamond \to X_nX_1\Diamond$ from the set of rules in O; note that $reversal(y)X_nX_1 = reversal(X_1X_ny)$.

Algorithm 11.3. Top-down parser.

Input: A CFG $I = (V_I, T_I, P_I, S_I)$.
Output: A PDA $O = (Q_O, \Delta_O, \Gamma_O, R_O, s_O, S_O, F_O)$ such that O works as an I-based top-down parser.
Method:

begin
  set $\Gamma_O = V_I, \Delta_O = T_I, S_O = S_I$
  set $R_O = \emptyset$
  for each $A \to x \in P_I$, where $A \in (V_I - T_I)$ and $x \in V_I^*$ **do**
    add $A\Diamond \to reversal(x)\Diamond$ to R_O (expansion rules)
  for each $a \in T_I$ **do**
    add $a\Diamond a \to \Diamond$ to R_O (popping rules)
end

Prove the following lemma as an exercise.

Lemma 11.1. *Algorithm 11.3 is correct.*

Left Parses. Consider the top-down parser O produced by Algorithm 11.3 from I_{CFG}. Observe that the expansion rules in O correspond to the grammatical rules in I according to the equivalence

$$A \to x \in P_I \text{ iff } A\Diamond \to reversal(x)\Diamond \in R_O.$$

Suppose that O simulates $S\ _{lm} \Rightarrow^* w\ [\rho]$ in I. To obtain ρ as the left parse of w corresponding to this derivation, record the grammatical rules corresponding to the expansion rules applied by O during this simulation. Specifically, if O makes an expansion by $A\Diamond \to reversal(x)\Diamond$ to simulate the application of $r : Ax \in P_I$ in I, write r. After the simulation of $S\ _{lm} \Rightarrow^* w$ is completed, the sequence of all rules recorded in this way is the left parse ρ. The following example describes O that produces left parses in this way.

Example 11.3. Take the grammar E (see Convention 11.6), defined as

$$1 : S \to S \vee S, \quad 2 : S \to S \wedge S, \quad 3 : S \to (S), \quad 4 : S \to i.$$

Consider E as the input grammar I of Algorithm 11.3. This algorithm turns I to a top-down parser for I, O, which has the expansion rules

$$S\Diamond \to S \vee S\Diamond, S\Diamond \to S \wedge S\Diamond, S\Diamond \to)S(\Diamond, S\Diamond \to i\Diamond.$$

Apart from these rules, the second **for** loop of the algorithm introduces the popping rules $a\Diamond a \to \Diamond$, for all $a \in \{\vee, \wedge, (,), i\}$.

For instance, consider $S\ _{lm} \Rightarrow^* i \vee i \wedge i\ [14244]$ in I. O parses $i \vee i \wedge i$, as described in Table 11.3, whose columns contain the following information:

Column 1: $S\ _{lm} \Rightarrow^* i \vee i \wedge i\ [14244]$ in I;

Column 2: the production of the left parse14244;

Column 3: $\triangleright S\Diamond i \vee i \wedge i \triangleleft \Rightarrow^* \triangleright\Diamond \triangleleft$ in O.

■

Table 11.3. Top-down parsing of $i \vee i \wedge i$.

Derivation in I	Left Parse	Computation in O
S		$\triangleright S \diamond i \vee i \wedge i \triangleleft$
$_{lm}\Rightarrow S \vee S$	1	$\Rightarrow \triangleright S \vee S \diamond i \vee i \wedge i \triangleleft$
$_{lm}\Rightarrow i \vee S$	4	$\Rightarrow \triangleright S \vee i \diamond i \vee i \wedge i \triangleleft$
		$\Rightarrow \triangleright S \vee \diamond \vee i \wedge i \triangleleft$
		$\Rightarrow \triangleright S \diamond i \wedge i \triangleleft$
$_{lm}\Rightarrow i \vee S \wedge S$	2	$\Rightarrow \triangleright S \wedge S \diamond i \wedge i \triangleleft$
$_{lm}\Rightarrow i \vee i \wedge S$	4	$\Rightarrow \triangleright S \wedge i \diamond i \wedge i \triangleleft$
		$\Rightarrow \triangleright S \wedge \diamond \wedge i \triangleleft$
		$\Rightarrow \triangleright S \diamond i \triangleleft$
$_{lm}\Rightarrow i \vee i \wedge i$	4	$\Rightarrow \triangleright i \wedge \diamond \wedge i \triangleleft$
		$\Rightarrow \triangleright \diamond \triangleleft$

Bottom-Up Parsing

Let $G = (V, T, P, S) \in$ **CFG**. Given $w \in T^*$, a G-based bottom-up parser works on w so that it reversely simulates a rightmost derivation of w in G. If there is no rightmost derivation of w in G, the parser rejects w because $w \notin L(G)$. However, if the parser successfully completes the reverse simulation of $S\ _{rm}\Rightarrow^*\ w\ [\rho]$ in G, the parser accepts w to express that $w \in L(G)$, and in addition, it usually produces the right parse reversal(ρ) as output.

Next, we give Algorithm 11.4 that turns any CFG $I = (V_I, T_I, P_I, S_I)$ to a PDA $O = (Q_O, \Delta_O, \Gamma_O, R_O, s_O, S_O, F_O)$, so O acts as an I-based bottom-up parser. To give an insight into this algorithm, consider any rightmost derivation $S_I\ _{rm}\Rightarrow^*\ w$ in I, where $w \in T_I*$. In essence, O simulates this derivation in reverse, proceeding from w toward S_I, and during all this simulation, it keeps its start symbol S_O as the deepest symbol, occurring right behind $\triangleright$. When it reaches the configuration $\triangleright S_O S_I \Diamond \triangleleft$, it moves to $\triangleright \Diamond \triangleleft$ and, thereby, successfully completes the parsing of w. More specifically, express $S_I\ _{rm}\Rightarrow^*\ w$ in I as $S_I\ _{rm}\Rightarrow^*\ zv\ _{rm}\Rightarrow^*\ tv$, where $t, v \in T_I^*, w = tv$, and $z \in V_I^*$. After reading t and making a sequence of moves corresponding to $zv\ _{rm}\Rightarrow^*\ tv, O$ uses its pushdown to record z and contains v as the remaining input to read. In brief, it occurs in the configuration $\triangleright S_O z \Diamond v \triangleleft$. To explain the next move of O that simulates the rightmost derivation step in I, express $S_I\ _{rm}\Rightarrow^*\ zv\ _{rm}\Rightarrow^*\ tv$ in greater detail as

$$\begin{aligned} S_I \ {}_{rm}\!\Rightarrow^* &\ yAv \\ {}_{rm}\!\Rightarrow &\ yxv \ [A \to x] \\ {}_{rm}\!\Rightarrow^* &\ tv, \end{aligned}$$

where $yx = z$ and $A \to x \in P_I$. From $\triangleright S_O yx\Diamond v\triangleleft$, which equals $\triangleright S_O z\Diamond v\triangleleft$, M simulates $yAv \ {}_{rm}\!\Rightarrow yxv \ [A \to x]$ in reverse as

$$\triangleright S_O yx\Diamond v\triangleleft \quad \Rightarrow \quad \triangleright S_O yA\Diamond v \triangleleft .$$

In addition, whenever needed, O can shift the first symbol of the remaining input onto the pushdown. In this way, step by step, O simulates $S_I \ {}_{rm}\!\Rightarrow^* zv \ {}_{rm}\!\Rightarrow^* tv$ in I until it reaches the configuration $\triangleright S_O S_I\Diamond\triangleleft$. To reach $\triangleright\Diamond\triangleleft$ and complete the acceptance of w, we add $S_O S_I\Diamond \to \Diamond$ to R_O.

Algorithm 11.4. Bottom-up parser.

Input: A CFG $I = (V_I, T_I, P_I, S_I)$.
Output: A PDA $O = (Q_O, \Delta_O, \Gamma_O, R_O, s_O, S_O, F_O)$ such that O works as an I-based bottom-up parser.
Method:

begin
  set $Q_O = \{\Diamond\}, \Delta_O = T_I, \Gamma_O = V_I \cup \{S_O\}$, where $S_O \notin V_I$
  set $R_O = \{S_O S_I\Diamond \to \Diamond\}$
  for each $r : A \to x \in P_I$ **do** (reducing rules)
    add $x\Diamond \to A\Diamond$ to R_O
  for each $a \in T_I$ **do**
    add $\Diamond a \to a\Diamond$ to R_O (shifting rules)
end

Next, we prove that $L(O) = L(I)$. In addition, in this proof, we demonstrate that O works as an I-based bottom-up parser in the sense that O simulates the construction of the rightmost derivations in I in reverse.

Lemma 11.2. *Algorithm 11.4 is correct, i.e., $L(O) = L(I)$, and O acts as an I-based bottom-up parser.*

Proof. To prove $L(I) = L(O)$, we first establish Claims A and B.
Claim A. Let

$$S_I \;{}_{rm}\!\Rightarrow^* xv$$
$$\qquad {}_{rm}\!\Rightarrow^* uv \; [\pi]$$

in I, where $v, u \in T_I^*, x \in V_I^*$, and $\pi \in P_I^*$. Then, O computes

$$\rhd S_O \Diamond uv \lhd \quad \Rightarrow^* \quad \rhd S_O x \Diamond v \lhd .$$

Proof of Claim A by induction on $|\pi| = \mathbf{0}$
Basis. Let $|\pi| = 0$. That is, $S_I \;{}_{rm}\!\Rightarrow^* xv \;{}_{rm}\!\Rightarrow^0 uv$ in I, so $u = x$. Observe that O computes $\rhd S_O \Diamond uv \lhd \Rightarrow^{|u|} \rhd S_O u \Diamond v \lhd$ by shifting u onto the pushdown by $|u|$ consecutive applications of shifting rules of the form $\Diamond a \rightarrow a\Diamond$, where $a \in T_I$ (see Algorithm 11.4). Thus, the basis holds true.

Induction Hypothesis. Suppose that the claim holds for each P_I^*, with $|\pi| \leq i$, for some $i \in {}_0\mathbb{N}$.

Induction step. Consider

$$S_I \;{}_{rm}\!\Rightarrow^* xv \;{}_{rm}\!\Rightarrow^* uv \; [p]$$

in I, where $\pi \in P_I^*, |\pi| = i$, and $p \in P_I$, so $|p| = i + 1$. Express $xv \;{}_{rm}\!\Rightarrow^* uv \; [p]$ as

$$xv \;{}_{rm}\!\Rightarrow^* yrhs(p)v \; [p] \;{}_{rm}\!\Rightarrow^* uv \; [\pi],$$

where $x = ylhs(p)$. By inspecting Algorithm 11.4, we see that $p \in P_I$ implies $rhs(p)\Diamond \rightarrow lhs(p)\Diamond R_O$. In the rest of this proof, we distinguish these two cases: (1) $yrhs(p)T_I^*$ and (2) $yrhs(p)T_I^*$.

(1) Let $yrhs(p) \in T_I^*$. Then, $yrhs(p) = u$. Construct $\rhd S_O \Diamond uv \lhd \Rightarrow^{|u|} \rhd S_O yrhs(p) \Diamond v \lhd$ in O by shifting u onto the pushdown, so

$$\rhd S_O \Diamond uv \lhd \Rightarrow^{|u|} \rhd S_O yrhs(p) \Diamond v \lhd$$
$$\Rightarrow \quad \rhd S_O ylhs(p) \Diamond v \lhd \quad [rhs(p)\Diamond \rightarrow lhs(p)\Diamond].$$

As $x = ylhs(p)$, we have just proved $\rhd S_O \Diamond uv \lhd \Rightarrow^* \rhd S_O x \Diamond v \lhd$ in O.

(2) Let $yrhs(p) \notin T_I^*$. Express $yrhs(p)$ as $yrhs(p) = zBt$, where $t \in T_I^*$, $z \in V_I^*$, and $B \in (V_I - T_I)$, so B is the rightmost nonterminal appearing in $yrhs(p)$. Consider

$$\begin{aligned} S_I \ {}_{rm}\Rightarrow^* &\ xv \\ {}_{rm}\Rightarrow &\ zBtv\ [p] \\ {}_{rm}\Rightarrow^* &\ uv\ [\pi]. \end{aligned}$$

Since $|\pi| = i, \triangleright S_O \lozenge uv \triangleleft \Rightarrow^* \triangleright S_O zB \lozenge tv \triangleleft$ by the inductive hypothesis. By shifting t onto the pushdown, we obtain $\triangleright S_O zB \lozenge tv \triangleleft \Rightarrow^* \triangleright S_O zBt \lozenge v \triangleleft$. As $yrhs(p) = zBt$, we have

$$\begin{aligned} \triangleright S_O \lozenge uv \triangleleft &\Rightarrow^{|t|} \triangleright S_O yrhs(p) \lozenge v \triangleleft \\ &\Rightarrow \triangleright S_O ylhs(p) \lozenge v \triangleleft\ [rhs(p)\lozenge \rightarrow lhs(p)\lozenge]. \end{aligned}$$

Therefore, $\triangleright S_O \lozenge uv \triangleleft \Rightarrow^* \triangleright S_O x \lozenge v \triangleleft$ in O because $x = ylhs(p)$, which completes (2).

Consequently, Claim A holds true.

Claim B. Let $\triangleright S_O \lozenge uv \triangleleft \Rightarrow^* \triangleright S_O x \lozenge v \triangleleft\ [\rho]$ in O, where $u, v \in \Delta_O^*$, $x \in V_I^*$, and $\rho \in (R_O - \{S_O S_I \lozenge \rightarrow \lozenge\})^*$. Then, $xv\ {}_{rm}\Rightarrow^* uv$ in I.

Proof of Claim B by induction on $|\rho| = 0$.

Basis. Let $|\rho| = 0$, so $u = x = \varepsilon$ and $\triangleright S_O \lozenge v \triangleleft \ {}^0\Rightarrow \triangleright S_O \lozenge v \triangleleft$ in O. Clearly, $v\ {}_{rm}\Rightarrow^0 v$ in I.

Induction Hypothesis. Suppose that the claim holds for each $\rho \in (R_O - \{S_O S_I \lozenge \rightarrow \lozenge\})^*$ with $|\rho| \leq i$, for some $i \geq 0$.

Induction Step. Consider any sequence of $i + 1$ moves of the form $\triangleright S_O \lozenge uv \triangleleft \Rightarrow^* \triangleright S_O x \lozenge v \triangleleft\ [\rho r]$ in O, where $u, v \in \Delta_O^*$, $x \in V_I^*$, $\rho \in (R_O - \{S_O S_I \lozenge \rightarrow \lozenge\})^*$, $|\rho| = i$, and $r \in R_O$. Express $\triangleright S_O \lozenge uv \triangleleft \Rightarrow^* \triangleright S_O x \lozenge v \triangleleft\ [\rho r]$ as

$$\begin{aligned} \triangleright S_O \lozenge uv \triangleleft &\Rightarrow^* \triangleright S_O y \lozenge t \triangleleft\ [\rho] \\ &\Rightarrow \triangleright S_O x \lozenge v \triangleleft\ [r], \end{aligned}$$

where $y \in V_I^*$ and $t \in \Delta_O^*$. As $r \neq S_O \lozenge \rightarrow \lozenge$, either $rhs(r) = A\lozenge$ with $A \in (V_I - T_I)$ or $rhs(r) = a\lozenge$ with $a\Delta_O$. Next, we distinguish these two cases: (1) $rhs(r) = A\lozenge$ with $A \in (V_I - T_I)$ and (2) $rhs(r) = a\lozenge$ with $a \in \Delta_O$.

(1) Let $rhs(r) = A\Diamond$ with $A \in (V_I - T_I)$, so $t = v, r$ is of the form $z\Diamond \to A\Diamond, x = hA$, and $y = hz$, for some $A \to z \in P_I$ and $h \in V_I^*$. By using $A \to z \in P_I, I$ makes $hAv\ {}_{rm}\Rightarrow hzv$. By the induction hypothesis, $yv\ {}_{rm}\Rightarrow^* uv$ in I. Thus, $xv\ {}_{rm}\Rightarrow^* uv$ in I.
(2) Let $rhs(r) = a\Diamond$ with $a \in \Delta_O$, so $t = av, r$ is of the form $\Diamond a \to a\Diamond, x = ya$. Thus, $xv = yt$. Recall that $\triangleright S_O\Diamond uv\triangleleft \Rightarrow^* \triangleleft S_O y\Diamond t\triangleleft$ $[\rho]$, with $|\rho| = i$. By the induction hypothesis, $yt\ {}_{rm}\Rightarrow^* uv$ in I. Since $xv = yt, xv\ {}_{rm}\Rightarrow^* uv$ in I.

Thus, Claim B holds true.

Consider Claim A for $v = \varepsilon$ and $x = S_I$. At this point, for all $u \in T_I^*, S_I\ {}_{rm}\Rightarrow^* u$ in I implies $\triangleright S_O\Diamond u\triangleleft \Rightarrow^* \triangleright S_O S_I\Diamond\triangleleft$ in O. By using $S_O S_I\Diamond \to \Diamond, O$ makes $\triangleright S_O S_I\Diamond\triangleleft \Rightarrow^* \triangleright\Diamond\triangleleft$ in O. Hence, $L(I) \subseteq L(O)$.

Consider Claim B for $v = \varepsilon$ and $x = S_I$. Under this consideration, if $\triangleright S_O\Diamond u\triangleleft \Rightarrow^* \triangleright S_O S_I\Diamond\triangleleft$ $[\rho]$ in O, then $S_I\ {}_{rm}\Rightarrow^* u$ in I, for all $u \in T_I^*$. During any acceptance of a string, $u \in T_I^*, O$ applies $S_O S_I\Diamond \to \Diamond$ precisely once, and this application occurs during the very last step in order to remove $S_O S_I$ and reach the configuration $\triangleright\Diamond\triangleleft$. Indeed, observe that any earlier application of this rule implies that subsequently O can never completely read u and, simultaneously, empty the pushdown; the details of this observation is left as an exercise. Thus, $\triangleright S_O\Diamond u\triangleleft \Rightarrow^* \triangleright\Diamond\triangleleft$ $[\rho]$ in O implies $S_I\ {}_{rm}\Rightarrow^* u$ in I, so $L(O) \subseteq L(I)$.

Consequently, $L(I) = L(O)$ because $L(I) \subseteq L(O)$ and $L(O) \subseteq L(I)$. As an exercise, examine the proof of Claim A to see that O works so that it simulates the rightmost derivations in I in reverse. Thus, O works as an I-based bottom-up parser, and Lemma 11.2 holds true. □

Right Parses. Consider the I-based top-down parser O produced by Algorithm 11.4 from $I \in$ **CFG**. Observe that the reducing rules in O correspond to the grammatical rules in I according to the equivalence

$$A \to x \in P_I \quad \text{iff } x\Diamond \to A\Diamond \in R_O.$$

Suppose that O reversely simulates $S_I\ {}_{rm}\Rightarrow^* w$ $[\rho]$ in I. To obtain $reversal(\rho)$ as the right parse of w corresponding to $S_I\ {}_{rm}\Rightarrow^* w$ $[\rho]$, record the reducing rules applied by O during this simulation. That is, if O makes a reduction by $x\Diamond \to A\Diamond$ to simulate the application of $A \to x \in P_I$ in I, write out r. When the simulation of $S_I\ {}_{rm}\Rightarrow^* w$ $[\rho]$

is completed, the corresponding right parse is obtained in this way. The following example illustrates O extended in this way.

We have described O in terms of the reverse simulation of the rightmost derivations in I. As noted at the beginning of this section, instead of this description, the way O works can also be described in terms of derivation trees constructed in a bottom-up way so that this construction starts from their frontiers and proceeds up toward the roots. Leaving a general version of this alternative description as an exercise, we include such a bottom-up construction of a derivation tree within the following example.

Example 11.4. Consider the CFG H (see Convention 11.6), defined as

$$1 : S \to S \vee A, \quad 2 : S \to A, \quad 3 : A \to A \wedge B, \quad 4 : A \to B,$$
$$5 : B \to (S), \quad 6 : B \to i.$$

Consider H as the input grammar I of Algorithm 11.4. This algorithm turns it to the H-based bottom-up parser O, which has the reducing rules

$$1 : S \vee A\Diamond \to S\Diamond, \quad 2 : A\Diamond \to S\Diamond, \quad 3 : A \wedge B\Diamond \to A\Diamond,$$
$$4 : B\Diamond \to A\Diamond, \quad 5 : (S)\Diamond \to B\Diamond, \quad 6 : B\Diamond \to i\Diamond.$$

We have labeled these rules by the labels that denote the grammatical rules from which they are constructed; for instance, $1 : S \vee A\Diamond \to S\Diamond$ is constructed from $1 : S \to S \vee A$. Apart from these six rules, the algorithm adds the shifting rules $\Diamond a \to a\Diamond$, for all $a \in \{\vee, \wedge, (,), i\}$. It also introduces $ZS\Diamond \to \Diamond$, where Z is declared as the start pushdown symbol of O.

For instance, take $i \vee i \wedge i$ as the input string. Consider Table 11.4. This table explains how O parses $i \vee i \wedge i$ by using its three columns, whose contents are described as follows:

Column 1: the bottom-up construction of $\clubsuit(S\ {}_{rm}\Rightarrow^* i \vee i \wedge i)$ in I;

Column 2: the construction of 64264631 as the right parse;

Column 3: $\triangleright Z\Diamond i \vee i \wedge i \triangleleft \Rightarrow^* \triangleright\Diamond\triangleleft$ in O.

Take the resulting right parse 64264631. Reverse it to obtain 13646246, according to which I makes the following rightmost

Table 11.4. Bottom-up parsing of $i \vee i \wedge i$.

Derivation Tree in I	Right Parse	Computation in O
$i \vee i \wedge i$		$\Rightarrow \triangleright Z\Diamond i \vee i \wedge i \triangleleft$
		$\Rightarrow \triangleright Zi\Diamond \vee i \wedge i \triangleleft$
$B\langle i\rangle \vee i \wedge i$	6	$\Rightarrow \triangleright ZB\Diamond \vee i \wedge i \triangleleft$
$A\langle B\langle i\rangle\rangle \vee i \wedge i$	4	$\Rightarrow \triangleright ZA\Diamond \vee i \wedge i \triangleleft$
$S\langle A\langle B\langle i\rangle\rangle\rangle \vee i \wedge i$	2	$\Rightarrow \triangleright ZS\Diamond \vee i \wedge i \triangleleft$
		$\Rightarrow \triangleright ZS \vee \Diamond i \wedge i \triangleleft$
		$\Rightarrow \triangleright ZS \vee i\Diamond \wedge i \triangleleft$
$S\langle A\langle B\langle i\rangle\rangle\rangle \vee B\langle i\rangle \wedge i$	6	$\Rightarrow \triangleright ZS \vee B\Diamond \wedge i \triangleleft$
$S\langle A\langle B\langle i\rangle\rangle\rangle \vee A\langle B\langle i\rangle\rangle \wedge i$	4	$\Rightarrow \triangleright ZS \vee A\Diamond \wedge i \triangleleft$
		$\Rightarrow \triangleright ZS \vee A \wedge \Diamond i \triangleleft$
		$\Rightarrow \triangleright ZS \vee A \wedge i\Diamond \triangleleft$
$S\langle A\langle B\langle i\rangle\rangle\rangle \vee A\langle B\langle i\rangle\rangle \wedge B\langle i\rangle$	6	$\Rightarrow \triangleright ZS \vee A \wedge B\Diamond \triangleleft$
$S\langle A\langle B\langle i\rangle\rangle\rangle \vee A\langle A\langle B\langle i\rangle\rangle \wedge B\langle i\rangle\rangle$	3	$\Rightarrow \triangleright ZS \vee A\Diamond \triangleleft$
$S\langle S\langle A\langle B\langle i\rangle\rangle\rangle \vee A\langle A\langle B\langle i\rangle\rangle \wedge B\langle i\rangle\rangle\rangle$	1	$\Rightarrow \triangleright ZS\Diamond \triangleleft$
		$\Rightarrow \triangleright\Diamond\triangleleft$

derivation:

$$\begin{aligned} S \;{}_{rm}\!\!\Rightarrow\;& S \vee A && [1] \\ {}_{rm}\!\!\Rightarrow\;& S \vee A \wedge B && [3] \\ {}_{rm}\!\!\Rightarrow\;& S \vee A \wedge i && [6] \\ {}_{rm}\!\!\Rightarrow\;& S \vee B \wedge i && [4] \\ {}_{rm}\!\!\Rightarrow\;& S \vee i \wedge i && [6] \\ {}_{rm}\!\!\Rightarrow\;& A \vee i \wedge i && [2] \\ {}_{rm}\!\!\Rightarrow\;& B \vee i \wedge i && [4] \\ {}_{rm}\!\!\Rightarrow\;& i \vee i \wedge i && [6]. \end{aligned}$$

In brief, $S \;{}_{rm}\!\!\Rightarrow^* i \vee i \wedge i$ [13646246]. ■

The algorithms and lemmas achieved earlier in this section have important theoretical consequences. Indeed, we have left a proof that $\mathbf{L_{CFG}} \subseteq \mathbf{L_{PDA}}$ as an exercise (see Theorem 6.17). Observe that this inclusion follows from Algorithm 11.3 and Lemma 11.1. Alternatively, Algorithm 11.4 and Lemma 11.2 imply $\mathbf{L_{CFG}} \subseteq \mathbf{L_{PDA}}$ too.

Corollary 11.1. *For every $I \in$ **CFG**, there exists an equivalent $O \in$ **PDA**, so $\mathbf{L_{CFG}} \subseteq \mathbf{L_{PDA}}$.* ■

The parsers constructed in Algorithms 11.3 and 11.4 represent a framework of most top-down and bottom-up parsers, respectively.

In general, however, they work in a non-deterministic way, and as such, they are difficult to implement and apply. Therefore, throughout the upcoming two application-oriented sections, we concentrate our attention solely on their deterministic versions, which are central to parsing in practice. Furthermore, until now, this chapter has still maintained the mathematical terminology used in the previous chapter. The following convention relaxes this strict formalism in order to make the upcoming parsers easy to implement.

Convention 11.7. Every parser discussed throughout the upcoming Sections 11.4 and 11.5 is deterministic. Rather than its strictly mathematical rule-based specification, we describe it as a parsing algorithm based on its parsing table. Although every parsing table always depends on the parser in question, its general format represents a two-dimensional array with rows and columns denoted by top pushdown symbols and input symbols, respectively. If A occurs on the pushdown top and ins is the current input symbol, the entry corresponding to A and ins specifies the proper parsing action to be performed in this configuration.

Consider the two special symbols, $\triangleright$ and $\triangleleft$, which always denote the pushdown bottom and the input end, respectively (see Convention 11.5). The parser finds out that the pushdown is empty when $\triangleright$ appears on the pushdown top. Similarly, when $\triangleleft$ occurs as the input symbol, all the input string has been read. Of course, O can never remove $\triangleright$ or $\triangleleft$. ■

11.4 Top-Down Parsing

Compared to Sections 11.2 and 11.3, the rest of this chapter becomes even less theoretical and more practical. Regarding top-down parsing, while the previous section sketched its basic methodology in general, the current section gives a more realistic insight into this approach to parsing because it restricts its attention to deterministic top-down parsers, which fulfill a central role in practice.

Take a CFG G and a G-based top-down parser working on an input string w (see Section 11.2). Recall that a top-down parser verifies that w is syntactically correct, so it simulates the construction of a derivation tree for w in a top-down way. That is, reading w

from left to right, the parser starts from the tree root and proceeds down toward the frontier denoted by w. To put it in derivation terms, it builds up the leftmost derivation of w, so it starts from S_G and proceeds toward w. If the parser works deterministically, the parser makes a completely deterministic selection of an applied rule during every single computational step.

Section 11.4 concentrates its attention on *predictive parsing*, which is the most frequently used deterministic top-down parsing method in practice. First, it defines and discusses predictive sets and LL grammars, where the first L stands for the *l*eft-to-right scan of the input string w, while the second L means that the parser simulates the *l*eftmost derivation of w. By using the predictive sets, the parser makes the simulation of the leftmost derivations in a completely deterministic way. Next, this section describes two fundamental versions of predictive parsing. First, it explains recursive descent parsing, which frees us from explicitly implementing a pushdown list. Then, it uses the LL grammars and predictive sets to construct predictive tables used by predictive table-driven parsing, which explicitly implements a pushdown list. In this version of predictive parsing, any grammatical change only leads to a modification of the table while its control procedure remains unchanged, which is obviously its great advantage. We also explain how predictive parsers handle syntax errors to recover from them.

Throughout this section, we make frequent use of the terminology introduced in Section 1.2, such as *suffixes*, *prefixes*, and *symbol*. In the examples, we illustrate the material under discussion in terms of the CFG J from Convention 11.6.

Predictive sets and LL grammars

Consider a CFG, $G = (V, T, P, S)$, and an input string, w. Let M be a G-based top-down parser constructed by Algorithm 11.3. Suppose that M has already found the beginning of the leftmost derivation for w, $S\ {}_{lm}\Rightarrow^* tAv$, where t is a prefix of w. More precisely, let $w = taz$, where a is the current input symbol, which follows t in w, and z is the suffix of w, which follows a. In tAv, A is the leftmost nonterminal to be rewritten in the following step. Suppose that there exist several different A-rules. Under the assumption that M works deterministically (see Convention 11.7), M has to select one of the A-rules to continue the parsing process, and it cannot revise this selection later on. If M represents a *predictive parser*, it selects the right rule by predicting whether its application gives rise to a leftmost derivation of a string starting with a. To make this prediction, every rule $r \in P$ is accompanied with its *predictive set* containing all terminals that can begin a string resulting from a derivation whose first step is made by r. If the A-rules have their predictive sets pairwise disjoint, M deterministically selects the rule whose predictive set contains a. To construct the predictive sets, we first need the $first$ and $follow$ sets, described as follows.

The predictive set corresponding to $r \in P$ obviously contains the terminals that occur as the first symbol in a string derived from $rhs(r)$, and this observation gives rise to the following useful definition.

Definition 11.1. Let $G = (V, T, P, S)$ be a CFG. For every string $x \in V^*$,

$$first(x) = \{a \mid x \Rightarrow^* w, \text{ where either } w \in T^+ \text{ with}$$
$$a = symbol(w, 1) \text{ or } w = \varepsilon = a\},$$

where $symbol(w, 1)$ denotes the leftmost symbol of w (see Section 1.2). ■

In general, $first$ is defined in terms of $\Rightarrow$, but we could rephrase this definition in terms of the leftmost derivations, which play a crucial role in top-down parsing, because for every $w \in T^*, x \Rightarrow^* w$ iff $x\ {}_{lm}\Rightarrow^* w$ (see Theorem 6.1). That is,

$$first(x) = \{a \mid x\ {}_{lm}\Rightarrow^* w, \text{ where either } w \in T^+ \text{ with}$$
$$a = symbol(w, 1) \text{ or } w = \varepsilon = a\}.$$

It is worth noting that if $x \Rightarrow^* \varepsilon$, where $x \in V^*$, then ε is in $first(x)$; as a special case, for $x = \varepsilon$, $first(\varepsilon) = \{\varepsilon\}$.

Next, we will construct the $first$ sets for all strings contained in $T \cup \{lhs(r) \mid r \in P\} \cup \{y \mid y \in suffixes(rhs(r)) \text{ with } r \in P\}$. We make use of some subsets of these $first$ sets later in this section (see Algorithm 11.6 and Definition 11.3).

Basic idea. To construct $first(x)$ for every $x \in T \cup \{lhs(r) \mid r \in P\} \cup \{y \mid y \in suffixes(rhs(r))$ with $r \in P\}$, we initially set $first(a)$ to $\{a\}$, for every $a \in T \cup \{\varepsilon\}$, because $a\ {}_{lm}\Rightarrow^0 a$ for these as. If $A \to uw \in P$ with $u \Rightarrow^* \varepsilon$, then $A \Rightarrow^* w$, so we add the symbols of $first(w)$ to $first(A)$ (by Algorithm 6.4, we can determine whether u is a string consisting of ε-nonterminals, in which case $u\ {}_{lm}\Rightarrow^* \varepsilon$). Keep extending all the $first$ sets in this way until no more symbols can be added to any of the $first$ sets.

The ε-rules deserve our special attention because they may give rise to derivations that erase substrings of sentential forms, and this possible erasure complicates the selection of the next applied rule. Indeed, consider a CFG, $G = (V, T, P, S)$, and suppose that $A \to x \in P$, with $x\ {}_{lm}\Rightarrow^* \varepsilon$. At this point, the parser needs to decide whether from A, it should make either $A\ {}_{lm}\Rightarrow x\ {}_{lm}\Rightarrow^* \varepsilon$ or a derivation that produces a non-empty string. To make this decision, we need to determine the set $follow(A)$ containing all terminals that can follow A in any sentential form; if this set contains the current input symbol that is out of $first(x)$, the parser simulates $A\ {}_{lm}\Rightarrow x\ {}_{lm}\Rightarrow^* \varepsilon$. In addition, to express that a sentential form ends with A, we include $\triangleleft$ into $follow(A)$ (see Convention 11.5 for $\triangleleft$).

Definition 11.2. Let $G = (V, T, P, S)$ be a CFG. For every $A \in V - T$,

$$follow(A) = \{a \in T \cup \{\triangleleft\} \mid Aa \in substrings(F(G)\{\triangleleft\})\},$$

where $F(G)$ denotes the set of sentential forms of G (see Definition 6.1). ■

Basic idea. We next construct $follow(A)$ for every $A \in (V - T)$. As S is a sentential form, we initialize $follow(S)$ with $\{\triangleleft\}$. Consider any $B \to uAv \in P$. If a is a terminal in $first(v)$, then $Aa \in substrings(F(G))$, so we add a to $follow(A)$. In addition, if ε is in $first(v)$, then $v\ {}_{lm}\Rightarrow^* \varepsilon$ and, consequently, $follow(B) \subseteq follow(A)$, so we add all symbols from $follow(B)$ to $follow(A)$. Keep extending all the $follow$ sets in this way until no symbol can be added to any of them.

Based on the $first$ and $follow$ sets, we define the predictive sets as follows.

Definition 11.3. For each $r \in P$, its *predictive set* is denoted by *predictive-set*(r) and defined as follows:

(I) if $\varepsilon \notin first(rhs(r))$, *predictive-set*$(r) = first(rhs(r))$;
(II) if $\varepsilon \in first(rhs(r))$, *predictive-set*$(r) = (first(rhs(r)) - \{\varepsilon\}) \cup follow(lhs(r))$. ■

LL grammars

Reconsider the discussion at the very beginning of this section. Recall that G, M, and w denote a CFG, a G-based top-down parser, and an input string w, respectively. Suppose that M has already simulated the beginning of the leftmost derivation $S\ _{lm} \Rightarrow^* tAv$ for an input string taz, where a is the current input symbol and $taz = w$, and it needs to select one of several different A-rules to rewrite A in tAv and, thereby, make another step. If for an A-rule $r, a \in$ *predictive-set*(r) and for any other A-rule $p, a \notin$ *predictive-set*(p), M deterministically selects r. This idea leads to the following definition of LL grammars.

Definition 11.4. A CFG $G = (V, T, P, S)$ is an *LL grammar* if for each $A \in (V - T)$, any two different A-rules, $p, q \in P$ and $p \neq q$, satisfy *predictive-set*$(p) \cap$ *predictive-set*$(q) = \emptyset$. ■

As already noted, in LL grammars, the first L stands for a *l*eft-to-right scan of symbols, and the other L stands for a *l*eftmost derivation. Sometimes, in greater detail, the literature refers to the LL grammars as *LL(1) grammars* to point out that the top-down parsers based on these grammars always look at one input symbol during each step of the parsing process. Indeed, these grammars represent a special case of *LL(k) grammars*, where $k \geq 1$, which underlie parsers that make a k-symbol look ahead. In this introductory textbook, however, we discuss only LL(1) grammars and simply refer to them as LL grammars for brevity.

Example 11.5. Consider the CFG J (see Convention 11.6), defined as

$$1: S \to AC, \quad 2: C \to \vee AC, \quad 3: C \to \varepsilon, \quad 4: A \to BD,$$
$$5: D \to \wedge BD, \quad 6: D \to \varepsilon, \quad 7: B \to (S), \quad 8: B \to i.$$

Algorithm 11.5 *first*. For each rule in J, we construct $first(u)$ for every $u \in \Delta \cup \{lhs(r)) \mid r \in P_J\} \cup \{y \mid y \in suffixes(rhs(r)),$ with $r \in P_J\}$ by Algorithm 11.5. First, we set $first(a) = \{a\}$ for every $a \in \{i, (,), \vee, \wedge\} \cup \{\varepsilon\}$. Consider $B \to i$. By the **repeat** loop of Algorithm 11.5, as $first(i) = \{i\}$, we include i in $first(B)$. As $i \in first(B)$ and $A \to BD \in P_J$, we add i to $first(BD)$ as well. Complete this construction as an exercise. However, to construct *predictive-set*(r) for each $r \in P_J$, we only need $\{first(rhs(r)) \mid r \in P_J\}$, which represents a subset of all the *first* sets constructed by Algorithm 11.5. The members of $\{first(rhs(r)) \mid r \in P_J\}$ are listed in the second column of Table 11.5.

Algorithm 11.5. *first*.

Input: A CFG $G = (V, T, P, S)$.
Output: Set $first(u)$ for every $u \in T \cup \{lhs(r) \mid r \in P\} \cup \{y \mid y \in suffixes(rhs(r))$ with $r \in P\}$.
Method:

begin
  set $first(a) = \{a\}$ for every $a \in T \cup \{\varepsilon\}$
  set all the other constructed *first* sets to $\emptyset$
  repeat
    if $r \in P, u \in prefixes(rhs(r))$ and $u\ _{lm}\Rightarrow^* \varepsilon$ **then**
      extend $first(suffix(rhs(r), |rhs(r)| - |u|))$ by $first(symbol(rhs(r), |u| + 1))$
      extend $first(lhs(r))$ by $first(suffix(rhs(r), |rhs(r)| - |u|))$
  until no change
end

Algorithm 11.6 *follow*. Consider $first(u)$, for each $u \in \{y \mid y \in suffixes(rhs(r))$, with $r \in P_J\}$. We construct $follow(X)$, for every $X \in (V_J - T_J)$ by Algorithm 11.6 as follows. We initially have $follow(S) = \{\lhd\}$. As $B \to (S) \in P_J$ and $) \in first())$, we add $)$

to $follow(S)$. Since (i) $S \to AC \in P_J$, (ii) $\varepsilon \in first(C)$, and (iii) $follow(S)$ contains) and $\lhd$, we add) and $\lhd$ to $follow(A)$. As $\vee \in first(C)$, we add $\vee$ to $follow(A)$ too. Complete this construction as an exercise. The third column of Table 11.5 contains $follow(lhs(r))$, for each $r \in P_J$.

Algorithm 11.6. *follow*.

Input: A CFG $G = (V, T, P, S)$, and $first(u)$ for every $u \in \{y \mid y \in suffixes(rhs(r))$ with $r \in P\}$ (see Algorithm 11.5).
Output: Sets $follow(A)$ for all $A \in (V - T)$.
Method:

begin
  set $follow(S) = \{\lhd\}$, and
  set all the other constructed $follow$ sets to $\emptyset$
  repeat
    if $r \in P$ **and** $Au \in suffixes(rhs(r))$, where $A \in (V - T)$,
      $u \in V^*$ **then**
      add the symbols in $(first(u) - \{\varepsilon\})$ to $follow(A)$
      if $\varepsilon \in first(u)$ **then**
        add the symbols in $follow(lhs(r))$ to $follow(A)$
  until **no change**
end

Predictive Sets (Definition 11.3). The fourth column of the Table 11.5 contains *predictive-set*(r), for each $r \in P_J$. Note that $follow(lhs(r))$ is needed to determine *predictive-set*(r) if $\varepsilon \in first(rhs(r))$ because at this point *predictive-set*$(r) = (first(rhs(r)) - \{\varepsilon\}) \cup$

Table 11.5. Predictive sets for rules in P_J.

Rule r	$first(rhs(r))$	$follow(lhs(r))$	*predictive-set*(r)
$S \to AC$	$i, ($	$), \lhd$	$i, ($
$C \to \vee AC$	$\vee$	$), \lhd$	$\vee$
$C \to \varepsilon$	ε	$), \lhd$	$), \lhd$
$A \to BD$	$i, ($	$\vee,), \lhd$	$i, ($
$D \to \wedge BD$	$\wedge$	$\vee,), \lhd$	$\wedge$
$D \to \varepsilon$	ε	$\vee,), \lhd$	$\vee,), \lhd$
$B \to (S)$	$($	$\wedge, \vee,), \lhd$	$($
$B \to i$	i	$\wedge, \vee,), \lhd$	i

$follow(lhs(r))$; if $\varepsilon \notin first(rhs(r))$, it is not needed because $predictive\text{-}set(r) = first(rhs(r))$ (see Definition 11.3). Take, for instance, $S \to AC \in P_J$, with $first(AC)) = \{i, (\}$. As $\varepsilon \notin first(rhs(r)), predictive\text{-}set(S \to AC) = first(AC) = \{i, (\}$. Consider $C \to \varepsilon \in P_J$, with $first(\varepsilon) = \{\varepsilon\}$ and $follow(C) = \{), \lhd\}$. As $\varepsilon \in first(rhs(C \to \varepsilon)), predictive\text{-}set(C \to \varepsilon) = (first(\varepsilon) - \{\varepsilon\}) \cup follow(C) = \emptyset \cup \{), \lhd\} = \{), \lhd\}$. Complete this construction as an exercise.

LL Grammar. Observe that $predictive\text{-}set(C \to \vee AC) \cap predictive\text{-}set(C \to \varepsilon) = \{\vee\} \cap \{), \lhd\} = \emptyset$. Analogously, $predictive\text{-}set(D \to \wedge BD) \cap predictive\text{-}set(D \to \varepsilon) = \emptyset$ and $predictive\text{-}set(B \to (S)) \cap predictive\text{-}set(B \to i) = \emptyset$. Thus, J is an LL CFG.

■

Predictive parsing

Next, we first describe the recursive-descent parsing method. Then, we use the LL grammars and their predictive sets to create predictive tables and deterministic top-down parsers driven by these tables. Finally, we explain how to handle errors in predictive parsing.

Predictive recursive-descent parsing

Given an LL grammar, $G = (V, T, P, S)$, we next explain how to construct a *G-based predictive recursive-descent parser*, which makes use of the programming routines **SUCCESS** and **ERROR**, described in the following convention.

Convention 11.8. SUCCESS announces a successful completion of the parsing process while **ERROR** announces a syntax error in the parsed program. ■

It is worth noting that the following parser moves back to the symbol preceding the current input symbol. As demonstrated shortly, it performs this return when a rule selection is made by using a symbol from the *follow* set; at this point, this symbol is not actually used up, so the parser needs to move back to the symbol preceding *ins*.

Reconsider Algorithm 11.3, which turns a CFG G to a G-based top-down parser M as a PDA, which requires, in strict terms, an implementation of a pushdown list. As a crucial advantage, *recursive descent*, i.e., a top-down parsing method discussed in the following, frees us from this implementation. Indeed, the pushdown list is invisible in this method because it is actually realized by the pushdown used to support recursion in the programming language in which we write the recursive-descent parser. As this method does not require an explicit manipulation with the pushdown list, it comes as no surprise that it is extremely popular in practice. Therefore, in its description, as follows, we pay special attention to its implementation.

Basic idea. Consider a programming language defined by an LL grammar G. Let $w = t_1 \cdots t_j t_{j+1} \cdots t_m$ be an input string. Like any top-down parser, a G-based recursive-descent parser simulates the construction of a derivation tree with its frontier equal to w by using the grammatical rules, so it starts from the root and works down to the leaves, reading w in a left-to-right way. In terms of derivations, the parser looks for the leftmost derivation of w. To find it, for each nonterminal Y, the parser has a Boolean function, *rd-function* Y, which simulates rewriting the leftmost nonterminal Y. More specifically, with the right-hand side of a Y-rule, $Y \to X_1 \cdots X_i X_{i+1} \cdots .X_n$, *rd-function* Y proceeds from X_1 to X_n. Assume that *rd-function* Y currently works with X_i and that t_j is the input symbol. At this point, depending on whether X_i is a terminal or a nonterminal, this function works as follows:

- If X_i is a terminal, *rd-function* Y matches X_i against t_j. If $X_i = t_j$, it reads t_j and proceeds to X_{i+1} and t_{j+1}. If $X_i \neq t_j$, a syntax error occurs, which the parser has to handle.
- If X_i is a nonterminal, the parser finds out whether there is an X_i-rule r satisfying $t_j \in$ *predictive-set*(r). If so, the parser calls *rd-function* X_i, which simulates rewriting X_i according to r. If not, a syntax error occurs, and the parser has to handle it.

The parser starts the parsing process from *rd-function* S, which corresponds to the start symbol of G and ends when it eventually returns to this function after completely reading w. If during this entire process no syntax error occurs, G-based recursive-descent parser has found the leftmost derivation of w, which thus represents a syntactically well-formed program; otherwise, w is syntactically incorrect.

As this method does not require explicitly manipulating a pushdown list, it is very popular in practice; in particular, it is suitable for parsing declarations and general program flow as the following example illustrates. This example describes every *rd-function* Y as a pseudocode

function Y
begin
⋮
end

A function like this returns a true–false value as a result.

Example 11.6. Reconsider the LL grammar J (see Convention 11.6 and Example 11.5). Table 11.6 repeats its rules together with the corresponding predictive sets.

Next, we construct a J-based recursive-descent parser as a collection of Boolean *rd-function* s corresponding to nonterminals S, A, B, C, and D. By using the predictive sets, we make this construction in such a way that the parser always selects the next applied rule deterministically.

Consider the start symbol S and the only S-rule $S \to AC$ with $predictive\text{-}set(S \to AC) = \{i, (\}$. As a result, *rd-function* S has the following form.

Table 11.6. Rules in P_J together with their predictive sets.

Rule r	$predictive\text{-}set(r)$
$S \to AC$	$i, ($
$C \to \vee AC$	$\vee$
$C \to \varepsilon$	$), \triangleleft$
$A \to BD$	$i, ($
$D \to \wedge BD$	$\wedge$
$D \to \varepsilon$	$\vee,), \triangleleft$
$B \to (S)$	$($
$B \to i$	i

```
function S
begin
  if ins ∈ {i, (} then
    if A then
      if C then
        return true
  return false
end
```

Consider the two C-rules that include $C \to \vee AC$, with *predictive-set*$(C \to \vee AC) = \{\vee\}$, and $C \to \varepsilon$, with *predictive-set*$(C \to \varepsilon) = \{), \lhd\}$. Therefore, the *rd-function* C selects $C \to \vee AC$ if the input symbol is $\vee$, and this function selects $C \to \varepsilon$ if this symbol is in $\{), \lhd\}$. If the input symbol differs from $\vee,)$, or $\lhd$, a syntax error occurs. Thus, the *rd-function* C has the following form.

```
function C
begin
  case ins of
    ∨:
      set ins to the next input symbol
      if A then
        if C then
          return true
    ), ◁:
      return true
  return false
end
```

There exists a single A-rule of the form $A \to BD$ in J, with *predictive-set*$(A \to BD) = \{i, (\}$. Its *rd-function* A, given as follows, is similar to the *rd-function* S.

```
function A
begin
  if ins ∈ {i, (} then
    if B then
      if D then
        return true
  return false
end
```

Consider the two D-rules $D \to \wedge BD$ with *predictive-set*$(D \to \wedge BD = \{\wedge\}$ and $D \to \varepsilon$, with *predictive-set*$(D \to \varepsilon = \{\vee,), \lhd\}$. Therefore, the *rd-function* D selects $D \to \wedge BD$ if the input symbol is $\wedge$, and it selects $D \to \varepsilon$ if this symbol is in $\{\vee,), \lhd\}$. The *rd-function* D has thus this form

```
function D
begin
  case ins of
    ∧:
      set ins to the next input symbol
      if B then
        if D then
          return true
    ∨, ), ◁:
      return true
  return false
end
```

Finally, consider the B-rules $B \to (S)$ with *predictive-set*$(B \to (S)) = \{(\}$ and $B \to i$ with *predictive-set*$(B \to i) = \{i\}$. Therefore, the *rd-function* B selects $B \to (S)$ if the input symbol is (and this function selects $B \to i$ if the input symbol is i. If the input symbol differs from (or i, the syntax error occurs. Thus, the *rd-function* B has the following form.

```
function B
begin
  case ins of
    (:
      set ins to the next input symbol
      if S then
        if ins =) then
          set ins to the next input symbol
          return true
    i:
      set ins to the next input symbol
      return true
  return false
end
```

Having these functions in the parser, its main body is based on the following simple **if** statement, which decides whether the source program is syntactically correct by the final Boolean value of the *rd-function S*:

```
begin
  if S then
    SUCCESS
  else
    ERROR
end
```

In the above functions, we frequently make an advance or a return within the string of input symbols. As an exercise, explain how to implement these actions.

■

Predictive table-driven parsing

In the predictive recursive-descent parsing, for every nonterminal A and the corresponding A-rules, there exists a specific Boolean function, so any grammatical change usually necessitates reprogramming several of these functions. Therefore, unless a change of this kind is ruled out, we often prefer an alternative *predictive table-driven parsing* based on a single general control procedure that is completely based on a predictive table. At this point, a change in the grammar only implies an adequate modification of the table, while the control procedure remains unchanged. As opposed to the predictive recursive-descent parsing, however, this parsing method maintains a pushdown explicitly, not implicitly via recursive calls like in predictive recursive-descent parsing.

Consider an LL CFG G. Like a general top-down parser (see Algorithm 11.3), a *G-based predictive table-driven parser* is underlain by a PDA, $M = (Q_M, \Delta_M, \Gamma_M, R_M, s_M, S_M, F_M)$ (see Definition 4.1). However, a strictly mathematical specification of M, including all its rules in R_M, would be somewhat tedious and lengthy from a practical point of view. Therefore, to make the parser easy to implement, we describe the parser in the way described in Convention 11.7; that is, M is specified as an algorithm together with a *G-based predictive table*, denoted by PT_G, by which the parser determines a move from every configuration. The rows and columns of PT_G are denoted by the members of $V_G - V_T$ and $T_G \cup \{\lhd\}$, respectively. Each of its entry contains a member of P_G, or it is blank. More precisely, for each $A \in V_G - T_G$ and each $t \in T_G$, if there exists $r \in P_G$ such that $lhs(r = A$ and $t \in predictive\text{-}set(r), PT_G[A, t] = r$; otherwise, $PT_G[A, t]$ is blank, which signalizes a syntax error. Making use of PT_G, M works with the input string, as described in the following. Before giving this description, we want to point out once again that G represents an LL CFG. Indeed, unless G is an LL CFG, we might place more than one rule into a single PT_G entry and, thereby, make M non-deterministic.

Let X be the *pd* top symbol. Initially, set *pd* to $\rhd S$. Perform one of actions in (I)–(V), given in the following.

(I) If $X \in T_G$ and $X = ins$, the *pd* top symbol is a terminal coinciding with ins, so M *pops* the pushdown by removing X from its top and, simultaneously, advances to the next input symbol;

(II) if $X \in T_G$ and $X \neq ins$, the pushdown top symbol is a terminal that differs from ins, so the parser announces an error by **ERROR**;

(III) if $X \in V_G - T_G$ and $PT_G[X, ins] = r$, with $r \in P_G$, where $lhs(r = X, M$ *expands* the pushdown by replacing X with $reversal(rhs(r))$;

(IV) if $X \in V_G - T_G$ and $PT_G[X, ins]$ is blank, M announces an error by **ERROR**;

(V) if $\rhd = X$ and $ins = \lhd$, the pushdown is empty, and the input string is completely read, so M halts and announces a successful completion of parsing by **SUCCESS**.

Throughout the rest of this section, we also often make use of the operations **EXPAND** and **POP**. We next define them in terms of the general version of a top-down parser constructed by Algorithm 11.3.

Definition 11.5. Let $G = (V_G, T_G, P_G, S_G)$ be a CFG, and let M be a PDA that represents a G-based top-down parser produced by Algorithm 11.3.

(I) Let $r : A \to x \in P_G$. **EXPAND**(r) applies $A\Diamond \to reversal(x)\Diamond$ in M.
(II) **POP** applies $a\Diamond a \to \Diamond$ in M. ■

Less formally, **EXPAND**($A \to x$) replaces the *pd* top with the reversal of x. If *ins* coincides with the *pd* top, **POP** removes the *pd* top and, simultaneously, advances to the next input symbol.

Algorithm 11.7. Predictive table-driven parser.

Input: An LL grammar G, its predictive table PT_G, and an input string, $w\lhd$, where $w \in T^*$.
Output: **SUCCESS** if $w \in L(G)$, and **ERROR** if $w \notin L(G)$.
Method:

```
begin
  set pd to ▷S
  repeat
    let X denote the current pd top symbol
  case X of
    X ∈ T:
      if  X = ins then
        POP
      else
        ERROR
    X ∈ V − T:
      if PT_G[X, ins] = r with r ∈ P then
        EXPAND(r)
      else
        ERROR
    X = ▷:
      if ins = ◁ then
        SUCCESS
      else
        ERROR
  until  SUCCESS or ERROR
end
```

As explained in Section 11.2, apart from deciding whether $w \in L(G)$, a G-based top-down parser frequently produces the left parse of w, i.e., the sequence of rules according to which the leftmost derivation

Table 11.7. PT_J.

	$\vee$	$\wedge$	(	)	i	$\lhd$
S			1		1	
C	2			3		3
A			4		4	
D	6	5		6		6
B			7		8	
$\rhd$						**OK**

of w is made in G provided that $w \in L(G)$. Consider the parser represented by Algorithm 11.7. To obtain left parses, extend this parser by writing out r whenever this algorithm performs **EXPAND**(r), where $r \in P$. In this way, we produce the left parse of an input string in the conclusion of the following example.

Example 11.7. Reconsider the LL grammar J (see Example 11.5). See Table 11.6 to recall its rules with the corresponding predictive sets.

By using the predictive sets corresponding to these rules, we construct the predictive table PT_J (see Table 11.7).

Take $i \wedge i \vee i$. Algorithm 11.7 parses this string as described in Table 11.8. The first column gives the configurations of the parser. The second column states the corresponding PT_J entries together with the rules they contain. The third column gives the action made by the parser. The fourth column gives the sentential forms derived in the leftmost derivation.

Suppose that the parser should produce the left parse of $i \wedge i \vee i$, i.e., the sequence of rules according to which the leftmost derivation of $i \wedge i \vee i$ is made in J (see (I) at the beginning of Section 11.2). The parser can easily obtain this parse by writing out the applied rules according to which the expansions are performed during the parsing process. Specifically, take the sequence of rules in the second column of Table 11.8 to obtain 14858624863 as the left parse of $i \wedge i \vee i$ in J. Indeed, $S\ _{lm}\Rightarrow^* i \wedge i \vee i$ [14858624863] in J.

Before closing this example, let us point out that from Algorithm 11.7 and PT_J, we could obtain a strictly mathematical specification of the parser as a PDA M. In essence, M has the form of the parser constructed in Algorithm 11.3 in Section 11.2. That is,

Table 11.8. Predictive table-driven parsing.

Configuration	Table Entry and Rule	Action	Sentential Form
			$\underline{S}$
$\triangleright S\Diamond i \wedge i \vee i \triangleleft$	$[S, i] = 1 : S \to AC$	**EXPAND**(1)	$_{lm}\Rightarrow \underline{A}C$ [1]
$\triangleright CA\Diamond i \wedge i \vee i \triangleleft$	$[A, i] = 4 : A \to BD$	**EXPAND**(4)	$_{lm}\Rightarrow \underline{B}DC$ [4]
$\triangleright CDB\Diamond i \wedge i \vee i \triangleleft$	$[B, i] = 8 : B \to i$	**EXPAND**(8)	$_{lm}\Rightarrow i\underline{D}C$ [8]
$\triangleright CDi\Diamond i \wedge i \vee i \triangleleft$		**POP**	
$\triangleright CD\Diamond \wedge i \vee i \triangleleft$	$[D, \wedge] = 5 : D \to \wedge BD$	**EXPAND**(5)	$_{lm}\Rightarrow i \wedge \underline{B}DC$ [5]
$\triangleright CDB \wedge \Diamond \wedge i \vee i \triangleleft$		**POP**	
$\triangleright CDB\Diamond i \vee i \triangleleft$	$[B, i] = 8 : B \to i$	**EXPAND**(8)	$_{lm}\Rightarrow i \wedge i\underline{D}C$ [8]
$\triangleright CDi\Diamond i \vee i \triangleleft$		**POP**	
$\triangleright CD\Diamond \vee i \triangleleft$	$[D, \vee] = 6 : D \to \varepsilon$	**EXPAND**(6)	$_{lm}\Rightarrow i \wedge i\underline{C}$[6]
$\triangleright C\Diamond \vee i \triangleleft$	$[C, \vee] = 2 : C \to \vee AC$	**EXPAND**(2)	$_{lm}\Rightarrow i \wedge i \vee \underline{A}C$ [2]
$\triangleright CA \vee \Diamond \vee i \triangleleft$		**POP**	
$\triangleright CA\Diamond i \triangleleft$	$[A, i] = 4 : A \to BD$	**EXPAND**(4)	$_{lm}\Rightarrow i \wedge i \vee \underline{B}DC$ [4]
$\triangleright CDB\Diamond i \triangleleft$	$[B, i] = 8 : B \to i$	**EXPAND**(8)	$_{lm}\Rightarrow i \wedge i \vee i\underline{D}C$ [8]
$\triangleright CDi\Diamond i \triangleleft$		**POP**	
$\triangleright CD\Diamond \triangleleft$	$[D, \triangleleft] = 6 : D \to \varepsilon$	**EXPAND**(6)	$_{lm}\Rightarrow i \wedge i \vee i\underline{C}$ [6]
$\triangleright C\Diamond \triangleleft$	$[C, \triangleleft] = 3 : C \to \varepsilon$	**EXPAND**(3)	$_{lm}\Rightarrow i \wedge i \vee i$ [3]
$\triangleright\Diamond\triangleleft$	$[\triangleright, \triangleleft] =$ **OK**	**SUCCESS**	

M makes **SUCCESS** by $\triangleright\Diamond\triangleleft \to \Diamond$. If a pair of the *pd* top X and the current input symbol b leads to **ERROR**, R_M has no rule with $X\Diamond b$ on its left-hand side. It performs **POP** by rules of the form $a\Diamond a \to \Diamond$, for each $a \in {}_J\Delta$. Finally, if a pair of the *pd* top and the current input symbol leads to **EXPAND**(r), where r is a rule from P_J, M makes this expansion according to r by $A\Diamond \to reversal(x)\Diamond$. For instance, as $PT_J[S, i] = 1$ and $1 : S \to AC \in P_J, R_M$ has $S\Diamond \to CA\Diamond$ to make an expansion according to 1. A completion of this mathematical specification of M is left as an exercise. However, not only is this completion a tedious task but also the resulting parser M specified in this way is difficult to understand what it actually does with its incredibly many rules. That is why, as already pointed out (see Convention 11.7), we always prefer the description of a parser as an algorithm together with a parsing table throughout the rest of this book. ■

Exclusion of Left Recursion

Deterministic top-down parsing places some non-trivial restrictions on the CFGs it is based on. Perhaps most importantly, no deterministic top-down parser can be based on a left-recursive CFG (see Section 6.1). Indeed, suppose that in order to simulate a left-most derivation step, a deterministic top-down parser would select a directly left-recursive rule of the form $A \rightarrow Ax$, where A is a nonterminal and x is a string (see Definition 6.15). Since the right-hand side starts with A, the parser would necessarily simulate the next leftmost derivation step according to the same rule, and it would loop in this way endlessly. As demonstrated in the Exercises, a general left recursion would also lead to an infinite loop like this. Therefore, deterministic top-down parsing is always underlain by non-left-recursive CFGs.

As demonstrated in Example 11.2, however, left-recursive CFGs specify some common programming-language syntactical constructions, such as conditions and expressions, in a very elegant and succinct way, so we want deterministic parsers based on them too. Fortunately, deterministic bottom-up parsers work with left-recursive CFGs perfectly well, which brings us to the topic of the following section — bottom-up parsing.

11.5 Bottom-Up Parsing

Given an input string, w, a standard bottom-up parser constructs a derivation tree with frontier w in a bottom-up way. That is, reading w from left to right, the parser starts from frontier w and proceeds up toward the root. To put it in terms of derivations, it builds up the rightmost derivation of w in reverse so that it starts from w and proceeds toward the start symbol. Each action during this parsing process represents a *shift* or a *reduction*. The former consists in shifting the current input symbol onto the pushdown. During a reduction, the parser selects a *handle*, i.e., an occurrence of the right-hand side of a rule in the current sentential form, and after this selection, it reduces the handle to the left-hand side of the rule so that this reduction, seen in reverse, represents a rightmost derivation step.

As already pointed out in Section 11.2, in practice, we are primarily interested in deterministic bottom-up parsing, which always precisely determines how to make every single step during the parsing process. That is why we narrow our attention to fundamental deterministic bottom-up parsing methods in this section. We describe two of them. First, we describe *precedence parsing*, which is a popular deterministic bottom-up parsing method for expressions whose operators and their priorities actually control the parsing process. Then, we describe *LR parsing*, where L stands for a *l*eft-to-right scan of the input string and R stands for a *r*ightmost derivation constructed by the parser. LR parsers are as powerful as deterministic PDAs, so they represent the strongest possible parsers that work in a deterministic way. That is probably why they are so often implemented in practice, and we discuss them in detail later in this section.

Both sections have a similar structure. First, they describe fundamental parsing algorithms together with parsing tables the algorithms are based on. Then, they explain how to construct these tables. Finally, they sketch how to handle syntax errors.

Throughout the section, we use the same notions and conventions as in Section 11.4, including **SUCCESS** and **ERROR** (see Convention 11.8).

Operator-precedence parsing

In practice, we almost always apply an operator-precedence parser to expressions, such as the logical expressions defined by the CFG E (see Convention 11.6), and we explain how to make this parser based on this CFG rather than give a general explanation. First, we explain how an operator-precedence parser based on E works, describe the construction of its parsing table, and sketch how to handle syntax errors. Then, we base this parser on other expression-generating grammars. Finally, we outline the advantages and disadvantages of operator-precedence parsing from a general viewpoint.

Recall that $E = (V_E, T_E, P_E, S_E)$ is defined as

$$1 : S \to S \vee S, \quad 2 : S \to S \wedge S, \quad 3 : S \to (S), \quad 4 : S \to i,$$

where $T_E = \{\vee, \wedge, (,), i\}$ and $V_E - T_E = \{S\}$. The operator-precedence parsing table of E, OP_E, has a relatively simple format. Indeed, this table has its rows and columns denoted by the members of $\Delta_E \cup \{\lhd\}$ and $\Delta_E \cup \{\lhd\}$, respectively. Each OP_E entry is a member of $\{\lfloor, \rfloor, \Box, \}$, where $\Box$ denotes a blank entry. Table 11.9 presents OP_E, whose construction is explained later in this section.

Table 11.9. OP_E.

	$\wedge$	$\vee$	i	(	)	$\lhd$
$\wedge$	$\rfloor$	$\rfloor$	$\lfloor$	$\lfloor$	$\rfloor$	$\rfloor$
$\vee$	$\lfloor$	$\rfloor$	$\lfloor$	$\lfloor$	$\rfloor$	$\rfloor$
i	$\rfloor$	$\rfloor$				
(	$\lfloor$	$\lfloor$	$\lfloor$	$\lfloor$		
)	$\rfloor$	$\rfloor$			$\rfloor$	$\rfloor$
$\rhd$	$\lfloor$	$\lfloor$	$\lfloor$	$\lfloor$		**OK**

Operator-precedence parser

An E-based operator-precedence parser makes shifts and reductions by operations **OP-REDUCE** and **OP-SHIFT**, respectively. We define both operations in the following, making use of the *pd* and *ins* notation introduced in Convention 11.5.

Definition 11.6. Let $E = (V_E, T_E, P_E, S_E)$ and OP_E have the same meanings as above. Operations **OP-REDUCE** and **OP-SHIFT** are defined in the following way.

(I) **OP-REDUCE** performs (1)–(3) given as follows.

 (a) Let $pd = y$, where $y \in \{\lhd\}V_E^*$, and let a be the topmost *pd* symbol such that $a \in \{\lhd\} \cup \Delta_E$ and $y = xaubv$, with $OP_E[a, b] = \lfloor$, where $x \in \{\lhd\}V_E^* \cup \{\varepsilon\}(a = \lhd \text{ iff } x = \varepsilon), b \in T_E, u \in N_E^*, v \in V_E^*$. Then, the current handle is ubv.

 (b) Select $r \in P_E$ with $rhs(r) = ubv$.

 (c) Change ubv to $lhs(r)$ on the *pd* top.

 To express, in a greater detail, that **OP-REDUCE** is performed according to r P_E, write **OP-REDUCE** (r).

(II) **OP-SHIFT** pushes *ins* onto the *pd* top and advances to the next input symbol. ■

To illustrate **OP-REDUCE**, suppose that the current configuration of the parser is of the form $\rhd(S)\Diamond\lhd$. In terms of (1) in Definition 11.6, $y = \rhd(S)$, and the topmost *pd* terminal a satisfying (1) is $\rhd$; to be quite precise, $x = \varepsilon, a = \rhd, u = \varepsilon, b = ($, and $v = S)$. Thus, (S) is the handle. In (2), select $3 : S \to (S)$. In (3), reduce (S) to S and, thereby, change $\rhd(S)\Diamond\lhd$ to $\rhd S\Diamond\lhd$; in symbols, perform **OP-REDUCE** (3). It is worth noting that the case when $a = \rhd$ is not ruled out, as illustrated by this example as well. To give another example, in a briefer manner, apply **OP-REDUCE** to the configuration $\triangleright(S \vee S\Diamond) \wedge S\triangleleft$. This application actually performs **OP-REDUCE** (1) and changes $\triangleright(S \vee S\Diamond) \wedge S\triangleleft$ to $\triangleright(S\Diamond) \wedge S\triangleleft$. Observe that **OP-REDUCE** is not applicable to any configuration. Indeed, consider $\triangleright()\Diamond\triangleleft$. As E contains no rule with the right-hand side equal to $()$, **OP-REDUCE** is inapplicable to $\triangleright()\Diamond\triangleleft$.

To illustrate the other operation, **OP-SHIFT**, take, for instance, $\triangleright S \vee \Diamond i\triangleleft$. Note that **OP-SHIFT** changes $\triangleright S \vee \Diamond i\triangleleft$ to $\triangleright S \vee i\Diamond\triangleleft$.

Basic idea. Let X be the *pd* topmost terminal. The parser determines each parsing step based on the entry $OP_E[X, ins]$. According to this entry, the parser performs one of the following actions (I)–(IV):

(I) If $OP_E[X, ins]$ contains $\lfloor$, the parser performs **OP-SHIFT**.
(II) If $OP_E[X, ins]$ is $\rfloor$, the parser performs **OP-REDUCE**.
(III) Let $OP_E[X, ins] = \Box$. If $X =$ (and $ins =$), the parser performs **OP-SHIFT** onto *pd* to prepare the performance of **OP-REDUCE** according to $3 : S \to (S)$ right after this shift; otherwise, i.e., if $X \neq$ (or $ins \neq$), the blank entry signalizes a syntax error, so the parser performs **ERROR**.
(IV) If $OP_E[X, ins] =$ **OK** the parser performs **SUCCESS** and successfully completes the parsing process.

Algorithm 11.8. Operator-precedence parser.

Input: E, OP_E, and $w\triangleleft$, where $w \in T_E^*$, where E and OP_E have the same meaning as above.
Output: If $w \in L(E)$, **SUCCESS**, and if $w \notin L(E)$, **ERROR**.
Method:

begin
  set *pd* to $\triangleleft$
  let X be the topmost terminal of *pd*
  case $OP_E[X, ins]$ **of**
    $\lfloor$:
      OP-SHIFT

```
        ⌋:
          OP-REDUCE
        □:
          if X = ( and ins =) then
            OP-SHIFT
          else
            ERROR
        OK:
          SUCCESS
    until SUCCESS or ERROR
end
```

As already explained in Section 11.2, apart from deciding whether $w \in L(E)$, we may want Algorithm 11.8 to produce the right parse of w, defined as the reverse sequence of rules according to which E makes this rightmost derivation (see (II) at the beginning of Section 11.2) if w belongs to $L(E)$. To obtain this parse, we extend the parser by writing out r whenever this algorithm performs **OP-REDUCE**(r), where $r \in P_E$. In this way, we produce the right parse of an input string in the conclusion of the following example.

Example 11.8. Let us take, for instance, $w = i \wedge (i \vee i)$ in Algorithm 11.8, which represents the E-based operator-precedence parser, whose parsing table OP_E is described in Table 11.9. In Table 11.10, we describe the parse of $i \wedge (i \vee i)$ by Algorithm 11.8. In the first column, we give the OP_E entries, and in the second, we present the actions taken by the parser according to these entries. In the third column, we give the parser configurations, which have the form $\triangleright x \Diamond y \triangleleft$, where $\triangleright x$ is the current *pd* and $y \triangleleft$ is the input suffix that remains to be parsed; the leftmost symbol of $y \triangleleft$ represents ins. We underline the topmost *pd* terminals in these configurations.

Suppose that the parser should produce the right parse of $i \wedge (i \vee i)$. The parser can easily obtain this parse by writing out the applied rules according to which the reductions are performed during the parsing process. Specifically, take the sequence of rules in the second column of Table 11.10 to obtain 444132 as the right parse of $i \wedge (i \vee i)$ in E. ■

Table 11.10. Operator-precedence parsing.

Table entry	Action	Configuration
		$\underline{\rhd}\Diamond i \wedge (i \vee i)\lhd$
$[\rhd, i] = \lfloor$	**OP-SHIFT**	$\rhd\underline{i}\Diamond \wedge (i \vee i)\lhd$
$[i, \wedge] = \rfloor$	**OP-REDUCE**(4)	$\underline{\rhd}S\Diamond \wedge (i \vee i)\lhd$
$[\rhd, \wedge] = \lfloor$	**OP-SHIFT**	$\rhd S\underline{\wedge}\Diamond(i \vee i)\lhd$
$[\wedge, (] = \lfloor$	**OP-SHIFT**	$\rhd S \wedge \underline{(}\Diamond i \vee i)\lhd$
$[(, i] = \lfloor$	**OP-SHIFT**	$\rhd S \wedge (\underline{i}\Diamond \vee i)\lhd$
$[i, \vee] = \rfloor$	**OP-REDUCE**(4)	$\rhd S \wedge \underline{(}S\Diamond \vee i)\lhd$
$[(, \vee] = \lfloor$	**OP-SHIFT**	$\rhd S \wedge (S\underline{\vee}\Diamond i)\lhd$
$[\vee, i] = \lfloor$	**OP-SHIFT**	$\rhd S \wedge (S \vee \underline{i}\Diamond)\lhd$
$[i,)] = \rfloor$	**OP-REDUCE**(4)	$\rhd S \wedge (S\underline{\vee}S\Diamond)\lhd$
$[\vee,)] = \rfloor$	**OP-REDUCE**(1)	$\rhd S \wedge \underline{(}S\Diamond)\lhd$
$[(,)] = \Box$	**OP-SHIFT**	$\rhd S \wedge (S\underline{)}\Diamond\lhd$
$[), \lhd] = \rfloor$	**OP-REDUCE**(3)	$\rhd S\underline{\wedge}S\Diamond\lhd$
$[\wedge, \lhd] = \rfloor$	**OP-REDUCE**(2)	$\underline{\rhd}S\Diamond\lhd$
$[\rhd, \lhd] =$ **OK**	**SUCCESS**	

Construction of operator-precedence parsing table

The parsing table OP_E (see Table 11.9) can be easily constructed by using common sense and elementary mathematical rules concerning the precedence and associativity of operators occurring in the expressions generated by E. This construction thus assumes that for every pair of two operators, their mutual precedence is stated, and in addition, for every operator, it is specified whether it is left-associative or right-associative.

Basic idea. Mathematically, $\lfloor$ and $\rfloor$ can be viewed as two binary relations over Δ, defined as follows. For any pair of operators a and b, a floor b means that a has a lower precedence than b, so a handle containing a is reduced after a handle containing b. Regarding the other relation, $a\rfloor b$ says that a has a precedence before b, meaning that a handle containing a is reduced before a handle containing b. To obtain the complete definitions of $\lfloor$ and $\rfloor$, perform (I)–(IV), in which $a \in T_E \cup \{\rhd\}$ and $b \in T_E \cup \{\lhd\}$:

(I) If a and b are operators such that a has a higher mathematical precedence than b, then $a\rfloor b$ and $b\lfloor a$.

(II) If a and b are left-associative operators of the same precedence, then $a\rfloor b$ and $b\rfloor a$. If a and b are right-associative operators of the same precedence, then $a\lfloor b$ and $b\lfloor a$.

(III) In T_E, consider i, which represents an identifier and occurs on the right-hand side of rule $4 : S \rightarrow i$ in P_E. If a is a terminal that can legally precede operand i, then $a\lfloor i$, and if a can legally follow i, then $i\rfloor a$.

(IV) If a is a terminal that can legally precede (, then $a\lfloor($. If a can legally follow (, then $(\lfloor a$. Similarly, if a can legally precede), then $a\rfloor)$, and if a can legally follow), then $)\rfloor a$.

Following (I)–(IV), we now construct OP_E by using the two equivalences

$$OP_E[a, b] = \lfloor \text{ iff } a\lfloor b, \text{ and } OP_E[a, b] = \rfloor \text{ iff } a\rfloor b,$$

for all $a \in T_E \cup \{\triangleright\}$ and all $b \in T_E \cup \{\triangleleft\}$. The other entries are blank. All these blank entries signalize syntax errors, except $OP_E[(,)]$; if the parser occurs in $OP_E[(,)]$, it shifts) onto *pd* in order to perform **OP-REDUCE** according to $3 : S \rightarrow (S)$ right after this shift.

Example 11.9. In E, suppose that $\vee$ and $\wedge$ satisfy the standard mathematical precedence and associative rules. That is, $\wedge$ has a precedence before $\vee$, and both operators are left-associative. From (I) above, as $\wedge$ has a precedence before $\vee$, $\wedge\rfloor\vee$ and $\vee\lfloor\wedge$. From (II), since $\wedge$ is left-associative, define $\wedge\rfloor\wedge$. Regarding i, as $\vee$ can legally precede i, define $\vee\lfloor i$ according to (III). Considering the parentheses, (IV) implies $\wedge\lfloor($. Complete the definitions of $\lfloor$ and $\rfloor$ and use them to construct OP_E (see Table 11.9). ■

Operator-precedence parsers for other expressions

Until now, we have based the explanation of operator-precedence parsing strictly on E. Of course, apart from the logical expressions generated by E, this parsing method elegantly handles other expressions. Next, we explain how this method handles unary operators. Then, we sketch how to adapt it for arithmetic expressions. Finally, we point out that it works well with both ambiguous and unambiguous grammars.

Table 11.11. Operator-precedence table with unary operator $\neg$.

	$\neg$	$\wedge$	$\vee$	i	(	)	$\lhd$
$\neg$	$\lfloor$	$\rfloor$	$\rfloor$	$\lfloor$	$\lfloor$	$\rfloor$	$\rfloor$
$\wedge$	$\lfloor$	$\rfloor$	$\rfloor$	$\lfloor$	$\lfloor$	$\rfloor$	$\rfloor$
$\vee$	$\lfloor$	$\lfloor$	$\rfloor$	$\lfloor$	$\lfloor$	$\rfloor$	$\rfloor$
i	$\rfloor$	$\rfloor$	$\rfloor$			$\rfloor$	$\rfloor$
(	$\lfloor$	$\lfloor$	$\lfloor$	$\lfloor$	$\lfloor$		
)		$\rfloor$	$\rfloor$			$\rfloor$	$\rfloor$
$\rhd$	$\lfloor$	$\lfloor$	$\lfloor$	$\lfloor$	$\lfloor$		**OK**

Unary operators. Consider $\neg$ as a unary operator that denotes a logical negation. To incorporate this operator, we extend the CFGE by adding a rule of the form $S \to \neg S$ to obtain the CFG defined as

$$S \to \neg S,\ S \to S \vee S,\ S \to S \wedge S,\ S \to (S),\ S \to i.$$

Assume that $\neg$ satisfies the standard precedence and associative rules used in logic. That is, $\neg$ is a right-associative operator having a higher precedence than $\wedge$ and $\vee$. Return to rules (I)–(IV) of the preceding Example 11.9 in this section. By using these rules, we easily obtain the table that includes this unary operator (see Table 11.11).

Arithmetic Expressions with Right-Associative Operators. Consider the CFG

$$S \to S{+}S, S \to SS, S \to S*S, S \to S/S, S \to S \uparrow S, S \to (S), S \to i$$

in which the operators have the standard meaning and satisfy the common arithmetic precedence and associative rules ($\uparrow$ denotes the operator of exponentiation). That is, $\uparrow$ has a precedence before * and /, which have a precedence before + and −. The exponentiation operator $\uparrow$ is right-associative, while the others are left-associative. The precedence table for this grammar is straightforwardly made by construction rules (I)–(IV) (see Table 11.12).

Expressions involving relational operators are handled analogously, so we leave their discussion as an exercise.

Table 11.12. Arithmetic operator-precedence table.

	$\uparrow$	$*$	$/$	$+$	$-$	i	$($	$)$	$\lhd$
$\uparrow$	$\lfloor$	$\rfloor$	$\rfloor$	$\rfloor$	$\rfloor$	$\lfloor$	$\lfloor$	$\rfloor$	$\rfloor$
$*$	$\lfloor$	$\rfloor$	$\rfloor$	$\rfloor$	$\rfloor$	$\lfloor$	$\lfloor$	$\rfloor$	$\rfloor$
$/$	$\lfloor$	$\rfloor$	$\rfloor$	$\rfloor$	$\rfloor$	$\lfloor$	$\lfloor$	$\rfloor$	$\rfloor$
$+$	$\lfloor$	$\lfloor$	$\lfloor$	$\rfloor$	$\rfloor$	$\lfloor$	$\lfloor$	$\rfloor$	$\rfloor$
$-$	$\lfloor$	$\lfloor$	$\lfloor$	$\rfloor$	$\rfloor$	$\lfloor$	$\lfloor$	$\rfloor$	$\rfloor$
i	$\rfloor$	$\rfloor$	$\rfloor$	$\rfloor$	$\rfloor$			$\rfloor$	$\rfloor$
$($	$\lfloor$	$\lfloor$	$\lfloor$	$\lfloor$	$\lfloor$	$\lfloor$	$\lfloor$		
$)$	$\rfloor$	$\rfloor$	$\rfloor$	$\rfloor$	$\rfloor$			$\rfloor$	$\rfloor$
$\rhd$	$\lfloor$	$\lfloor$	$\lfloor$	$\lfloor$	$\lfloor$	$\lfloor$	$\lfloor$		**OK**

Ambiguity. As opposed to most top-down parsers, such as the predictive parsers (see Section 11.4), the precedence parsers work with ambiguous grammars without any problems. In fact, all the previous precedence parsers discussed in this section are based on ambiguous grammars, such as E. As is obvious, these parsers can be based on unambiguous grammars too. To illustrate, consider the unambiguous grammar H (see Convention 11.6), defined as

$$S \to S \vee A, S \to A, A \to A \wedge B, A \to B, B \to (S), B \to i.$$

Suppose that all the operators satisfy the same precedence and associative rules as in the equivalent ambiguous grammar above. As an exercise, by rules (I)–(IV), construct OP_H, and observe that this table coincides with the table given in Table 11.9.

On the one hand, as demonstrated earlier in this section, operator-precedence parsers work nicely for the CFGs that generate expressions, even if these grammars are ambiguous. On the other hand, they place several strict restrictions on the CFGs they are based on; perhaps most significantly, they exclude any involvement of ε-rules or rules having the same right-hand side but different left-hand sides. As a result, in practice, these parsers are frequently used in combination with predictive parsers discussed in Section 11.4. Combined in this way, the precedence parsers handle the syntax of expressions, while the other parsers handle the rest. Alternatively, bottom-up parsers are designed as LR parsers, discussed as follows.

LR parsing

This section discusses the *LR parsers* (L stands for the *l*eft-to-right scan of symbols, and R stands for the *r*ightmost derivation, which the bottom-up parsers construct in reverse, as already explained in Section 11.2). LR parsers are based on LR tables constructed from LR CFGs, i.e., the CFGs for which LR tables can be constructed; let us point out that there are non-LR CFGs for which these tables cannot be built up.

In practice, LR parsers are among the most popular parsers for their several indisputable advantages. First, they work fast. Furthermore, they easily and elegantly handle syntax errors because they never shift an erroneous input symbol onto the pushdown, and this property obviously simplifies the error recovery process. Most importantly, out of all deterministic parsers, they are ultimately powerful because *LR CFGs* generate the language family coinciding with $\mathbf{L_{dPDA\text{-}f}}$, i.e., the language family accepted by deterministic pushdown automata by final state (see Definition 4.3 and Corollary 4.1).

In this section, we first describe the fundamental LR parsing algorithm. Then, we explain how to construct the LR tables, which the algorithm makes use of.

LR parsing algorithm

Consider an LR grammar, $G = (V, T, P, S)$. Its *G-based LR table* consists of the *G-based action part* and the *G-based goto part*, denoted by $action_G$ and $goto_G$, respectively. Both parts have their rows denoted by members of the set $\Theta_G = \{\theta_1, \ldots, \theta_m\}$, whose construction is described later in this section. The columns of $action_G$ and $goto_G$ are denoted by the symbols of T and N, respectively; recall that N and T denote G's alphabets of terminals and nonterminals, respectively. For each $\theta_j \in \Theta_G$ and $t_i \in T \cup \{\triangleleft\}$, $action_G[\theta_j, t_i]$ entry is either a member of $\Theta_G \cup P \cup \{\mathbf{OK}\}$ or a blank entry (see Table 11.13). Frequently, the rules of P are labeled throughout this section, and instead of the rules themselves, only their labels are written in $action_G$ for brevity. For each $\theta_j \in \Theta_G$ and $A_i \in N_G$, $goto_G[\theta_j, A_i]$ is either a member of Θ_G or a blank entry (see Table 11.14).

Table 11.13. *action*.

	t_1	$\ldots$	t_i	$\ldots$	t_n
θ_1					
$\vdots$					
θ_j			$action[\theta_j, t_i]$		
$\vdots$					
θ_m					

Table 11.14. *goto*.

	A_1	$\ldots$	A_i	$\ldots$	A_n
θ_1					
$\vdots$					
θ_j			$goto[\theta_j, A_i]$		
$\vdots$					
θ_m					

Convention 11.9. As usual, whenever there is no danger of confusion, we omit G in the denotation above, so we simplify $\Theta_G, action_G, goto_G$, and Δ_G to $\Theta, action, goto$, and Δ, respectively. ■

Basic idea. Like an operator-precedence parser (see Algorithm 11.8), an LR parser scans w from left to right, and during this scan, it makes shifts and reductions. If $w \in L(G)$, it accepts w; otherwise, it rejects w. During the parsing process, every configuration of the parser is of the form $\triangleright q_0 Y_1 q_1 \cdots Y_{m-1} q_{m-1} Y_m q_m \Diamond v \triangleleft$, where the qs and the Ys are in Θ_G and V_G, respectively, and $v \in \mathit{suffixes}(w)$. Recall that according to Definition 4.1, the pushdown is written in the right-to-left way, so q_m is the topmost *pd* symbol, Y_m occurs as the second *pd* symbol, and so on up to the *pd* bottom, $\triangleright$. As a result, a member of Θ_G always appears as the topmost *pd* symbol. The LR parser makes shifts and reductions in a specific LR way, though. Next, we describe the operations **LR-REDUCE** and **LR-SHIFT** that denote the actions by which the LR parser makes its reductions and shifts, respectively (as usual, in the definition of **LR-REDUCE** and **LR-SHIFT**, we make use of the *pd* and *ins* notations introduced in Convention 11.5).

Definition 11.7. Operations **LR-REDUCE** and **LR-SHIFT**. Let $G = (V, T, P, S)$, Θ, *action*, and *goto* have the same meaning as above. In a G-based LR parser, we use operations **LR-REDUCE** and **LR-SHIFT** defined as follows:

LR-REDUCE. If $p : A \to X_1X_2 \cdots X_n \in P, X_j \in V, 1 \leq j \leq n$, for some $n \geq 0$ ($n = 0$ means $X_1X_2 \cdots X_n = \varepsilon$), $o_0X_1o_1X_2o_2 \cdots o_{n-1}X_no_n$ occurs as the $(2n + 1)$-symbol *pd* top, i.e., the current configuration of the parser is of the form $\triangleright \cdots o_0X_1o_1X_2o_2 \cdots o_{n-1}X_no_n\Diamond \cdots \triangleleft$, with o_n as the topmost *pd* symbol, where $o_k \in \Theta, 0 \leq k \leq n$, then **LR-REDUCE**$(p)$ replaces $X_1o_1X_2o_2 \cdots o_{n-1}X_no_n$ with Ah on the pushdown top, where $h \in \Theta_G$ is defined as $h = goto_G[o_0, A]$; otherwise, **ERROR**.

LR-SHIFT. Let t and q denote *ins* and the *pd* top symbol, respectively. Let $action[q, t] = o$, where $o \in \Theta_G$. At this point, **LR-SHIFT** extends *pd* by to, so o is the topmost pushdown symbol after this extension. In addition, it sets *ins* to the input symbol that follows t and, thereby, advances to the next input symbol. ■

Note that the following LR parser has always its *pd* top symbol from Θ_G, as follows from Definition 11.7.

Algorithm 11.9. LR Parser.

Input: An LR grammar $G = (V, T, P, S)$, an input string, w, with $w \in T_G^*$, and a G-based LR table consisting of *action* and *goto*.
Output: **SUCCESS** if $w \in L(G)$, or **ERROR** if $w \notin L(G)$.
Method:

```
begin
  set pd = ▷θ1
  repeat
    let q denote the pd topmost symbol
    case action[q, ins] of
      action[q, ins] ∈ Θ_G:
        LR-SHIFT
```

$action[q, ins] \in P$:
    LR-REDUCE(r) with $r = action[q, ins]$
$action[q, ins] = \square$:
    ERROR
$action[q, ins] =$ **OK**:
    SUCCESS
until **SUCCESS** or **ERROR**
end

To obtain the right parse of w, extend Algorithm 11.9 by writing out r whenever **LR-REDUCE** (r) occurs, where $r \in P$. This straightforward extension is left as an exercise.

Example 11.10. Consider the CFG H (see Convention 11.6), whose rules are

$$1 : S \to S \vee A, \quad 2 : S \to A, \quad 3 : A \to A \wedge B, \quad 4 : A \to B,$$
$$5 : B \to (S), \quad 6 : B \to i,$$

where S is the start symbol. This grammar has its two-part LR table, consisting of *action* and *goto*, depicted in Tables 11.15 and 11.16, respectively. Both *action* and *goto* have their rows denoted by the members of $\Theta_H = \{\theta_1, \theta_2, \ldots, \theta_{12}\}$. The columns of *action* are denoted by terminals $\wedge, \vee, (,), i$, and $\lhd$. The columns of *goto* are denoted by nonterminals S, A, and B.

Table 11.15. $action_H$.

	$\wedge$	$\vee$	i	(	)	$\lhd$
θ_1			θ_6	θ_5		
θ_2		θ_7				**OK**
θ_3	θ_8	2			2	2
θ_4	4	4			4	4
θ_5			θ_6	θ_5		
θ_6	6	6			6	6
θ_7			θ_6	θ_5		
θ_8			θ_6	θ_5		
θ_9		θ_7			θ_{12}	
θ_{10}	θ_8	1			1	1
θ_{11}	3	3			3	3
θ_{12}	5	5			5	5

Table 11.16. $goto_H$.

	S	A	B
θ_1	θ_2	θ_3	θ_4
θ_2			
θ_3			
θ_4			
θ_5	θ_9	θ_3	θ_4
θ_6			
θ_7		θ_{10}	θ_4
θ_8			θ_{11}
θ_9			
θ_{10}			
θ_{11}			
θ_{12}			

Table 11.17. LR parsing.

Configuration	Table Entry	Parsing Action
$\rhd\theta_1\lozenge i \wedge i \vee i\lhd$	$action[\theta_1, i] = \theta_6$	**LR-SHIFT**(i)
$\rhd\theta_1 i\theta_6\lozenge \wedge i \vee i\lhd$	$action[\theta_6, \wedge] = 6, goto[\theta_1, B] = \theta_4$	**LR-REDUCE**(6)
$\rhd\theta_1 B\theta_4\lozenge \wedge i \vee i\lhd$	$action[\theta_4, \wedge] = 4, goto[\theta_1, A] = \theta_3$	**LR-REDUCE**(4)
$\rhd\theta_1 A\theta_3\lozenge \wedge i \vee i\lhd$	$action[\theta_3, \wedge] = \theta_8$	**LR-SHIFT**($\wedge$)
$\rhd\theta_1 A\theta_3 \wedge \theta_8\lozenge i \vee i\lhd$	$action[\theta_8, i] = \theta_6$	**LR-SHIFT**(i)
$\rhd\theta_1 A\theta_3 \wedge \theta_8 i\theta_6\lozenge \vee i\lhd$	$action[\theta_6, \vee] = 6, goto[\theta_8, B] = \theta_{11}$	**LR-REDUCE**(6)
$\rhd\theta_1 A\theta_3 \wedge \theta_8 B\theta_{11}\lozenge \vee i\lhd$	$action[\theta_{11}, \vee] = 3, goto[\theta_1, A] = \theta_3$	**LR-REDUCE**(3)
$\rhd\theta_1 A\theta_3\lozenge \vee i\lhd$	$action[\theta_3, \vee] = 2, goto[\theta_1, S] = \theta_2$	**LR-REDUCE**(2)
$\rhd\theta_1 S\theta_2\lozenge \vee i\lhd$	$action[\theta_2, \vee] = \theta_7$	**LR-SHIFT**($\vee$)
$\rhd\theta_1 S\theta_2 \vee \theta_7\lozenge i\lhd$	$action[\theta_7, i] = \theta_6$	**LR-SHIFT**(i)
$\rhd\theta_1 S\theta_2 \vee \theta_7 i\theta_6\lozenge\lhd$	$action[\theta_6, \lhd] = 6, goto[\theta_7, B] = \theta_4$	**LR-REDUCE**(6)
$\rhd\theta_1 S\theta_2 \vee \theta_7 B\theta_4\lozenge\lhd$	$action[\theta_4, \lhd] = 4, goto[\theta_7, A] = \theta_{10}$	**LR-REDUCE**(4)
$\rhd\theta_1 S\theta_2 \vee \theta_7 A\theta_{10}\lozenge\lhd$	$action[\theta_{10}, \lhd] = 1, goto[\theta_1, S] = \theta_2$	**LR-REDUCE**(1)
$\rhd\theta_1 S\theta_2\lhd$	$action[\theta_2, \lhd] =$ **OK**	**SUCCESS**

With $i \wedge i \vee i \in L(H)$ as the expression, Algorithm 11.9 works as described in Table 11.17. The algorithm makes the successful parsing of $i \wedge i \vee i$ by the sequence of configurations given in the first column of this table. The second column gives the relevant entries of *action* and *goto*, and the third column specifies the actions made by the algorithm. Note that *goto* is relevant only when a reduction is performed; regarding a shift, it is not needed at all.

By writing out the rules according to which the LR parser makes each reduction, we obtain 64632641 as the right parse of $i \wedge i \vee i$ in H. ■

Construction of LR table

The parsing theory has developed many sophisticated methods for constructing the LR tables. Most of them are too complicated to include in this introductory text, however. Therefore, we just restrict our explanation of this construction to the fundamental ideas underlying a *simple LR table construction.* As its name indicates, it is simpler than the other constructions of LR tables, and this aspect represents its principal advantage. On the other hand, there exist LR CFGs for which this construction does not work. In other words, the LR parsers based on tables constructed in this way are slightly less powerful than the LR parsers based on tables produced by more complicated methods, which are equivalent to LR CFGs and, therefore, characterize $\mathbf{L_{dPDA\text{-}f}}$, as already pointed out at the beginning of this section. However, these complicated LR-table constructions are too complex to be included in this introductory text.

As a matter of fact, even if we restrict our attention to the simple LR-table construction, this construction still belongs to the most complicated topics of this introductory book. Therefore, we only give its gist while reducing the formalism concerning this construction as much as possible.

Items. Let G be an LR CFG. In every configuration, the *pd* top contains a handle prefix, which the G-based LR parser tries to extend so that a complete handle occurs on the *pd* top. As soon as the *pd* top contains a complete handle, the parser can make a reduction according to a rule r, with $rhs(r)$ equal to the handle. To express that a handle prefix appears as the *pd* top, we introduce an *item* of the form

$$A \rightarrow x|y,$$

for each rule $A \rightarrow z \in P_G$ and any two strings x and y such that $z = xy$. Intuitively, $A \rightarrow x|y$ means that if x occurs as the *pd* top and the parser makes a sequence of moves resulting in producing y right behind x on the *pd* top, then the parser gets z as the handle on the *pd* top, which can be reduced to A according to $A \to z$. An item of the form $A \to |z$ is called a start item, while an item of the form $A \to z|$ is called an end item.

Example 11.11. Reconsider the six-rule CFG H(see Convention 11.6), defined as

$$S \to S \vee A, S \to A, A \to A \wedge B, A \to B, B \to (S), B \to i.$$

From $S \to S \vee A$, we obtain the four items

$$S \to |S \vee A, S \to S| \vee A, S \to S \vee |A, \text{ and } S \to S \vee A|$$

in which $S \to |S \vee A$ and $S \to S \vee A|$ represent a start item and an end item, respectively. Consider $S \to S| \vee A$. In essence, this item says that if S currently appears as the *pd* top and a prefix of the input is reduced to $\vee A$, the parser obtains $S \vee A$ as the handle, which can be reduced to S by using $S \to S \vee A$.

Before going any further, note that several different items may be relevant to the same *pd* top string. To illustrate, take $S \to S \vee A|$ and $A \to A| \wedge B$. Note that both items have to be taken into account whenever the LR parser occurs in a configuration with the three-symbol string $S \vee A$ as the *pd* top. Consider, for instance, $\rhd S \vee A\Diamond\lhd$ and $\rhd S \vee A\Diamond \wedge i\lhd$. From $\rhd S \vee A\Diamond\lhd$, the H-based LR parser makes a reduction according to $S \to S \vee A$ and, thereby, successfully completes the parsing process. More formally, the parser performs

$$\rhd S \vee A\Diamond\lhd \quad \Rightarrow \quad \rhd S\Diamond \lhd.$$

In $\rhd S \vee A\Diamond \wedge i\lhd$, the parser has actually A as a prefix of the handle $A \wedge B$ on the pushdown. As a result, from $\rhd S \vee A\Diamond \wedge i\lhd$, it first makes several shifts and reductions before it obtains $A \wedge B$ as the handle on the pushdown top. Then, it reduces $A \wedge B$ to A according to $A \to A \wedge B$, after which it obtains $S \vee A$ as the handle, makes a reduction according to $S \to S \vee A$ and, thereby, completes the parsing process. To summarize this part of the parsing process formally, from $\rhd S \vee A\Diamond \wedge i\lhd$, the H-based LR parser computes

$$\begin{aligned} \rhd S \vee A\Diamond \wedge i\lhd \quad &\Rightarrow \quad \rhd S \vee A \wedge \Diamond i\lhd \\ &\Rightarrow \quad \rhd S \vee A \wedge i\Diamond\lhd \\ &\Rightarrow \quad \rhd S \vee A \wedge B\Diamond\lhd \end{aligned}$$

$$\Rightarrow \quad \rhd S \vee A\Diamond\lhd$$
$$\Rightarrow \quad \rhd S\Diamond \lhd .$$

■

Convention 11.10. Throughout the rest of this chapter, for a LR CFG G, I_G, I_G^{start}, and I_G^{end} denote the set of all its items, the set of start items, and the set of end items, respectively, so $I_G^{start} \subseteq I_G$ as well as $I_G^{end} \subseteq I_G$. Furthermore, $\Omega_G = power(I_G)$, i.e., Ω_G denotes the power set of I_G, defined as the set of all subsets of I_G (see Section 1.2). As usual, we omit the subscript G in this notation if no confusion exists; for instance, we often write I instead of I_G if G is understood. ■

As sketched in the conclusion of the previous example, several items are usually related to a single prefix of the right-hand side of some rules when the prefix occurs on the *pd* top in order to determine the next LR parsing step. Next, we construct the item sets corresponding to all prefixes of the right-hand sides of rules in P, and these sets are then used as members of Θ, so $\Theta \subseteq \Omega$. By using the members of Θ, we then construct the G-based LR table.

Construction of Θ. Initially, we change the start symbol S to a new start symbol Z in G, and add a dummy rule of the form $Z \to S$. As a result, we can be sure that in G, every derivation that generates a sentence in $L(G)$ starts by applying $Z \to S$. Apart from Θ, we introduce an auxiliary item set W. Initially, we set Θ and W to $\emptyset$ and $\{Z \to |S\}$, respectively. We repeat extensions (I) and (II), described in the following, until no new item set can be included in W in order to obtain all item sets in Θ. Let us note that during the computation of (I) and (II), Θ and W always represent subsets of Ω:

(I) Let $I \in W$. Suppose that u appears on the *pd* top, and let $A \to uBv \in P$, where $A, B \in \Delta$, and $u, v \in V^*$. Observe that if $A \to u|Bv \in I$ and $B \to |y \in I^{start}$, then by using $B \to y$, the G-based LR parser can reduce y to B, and this reduction does not affect u appearing on the *pd* top at all because $B \to |y$ is a start item. Thus, extend I by adding $B \to |y$ to it. Repeat this extension until I can no longer be extended in this way. Take the resulting I, and add it to Θ (if I was already there, Θ remains unchanged). To summarize this extension as a pseudocode, for $I \in W$, perform the following.

repeat
  if $A \to u|Bv \in I$ **and** $B \to z \in P$ **then**
    include $B \to |z$ into I
until no change
include I into Θ

(II) This extension is based on the relation $\circlearrowleft_G$ from $\Omega \times V$ to Ω, defined as follows. Intuitively, for $I \in \Omega$ and $X \in V, \circlearrowleft_G (I, X)$ specifies the set of items related to the configuration that M enters from I by pushing X on the *pd* top. Formally, for all $I \in \Omega$ and all $X \in V$,

$$\circlearrowleft_G (I, X) = \{A \to uX|v | A \to u|Xv \in I, A \in \Delta, u, v \in V^*\}.$$

Let $I \in W$ and $A \to uX|v \in I$, where $A \in \Delta, u, v \in V^*$, and $X \in V$. Consider a part of a rightmost derivation in G in reverse, during which a portion of the input string is reduced to X. Simulating this derivation part, the G-based LR parser actually obtains X on the *pd*. As a result, for every $I \in W$ and $X \in V$, the following **for loop** extends W by $\circlearrowleft_G (I, X)$ unless $\circlearrowleft_G (I, X)$ is empty.

for each $X \in V$ with $\circlearrowleft_G (I, X) \neq \emptyset$ **do**
  include $\circlearrowleft_G (I, X)$ into W

Based on (I) and (II), we next construct Θ.

Algorithm 11.10. Construction of Θ.

Input: An LR grammar $G = (V, T, P, S)$, extended by the dummy rule $Z \to S$, where Z is the new start symbol.
Output: Θ
Note: Apart from Θ, an auxiliary set $W \subseteq \Omega$ is used.
Method:

begin
  set $W = \{\{Z \to |S\}\}$
  set $\Theta = \emptyset$
  repeat

```
    for each I ∈ W do
      repeat
        if A → u|Bv ∈ I  and  B → z ∈ P then
          include B → |z into I
      until no change
      include I into Θ
    for each X ∈ V with ↻_G (I, X) ≠ ∅ do
      include ↻_G (I, X) into W
  until  no change
end
```

Example 11.12. Consider again H (see Convention 11.6). Add a dummy rule of the form $Z \to S$ to its rules, and define Z as the start symbol. The resulting LR CFG is defined as

$$Z \to S, S \to S \vee A, S \to A, A \to A \wedge B, A \to B, B \to (S), B \to i.$$

Next, apply Algorithm 11.10 with H as its input. At the beginning, set $\Theta_H = \emptyset$ and $W = \{\{Z \to |S\}\}$. By extending I, the algorithm extends $\{Z \to |S\} \in W_H$ to

$$\{Z \to |S, S \to |S \vee A, S \to |A, A \to |A \wedge B, A \to |B, B \to |(S), B \to |i\}.$$

Include this item set in Θ_H. Note that this new item set I contains $Z \to |S, S \to |S \vee A$, and for $I = \{Z \to |S, S \to |S \vee A\}$, we have $\circlearrowleft_H (I, S) = \{Z \to S|, S \to S| \vee A\}$. Thus, by performing extension (II), the algorithm includes $\{Z \to S|, S \to S|\vee A\}$ in $_HW$, after which it performs the second iteration of (I) and (II), and so on. Continuing in this way, this algorithm eventually produces the 12 item sets listed in the second column of Table 11.18. For brevity, these 12 item sets are referred to as θ_1–θ_{12} according to the first column of this table. ■

Construction of LR Table. Making use of Θ, we construct the *action* and *goto* parts of LR table by performing (I)–(III), given as follows, in which we automatically suppose that θ_i and θ_j belong to Θ. Concerning (I) and (II), given as follows, it is important to realize that for all $\theta_i \in \Omega$ and all $X \in V$, $\circlearrowleft_G (\theta_i, X)$ and I^{start} (see Convention 11.10) are necessarily disjoint.

Table 11.18. Θ_H.

State	Item Sets
θ_1	$\{Z \to \|S, S \to \|S \vee A, S \to \|A, A \to \|A \wedge B, A \to \|B, B \to \|(S), B \to \|i\}$
θ_2	$\{Z \to S\|, S \to S\| \vee A\}$
θ_3	$\{S \to A\|, A \to A\| \wedge B\}$
θ_4	$\{A \to B\|\}$
θ_5	$\{B \to (\|S), S \to \|S \vee A, S \to \|A, A \to \|A \wedge B, A \to \|B, B \to \|(S), B \to \|i\}$
θ_6	$\{B \to i\|\}$
θ_7	$\{S \to S \vee \|A, A \to \|A \wedge B, A \to \|B, B \to \|(S), B \to \|i\}$
θ_8	$\{A \to A \wedge \|B, B \to \|(S), B \to \|i\}$
θ_9	$\{B \to (S\|), S \to S\| \vee A\}$
θ_{10}	$\{S \to S \vee A\|, A \to A\| \wedge B\}$
θ_{11}	$\{A \to A \wedge B\|\}$
θ_{12}	$\{B \to (S)\|\}$

(I) To explain the construction of *goto*, consider item $A \to u|Bv \in I$, where $I \in \Theta, A, B \in N$ and $u, v \in V^*$. At this point, after reducing a portion of the input string to B, M actually extends the prefix u of the handle by B, so uB occurs on the *pd* top, which leads to the following extension.

if $\circlearrowleft_G (\theta_i, B) = \theta_j - I^{start}$, where $B \in N$ **then**
$goto[\theta_i, B] = \Theta_j$

(II) By analogy with the construction of the *goto* entries in (I), we obtain the *action* shift entries in the following way.

if $\circlearrowleft_G (\theta_i, b) = \theta_j - I^{start}$, where $b \in T$ **then**
$action[\theta_i, b] = \Theta_j$

(III) To explain the construction of the reduction entries in *action*, consider a rule $r : A \to u \in P$ and $A \to u| \in I^{end}$ (see Convention 11.10), which means that a complete handle u occurs on the pushdown top. At this point, the parser reduces u to A according to this rule, provided that after this reduction, A is followed by a terminal a that may legally occur after A in a sentential form. As a result, we obtain the following.

if $A \to u| \in \theta_i, a \in follow(A), r : A \to u \in P$ **then**
$action[\theta_i, a] = r$

(see Definition 11.2 and Algorithm 11.6 for the definition and construction of $follow$, respectively).

Recall that G starts every derivation by applying $0 : Z \to S$, and the LR parser works so that it simulates the rightmost derivations in G in reverse. Furthermore, note that when the input symbol equals $\lhd$, all the input has been read by the parser. Consequently, if $Z \to S| \in \theta_i$, we set $action[\theta_i, \lhd] =$ **OK** to signalize that the parsing process has been successfully completed.

Therefore,

if $Z \to S| \in \theta_i$ **then**
  $action[\theta_i, \lhd] =$ **OK**

Algorithm 11.11. LR Table.

Input: An LR CFG $G = (V, T, P, S)$, in which Z and $0 : Z \to S$ have the same meaning as in Algorithm 11.10, and Θ, which is constructed by Algorithm 11.10.
Output: A G-based LR table, consisting of the *action* and *goto* parts.
Note: We suppose that $A, B \in N, b \in T$, and $u, v \in V^*$ in this algorithm.
Method:

```
begin
  denote the rows of action and goto with the members of Θ
  denote the columns of action and goto with the members of
T and N
  repeat
    for all θi, θj ∈ Θ do
      if ↻G (θi, B) = θj − I^start, where B ∈ N then
        goto[θi, B] = θj
      if ↻G (θi, b) = θj − I^start, where b ∈ T then
        action[θi, b] = θj
      if A → u| ∈ θi ∩ I^end, a ∈ follow(A), j : A → u ∈ PG then
        action[θi, a] = j
  until no change
  if Z → S| ∈ θi then
    action[θi, ◁] = OK
  all the other entries remain blank and, thereby, signalize a syntax
  error
end
```

Example 11.13. Consider the same grammar as in Example 11.12. Number its rules to obtain

$$0: Z \to S, 1: S \to S \vee A, 2: S \to A, 3: A \to A \wedge B, 4: A \to B,$$
$$5: B \to (S), 6: B \to i.$$

Consider $\Theta_H = \{\theta_1, \theta_2, \ldots, \theta_{12}\}$, obtained in Example 11.12 (see Table 11.18). Denote the rows of *action* and *goto* with the members of Θ_H. Denote the columns of *action* and *goto* with the members of $\{\vee, \wedge, (,), i, \lhd\}$ and $\{S, A, B\}$, respectively.

According to the first **if** statement in Algorithm 11.11, $goto[\theta_1, S] = \theta_2$ because $S \to |S \vee A \in \theta_1$ and $S \to S| \vee A \in \theta_2$. By the second if statement, $action[\theta_2, \vee] = \theta_7$ because $S \to S| \vee A \in \theta_2$ and $S \to S \vee |A \in \theta_7$. By the third **if** statement, $action[\theta_3, \vee] = 2$ because $2: S \to A| \in \theta_3$ and $\vee \in follow(A)$. As an exercise, perform a complete execution of the **repeat** loop, containing the three **if** statements. After this, according to the conclusion of Algorithm 11.11, set $action[\theta_2, \lhd] =$ **OK** because θ_2 contains $Z \to S|$. The resulting table produced by this algorithm is given in Tables 11.15 and 11.16. ■

Out of the existing constructions of LR tables, the one described in this section belongs to the simplest methods of obtaining LR tables. Unfortunately, there exist LR grammars for which this construction does not work. In fact, it breaks down even when some quite common grammatical phenomena occur. Specifically, it cannot handle *reduction-shift conflict*, whose decision requires figuring out whether the parser should perform a shift or a reduction when both actions are possible. Furthermore, if two or more different reductions can be made in the given configuration, it cannot decide which of them it should select. To give an insight into this latter *reduction-reduction conflict*, suppose that the same set in Θ_H contains two end items, $A \to u|$ and $B \to v|$, but $follow(A) \cap follow(B) \neq \emptyset$. At this point, Algorithm 11.11 would illegally place both rules, $A \to u$ and $B \to v$, into the same entry in the action part of the LR table. There exists a number of complicated constructions that resolve these conflicts. However, these constructions are too complex to be included in this introductory text, as already pointed out.

Chapter 12

Applications of Grammars

In practice, most applications of automata are supported by suitable grammatical models. For instance, as demonstrated in the previous chapter, context-free grammars are the most widely used specification tools for the syntactic structure of programming languages, whose parsers are based on pushdown automata (see Section 11.2). As specification tools, computer science makes use of grammars in a much larger variety of its application fields, ranging from artificial intelligence to text mining, speech recognition, and computational biology up to computer art. Since grammars represent the principal counterparts of automata and since they are so hugely used in practice, the current chapter sketches some more grammatical applications apart from parsing. By no means, however, is it intended to provide comprehensive or even exhaustive coverage of these applications. In fact, it narrows its coverage only to a single grammatical model based on context-free rules — scattered context grammars (see Section 6.3).

As follows from their concept, scattered context grammars are applicable in every scientific field that formalizes its results by strings in which there exist some scattered context dependencies spread over them. Since numerous scientific areas formalize and study their results by using strings involving dependencies of this kind, any attempt at sketching applications of scattered context grammars in all these areas would be unbearably sketchy and, therefore, inappropriate. Instead of an encyclopedic approach like this, we devote this two-section chapter only to two scientific fields: natural language processing (Section 12.1) and musicology (Section 12.2). While the former represents a well-known and classical application area of these grammars, the latter is less frequently considered in this way. The chapter pays special attention to the formalization of structures that can be properly and elegantly described by scattered context rules, although they are indescribable by ordinary context-free rules. To put it in terms of this book, these structures represent languages in $\mathbf{L}_{\mathbf{CSG}} - \mathbf{L}_{\mathbf{CFG}}$.

12.1 Natural Language Processing

As opposed to a rather theoretical approach to scattered context grammars in Section 6.3, we now describe some of their linguistically oriented applications, with a focus on natural language processing. We narrow our attention to the investigation of English syntax because the reader is surely familiar with it very well. We explore several common linguistic phenomena involving scattered context in this syntax and explain quite clearly how to handle them by scattered context grammars. More precisely, we use these grammars to describe and transform English sentences involving these phenomena so that all these sentences are syntactically correct in terms of the English language. From a purely linguistic viewpoint, it is worth pointing out that scattered context grammars can be similarly applied to most members of other language families, including Indo-European, Sino-Tibetan, Niger-Congo, Afro-Asiatic, Altaic, and Japonic families of languages.

Syntax and related linguistic terminology

In the linguistic study concerning English syntax, we discuss and describe the principles and rules according to which we correctly construct and transform grammatical English sentences. To give an insight into the discussion of English syntax, we open this section with some simple examples that illustrate how we connect the theoretically oriented discussion of scattered context grammars in the previous chapters with the application-oriented discussion of English syntax in the current chapter. Then, we introduce the basic terminology used in syntax-oriented linguistics.

Introduction through examples

Observe that many common English sentences contain expressions and words that mutually depend on each other, although they are not adjacent to each other in the sentences. For example, consider the following sentence:

He usually goes to work early.

The subject (*he*) and the predicator (*goes*) are related; sentences

**He usually go to work early.*

and

**I usually goes to work early.*

are ungrammatical because the form of the predicator depends on the form of the subject, according to which the combinations **he... go* and **I... goes* are illegal (throughout this chapter, * denotes ungrammatical sentences or their parts). Clearly, any change in the subject implies a corresponding change in the predicator as well. Linguistic dependencies of this kind can be easily and elegantly captured by scattered context grammars. Let us construct a scattered context grammar that contains this production:

$$(\text{He}, \text{goes}) \rightarrow (\text{We}, \text{go}).$$

This production checks whether the subject is the pronoun *he* and whether the verb *go* is in the third-person singular. If the sentence satisfies this property, it can be transformed into the grammatically correct sentence

We usually go to work early.

Observe that the related words may occur far away from each other in the sentence in question. In the above example, the word *usually* occurs between the subject and the predicator. While it is fairly easy to use context-sensitive grammars to model context dependencies where only one word occurs between the related words, note that the number of words appearing between the subject and the predicator can be virtually unlimited. We can say

He almost regularly goes to work early.

but also

He usually, but not always, goes to work early.

and many more grammatical sentences like this. To model these context dependencies by ordinary context-sensitive grammars, many auxiliary productions have to be introduced to send the information concerning the form of a word to another word, which may occur at the opposite end of the sentence. As opposed to this awkward and tedious description, the single scattered context production above is needed to perform the same job regardless of the number of words appearing between the subject and the predicator.

We next give another example that illustrates the advantage of scattered context grammars over classical context-sensitive grammars under some circumstances. Consider these two sentences:

John recommended it.

and

Did John recommend it?

There exists a relation between the basic clause and its interrogative counterpart. Indeed, we obtain the second, interrogative clause by adding *did* in front of *John* and by changing *recommended* to *recommend* while keeping the rest of the sentence unchanged. In terms of scattered context grammars, this transformation can be described by the scattered context production

$$(\text{John, recommended}) \rightarrow (\text{Did John, recommend});$$

clearly, when applied to the first sentence, this production performs exactly the same transformation as we have just described. Although this transformation is possible by using an ordinary context production, the inverse transformation is much more difficult to achieve.

The inverse transformation can be performed by a scattered context production

$$(\text{Did}, \text{recommend}) \rightarrow (\varepsilon, \text{recommended});$$

obviously, by erasing *did* and changing *recommend* to *recommended*, we obtain the first sentence from the second one. Again, instead of *John*, the subject may consist of a noun phrase containing several words, which makes it difficult to capture this context dependency by ordinary context-sensitive grammars.

Considering the examples above, the advantage of scattered context grammars is more than obvious: Scattered context grammars allow us to change only some words during the transformation while keeping the others unchanged. On the other hand, context-sensitive grammars are inconvenient to perform transformations of this kind. A typical context-sensitive grammar that performs this job usually needs many more context-sensitive productions by which it repeatedly traverses the transformed sentence in question just to change very few context-dependent words broadly spread across the sentence.

Terminology

Taking into account the intuitive insight given above, we see that there are structural rules and regularities underlying syntactically well-formed English sentences and their transformations. Although we have already used some common linguistic notions, such as subject or predicator, we now introduce this elementary linguistic terminology more systematically so we can express these English sentences in terms of their syntactic structure in a more exact and general way. However, we restrict this introduction only to the very basic linguistic notions, most of which are taken from Huddleston and Pullum (2002, 2005). In the following chapter, which closes this book, we recommend several additional excellent linguistic treatments closely related to the discussion of this chapter.

Throughout the rest of this section, we narrow our discussion primarily to verbs and personal pronouns, whose proper use depends on the context in which they occur. For instance, *is, are, was*, and *been* are different forms of the same verb *be*, and their proper use depends on the context in which they appear. We say that words in these categories *inflect* and call this property *inflection*. Verbs and personal pronouns often represent the key elements of a clause — the *subject* and the *predicate*. In simple clauses like

She loves him.

, we can understand the notion of the subject and the predicate so that some information is "predicated of" the subject (*she*) by the predicate (*loves him*). In more complicated clauses, the best way to determine the subject and the predicate is the examination of their syntactic properties (see Huddleston and Pullum, 2002, for more details). The predicate is formed by a *verb phrase* — the most important word of this phrase is the verb, also known as the *predicator*. In some verb phrases, there occur several verbs. For example, in the sentence

He has been working for hours.,

the verb phrase contains three verbs: *has, been*, and *working*. The predicator is, however, always the first verb of a verb phrase (*has* in the above example). In this study, we focus on the most elementary clauses — *canonical clauses*. In these clauses, the subject always precedes the predicate, and these clauses are positive, declarative, and without subordinate or coordinate clauses.

Next, we describe the basic categorization of verbs and personal pronouns, and further characterize their inflectional forms in greater detail.

Verbs

We distinguish several kinds of verbs based on their grammatical properties. The set of all verbs is divided into two subsets: the set of *auxiliary verbs* and the set of *lexical verbs*. Furthermore, the set of auxiliary verbs consists of *modal verbs* and *non-modal verbs*. The set of modal verbs includes the following verbs: *can, may, must, will, shall, ought, need, dare*; the verbs *be, have*, and *do* are non-modal. All the remaining verbs are lexical. In reality, the above-defined classes overlap in certain situations; for example, there are sentences where *do* appears as an auxiliary verb, and in different situations, *do* behaves as a lexical verb. For simplicity, we do not take into account these special cases in what follows.

Table 12.1. Paradigms of English verbs.

Form	Paradigm	Person	Example
Primary	Present	3rd sg	*She* walks *home.*
		Other	*They* walk *home.*
	Preterite		*She* walked *home.*
Secondary	Plain form		*They should* walk *home.*
	Gerund-participle		*She is* walking *home.*
	Past participle		*She has* walked *home.*

Inflectional forms of verbs are called *paradigms*. In English, every verb, except for the verb *be*, may appear in each of the six paradigms described in Table 12.1 (see Huddleston and Pullum, 2002). Verbs in *primary form* may occur as the only verb in a clause and form the head of its verb phrase (predicator); on the other hand, verbs in *secondary form* have to be accompanied by a verb in primary form.

The verb *be* has nine paradigms in its neutral form. All primary forms have, in addition, their negative contracted counterparts. Compared to other verbs, there is one more verb paradigm called *irrealis*. The irrealis form *were* (and *weren't*) is used in sentences of an unrealistic nature, such as

I wish I were rich.

All these paradigms are presented in Table 12.2.

Personal pronouns

Personal pronouns exhibit a great amount of inflectional variation as well. Table 12.3 summarizes all their inflectional forms. The most important for us is the class of pronouns in *nominative* because these pronouns often appear as the subject of a clause.

Transformational scattered context grammars

As we have already mentioned, in this chapter, we primarily apply scattered context grammars to transform grammatical English sentences into other grammatical English sentences. To do so, we next

Table 12.2. Paradigms of the verb *be*.

Form	Paradigm	Person	Neutral	Negative
Primary	Present	1st sg	*am*	*aren't*
		3rd sg	*is*	*isn't*
		Other	*are*	*aren't*
	Preterite	1st sg, 3rd sg	*was*	*wasn't*
		Other	*were*	*weren't*
Secondary	Irrealis	1st sg, 3rd sg	*were*	*weren't*
	Plain form		*be*	—
	Gerund-participle		*being*	—
	Past participle		*been*	—

Table 12.3. Personal pronouns.

Non-reflexive				
Nominative	Accusative	Genitive		
Plain		Dependent	Independent	Reflexive
I	*me*	*my*	*mine*	*myself*
you	*you*	*your*	*yours*	*yourself*
he	*him*	*his*	*his*	*himself*
she	*her*	*her*	*hers*	*herself*
it	*it*	*its*	*its*	*itself*
we	*us*	*our*	*ours*	*ourselves*
you	*you*	*your*	*yours*	*yourselves*
they	*them*	*their*	*theirs*	*themselves*

slightly modify scattered context grammars so that they start their derivations from a language rather than a single start symbol. Even more importantly, these grammars define transformations of languages, not just their generation.

Definition 12.1. A *transformational scattered context grammar* is a quadruple

$$G = (V, T, P, I),$$

where

- V is the *total vocabulary*;
- $T \subset V$ is the set of terminals (or the *output vocabulary*);
- P is a finite set of scattered context productions;
- $I \subset V$ is the *input vocabulary*.

The derivation step is defined as in scattered context grammars (see Definition 6.26). The *transformation T that G defines from $K \subseteq I^*$* is denoted by $T(G, K)$ and defined as

$$T(G, K) = \{(x, y) \mid x \Rightarrow_G^* y, x \in K, y \in T^*\}.$$

If $(x, y) \in T(G, K)$, we say that *x is transformed to y* by G; x and y are called the *input* and *output sentences*, respectively. ■

As already pointed out, while scattered context grammars generate strings, transformational scattered context grammars translate them. In a sense, however, the language generated by any scattered context grammar $G = (V, T, P, S)$ can be expressed by using a transformational scattered context grammar $H = \big(V, T, P, \{S\}\big)$ as well. Observe that

$$L(G) = \Big\{y \mid (S, y) \in T\big(H, \{S\}\big)\Big\}.$$

Before we make use of transformational scattered context grammars in terms of English syntax in the following section, we give two examples to demonstrate the close relation of these grammars to the theoretically oriented studies given previously in this book. To link the theoretical discussions given in the previous chapters of this book to the present chapter, the first example presents a transformational scattered context grammar that works with purely abstract languages. In the second example, we discuss a transformational scattered context grammar that is somewhat more linguistically oriented.

Example 12.1. Define the transformational scattered context grammar as

$$G = (V, T, P, I),$$

where $V = \{A, B, C, a, b, c\}$, $T = \{a, b, c\}$, $I = \{A, B, C\}$, and

$$P = \{(A, B, C) \to (a, bb, c)\}.$$

For example, for the input sentence $AABBCC$,

$$AABBCC \Rightarrow_G aABbbcC \Rightarrow_G aabbbbcc.$$

Therefore, the input sentence $AABBCC \in I^*$ is transformed into the output sentence $aabbbbcc \in T^*$, and

$$(AABBCC, aabbbbcc) \in T(G, I^*).$$

If we restrict the input sentences to the language $L = \{A^n B^n C^n \mid n \geq 1\}$, we get

$$T(G, L) = \{(A^n B^n C^n, a^n b^{2n} c^n) \mid n \geq 1\},$$

so every $A^n B^n C^n$, where $n \geq 1$, is transformed into $a^n b^{2n} c^n$. ■

In the following example, we modify strings consisting of English letters by a transformational scattered context grammar, and in this way, we relate these grammars to lexically oriented linguistics, i.e., the area of linguistics that concentrates its study on vocabulary analysis and dictionary design.

Example 12.2. We demonstrate how to lexicographically order alphabetic strings and, simultaneously, convert them from their uppercase versions to their lowercase versions. More specifically, we describe a transformational scattered context grammar G that takes any alphabetic strings that consist of English uppercase letters enclosed in angle brackets, lexicographically orders the letters, and converts them into their corresponding lowercases. For instance, G transforms $\langle XXUY \rangle$ to $uxxy$.

More precisely, let J and T be alphabets of English uppercases and English lowercases, respectively. Let $\prec$ denote *lexical order* over J, i.e., $A \prec B \prec \cdots \prec Z$. Let h be the function that maps the uppercases to the corresponding lowercases, i.e., $h(A) = a$, $h(B) = b$, $\ldots$, $h(Z) = z$. Let i denote the inverse of h, so $i(a) = A$, $i(b) = B$, $\ldots$, $i(z) = Z$. Let $N = \{\hat{a} \mid a \in T\}$. We define the transformational scattered context grammar $G = (V, T, P, I)$, where T is defined as above, $I = J \cup \{\langle, \rangle\}$, $V = I \cup N \cup T$, and P is constructed as follows:

(1) For each A, $B \in I$, where $A \prec B$, add
$(B, A) \rightarrow (A, B)$ to P.

(2) For each $a \in T$, add
 $(\langle) \to (\hat{a})$ to P.
(3) For each $a \in T$ and $A \in J$, where $i(a) = A$, add
 $(\hat{a}, A) \to (a, \hat{a})$ to P.
(4) For each $a, b \in T$, where $i(a) \prec i(b)$, add
 $(\hat{a}) \to (\hat{b})$ to P.
(5) For each $a \in T$, add
 $(\hat{a}, \rangle) \to (\varepsilon, \varepsilon)$ to P.

Set $K = \{\langle\}J^*\{\rangle\}$. For instance, G transforms $\langle ORDER\rangle \in K$ to $deorr \in T^*$ as

$$\begin{aligned}\langle ORDER\rangle &\Rightarrow_G \langle OEDRR\rangle \Rightarrow_G \langle DEORR\rangle \\ &\Rightarrow_G \hat{d}DEORR\rangle \Rightarrow_G d\hat{d}EORR\rangle \Rightarrow_G d\hat{e}EORR\rangle \Rightarrow_G de\hat{e}ORR\rangle \\ &\Rightarrow_G de\hat{o}ORR\rangle \Rightarrow_G deo\hat{o}RR\rangle \Rightarrow_G deo\hat{r}RR\rangle \Rightarrow_G deor\hat{r}R\rangle \\ &\Rightarrow_G deorr\hat{r}\rangle \Rightarrow_G deorr,\end{aligned}$$

so $(\langle ORDER\rangle, deorr) \in T(G, K)$. Clearly, G can make the same transformation in many more ways; on the other hand, note that the set of all transformations of $\langle ORDER\rangle$ to $deorr$ is finite.

More formally, we claim that G transforms every $\langle A_1 \ldots A_n\rangle \in K$ into $b_1 \ldots b_n \in T^*$, for some $n \geq 0$, so that $i(b_1) \ldots i(b_n)$ represents a permutation of $A_1 \ldots A_n$, and for all $1 \leq j \leq n-1$, $i(b_j) \prec i(b_{j+1})$ (the case when $n = 0$ means that $A_1 \ldots A_n = b_1 \ldots b_n = \varepsilon$). To see why this claim holds, note that $T \cap I = \emptyset$, so every successful transformation of a string from K into a string from T^* is performed so that all symbols are rewritten during the computation. Through productions introduced in (1), G lexicographically orders the input uppercases. Through a production of the form $(\langle) \to (\hat{a})$ introduced in (2), G changes the leftmost symbol $\langle$ to $\hat{a}$. Through productions introduced in (3) and (4), G verifies that the alphabetic string is properly ordered and, simultaneously, converts its uppercase symbols into the corresponding lowercases in a strictly left-to-right one-by-one way. Observe that a production introduced in (2) is applied precisely once during every successful transformation because the left-to-right conversion necessitates its application, and on the other hand, no production can produce $\langle$. By a production from (5), G completes the transformation; note that if this completion is performed prematurely with some uppercases left, the transformation is necessarily unsuccessful because the uppercases cannot be turned to the corresponding lowercases. Based on these observations, it should be obvious that G performs the desired transformation. ■

Having illustrated the lexically oriented application, we devote the following section solely to the applications of transformational scattered context grammars in English syntax.

Scattered context in English syntax

In this section, we apply transformational scattered context grammars to English syntax. Before opening this topic, let us make an assumption regarding the set of all English words. We assume that this set, denoted by T, is finite and fixed. From a practical point of view, this is obviously a reasonable assumption because we all commonly use a finite and fixed vocabulary of words in everyday English (purely hypothetically, however, this may not be the case, as illustrated by the study that closes this section). Next, we subdivide this set into subsets with respect to the classification of verbs and pronouns described earlier in this chapter:

- T is the set of all words, including all their inflectional forms;
- $T_V \subset T$ is the set of all verbs, including all their inflectional forms;
- $T_{VA} \subset T_V$ is the set of all auxiliary verbs, including all their inflectional forms;
- $T_{Vpl} \subset T_V$ is the set of all verbs in plain form;
- $T_{PPn} \subset T$ is the set of all personal pronouns in the nominative.

To describe all possible paradigms of the verb $v \in T_{Vpl}$, we use the following notation:

- $\pi_{3rd}(v)$ is the verb v in third-person singular present;
- $\pi_{pres}(v)$ is the verb v in present (other than third-person singular);
- $\pi_{pret}(v)$ is the verb v in preterite.

There are several conventions we use throughout this section in order to simplify the presented case studies:

- We do not take into account capitalization and punctuation. Therefore, according to this convention,

He is your best friend.

and

he is your best friend

are equivalent.

- To make the following studies as simple and readable as possible, we expect every input sentence to be a canonical clause. In some examples, however, we make slight exceptions to this rule; for instance, sometimes we permit the input sentence to be negative. The first and last examples also demonstrate a simple type of coordinated canonical clauses.
- The input vocabulary is the set $I = \{\langle x \rangle \mid x \in T\}$, where T is the set of all English words, as stated above. As a result, every transformational scattered context grammar in this section takes an input sentence over I and transforms it into an output sentence over T. For instance, in the case of the declarative-to-interrogative transformation,

  $$\langle \text{he} \rangle \langle \text{is} \rangle \langle \text{your} \rangle \langle \text{best} \rangle \langle \text{friend} \rangle$$

  is transformed into

  is he your best friend

  As we have already mentioned, we omit punctuation and capitalization, so the above sentence corresponds to

  Is he your best friend?

Next, we conduct several studies that describe how to transform various kinds of grammatical sentences into other grammatical sentences by using transformational scattered context grammars.

Clauses with *neither* and *nor*

The first example shows how to use transformational scattered context grammars to negate clauses that contain the pair of the words *neither* and *nor*, such as

Neither Thomas nor his wife went to the party.

Clearly, the words *neither* and *nor* are related, but there is no explicit limit to the number of words appearing between them. The following transformational scattered context grammar G converts the above sentence to

Both Thomas and his wife went to the party.

In fact, the constructed grammar G is general enough to negate every grammatical clause that contains the pair of words *neither* and *nor*.

Set $G = (V, T, P, I)$, where $V = T \cup I$, and P is defined as follows:

$$P = \Big\{ (\langle \text{neither} \rangle, \langle \text{nor} \rangle) \to (\text{both}, \text{and}) \Big\} \\ \cup \Big\{ (\langle x \rangle) \to (x) \mid x \in T - \{\text{neither, nor}\} \Big\}.$$

For example, for the above sentence, the transformation can proceed in this way:

$$\begin{aligned} &\langle \text{neither} \rangle \langle \text{thomas} \rangle \langle \text{nor} \rangle \langle \text{his} \rangle \langle \text{wife} \rangle \langle \text{went} \rangle \langle \text{to} \rangle \langle \text{the} \rangle \langle \text{party} \rangle \\ \Rightarrow_G &\text{both } \langle \text{thomas} \rangle \text{ and } \langle \text{his} \rangle \langle \text{wife} \rangle \langle \text{went} \rangle \langle \text{to} \rangle \langle \text{the} \rangle \langle \text{party} \rangle \\ \Rightarrow_G &\text{both thomas and } \langle \text{his} \rangle \langle \text{wife} \rangle \langle \text{went} \rangle \langle \text{to} \rangle \langle \text{the} \rangle \langle \text{party} \rangle \\ \Rightarrow_G &\text{both thomas and his } \langle \text{wife} \rangle \langle \text{went} \rangle \langle \text{to} \rangle \langle \text{the} \rangle \langle \text{party} \rangle \\ \Rightarrow_G^5 &\text{both thomas and his wife went to the party.} \end{aligned}$$

The production

$$(\langle \text{neither} \rangle, \langle \text{nor} \rangle) \to (\text{both}, \text{and})$$

replaces *neither* and *nor* with *both* and *and*, respectively. Every other word $\langle w \rangle \in I$ is changed to $w \in T$. Therefore, if we denote all possible input sentences, described in the introduction of this example, by K, $T(G, K)$ represents the set of all negated sentences from K, and

$$\big(\langle \text{neither} \rangle \langle \text{thomas} \rangle \langle \text{nor} \rangle \langle \text{his} \rangle \langle \text{wife} \rangle \langle \text{went} \rangle \langle \text{to} \rangle \langle \text{the} \rangle \langle \text{party} \rangle, \\ \text{both thomas and his wife went to the party} \big) \in T(G, K).$$

Existential clauses

In English, clauses that indicate existence are called *existential*. These clauses are usually formed by the dummy subject *there*; for example,

There was a nurse present.

However, this dummy subject is not mandatory in all situations. For instance, the above example can be rephrased as

A nurse was present.

We construct a transformational scattered context grammar G that converts any canonical existential clause without the dummy subject *there* to an equivalent existential clause with *there*.

Set $G = (V, T, P, I)$, where $V = T \cup I \cup \{X\}$ (X is a new symbol such that $X \notin T \cup I$), and P is defined as follows:

$$\begin{aligned} P = \Big\{ & (\langle x\rangle, \langle \text{is}\rangle) \rightarrow (\text{there is } xX, \varepsilon), \\ & (\langle x\rangle, \langle \text{are}\rangle) \rightarrow (\text{there are } xX, \varepsilon), \\ & (\langle x\rangle, \langle \text{was}\rangle) \rightarrow (\text{there was } xX, \varepsilon), \\ & (\langle x\rangle, \langle \text{were}\rangle) \rightarrow (\text{there were } xX, \varepsilon) \mid x \in T \Big\} \\ \cup & \Big\{ (X, \langle x\rangle) \rightarrow (X, x) \mid x \in T \Big\} \\ \cup & \big\{ (X) \rightarrow (\varepsilon) \big\}. \end{aligned}$$

For the above sample sentence, we get the following derivation:

$$\begin{aligned} & \langle \text{a}\rangle\langle \text{nurse}\rangle\langle \text{was}\rangle\langle \text{present}\rangle \\ \Rightarrow_G & \text{there was a } X\langle \text{nurse}\rangle\langle \text{present}\rangle \\ \Rightarrow_G & \text{there was a } X \text{ nurse } \langle \text{present}\rangle \\ \Rightarrow_G & \text{there was a } X \text{ nurse present} \\ \Rightarrow_G & \text{there was a nurse present.} \end{aligned}$$

A production from the first set has to be applied first because initially there is no symbol X in the sentential form, and all other productions require X to be present in the sentential form. In our case, the production

$$(\langle \text{a}\rangle, \langle \text{was}\rangle) \rightarrow (\text{there was a } X, \varepsilon)$$

is applied; the use of other productions from this set depends on what tense is used in the input sentence and whether the subject is singular or plural. The production non-deterministically selects the first word of the sentence, puts *there was* in front of it, and the symbol X behind it; in addition, it erases *was* in the middle of the sentence. Next, all words $\langle w \rangle \in I$ are replaced with $w \in T$ by productions from the second set. These productions also verify that the previous non-deterministic selection was made at the beginning of the sentence; if not, there remains a word $\langle w \rangle \in I$ in front of X that cannot be rewritten. Finally, the derivation ends by erasing X from the sentential form.

This form of the derivation implies that if we denote the input existential clauses described in the introduction of this example by K, $T(G, K)$ represents the set of these clauses with the dummy subject *there*. As a result,

$$(\langle \text{a} \rangle \langle \text{nurse} \rangle \langle \text{was} \rangle \langle \text{present} \rangle, \text{there was a nurse present}) \in T(G, K).$$

Interrogative clauses

In English, there are two ways of transforming declarative clauses into interrogative clauses, depending on the predicator. If the predicator is an auxiliary verb, the interrogative clause is formed simply by swapping the subject and the predicator. For example, we get the interrogative clause

Is he mowing the lawn?

by swapping *he*, which is the subject, and *is*, which is the predicator, in

He is mowing the lawn.

On the other hand, if the predicator is a lexical verb, the interrogative clause is formed by adding the dummy *do* to the beginning of the declarative clause. The dummy *do* has to be of the same paradigm as the predicator in the declarative clause, and the predicator itself is converted to its plain form. For instance,

She usually gets up early.

is a declarative clause with the predicator *gets*, which is in the third-person singular, and the subject *she*. By inserting *do* in third-person singular to the beginning of the sentence and converting *gets* to its plain form, we obtain

Does she usually get up early?

To simplify the following transformational scattered context grammar G, which performs this conversion, we assume that the subject is a personal pronoun in the nominative.

Set $G = (V, T, P, I)$, where $V = T \cup I \cup \{X\}$ (X is a new symbol such that $X \notin T \cup I$), and P is defined as follows:

$$
\begin{aligned}
P = & \Big\{ (\langle p \rangle, \langle v \rangle) \to (vp, X) \mid v \in T_{\text{VA}}, p \in T_{\text{PPn}} \Big\} \\
\cup & \Big\{ (\langle p \rangle, \langle \pi_{\text{pret}}(v) \rangle) \to (\text{did } p, vX), \\
& \quad (\langle p \rangle, \langle \pi_{\text{3rd}}(v) \rangle) \to (\text{does } p, vX), \\
& \quad (\langle p \rangle, \langle \pi_{\text{pres}}(v) \rangle) \to (\text{do } p, vX) \mid v \in T_{\text{V}pl} - T_{\text{VA}}, p \in T_{\text{PPn}} \Big\} \\
\cup & \Big\{ (\langle x \rangle, X) \to (x, X), \\
& \quad (X, \langle y \rangle) \to (X, y) \mid x \in T - T_{\text{V}}, y \in T \Big\} \\
\cup & \big\{ (X) \to (\varepsilon) \big\}.
\end{aligned}
$$

For sentences whose predicator is an auxiliary verb, the transformation made by G proceeds as follows:

$$
\begin{aligned}
& \langle \text{he} \rangle \langle \text{is} \rangle \langle \text{mowing} \rangle \langle \text{the} \rangle \langle \text{lawn} \rangle \\
\Rightarrow_G \; & \text{is he } X \langle \text{mowing} \rangle \langle \text{the} \rangle \langle \text{lawn} \rangle \\
\Rightarrow_G \; & \text{is he } X \text{ mowing } \langle \text{the} \rangle \langle \text{lawn} \rangle \\
\Rightarrow_G \; & \text{is he } X \text{ mowing the } \langle \text{lawn} \rangle \\
\Rightarrow_G \; & \text{is he } X \text{ mowing the lawn} \\
\Rightarrow_G \; & \text{is he mowing the lawn.}
\end{aligned}
$$

The derivation starts by the application of a production from the first set, which swaps the subject and the predicator, and puts X behind them. Next, productions from the third set rewrite every word $\langle w \rangle \in I$ to $w \in T$. Finally, X is removed from the sentential form.

The transformation of the sentences in which the predicator is a lexical verb is more complicated:

$$\begin{aligned}
&\langle \text{she} \rangle \langle \text{usually} \rangle \langle \text{gets} \rangle \langle \text{up} \rangle \langle \text{early} \rangle \\
\Rightarrow_G\ &\text{does she } \langle \text{usually} \rangle \text{ get}X \langle \text{up} \rangle \langle \text{early} \rangle \\
\Rightarrow_G\ &\text{does she usually get}X \langle \text{up} \rangle \langle \text{early} \rangle \\
\Rightarrow_G\ &\text{does she usually get}X \text{ up } \langle \text{early} \rangle \\
\Rightarrow_G\ &\text{does she usually get}X \text{ up early} \\
\Rightarrow_G\ &\text{does she usually get up early.}
\end{aligned}$$

As the predicator is in the third-person singular, a production from

$$\left\{ (\langle p \rangle, \langle \pi_{3rd}(v) \rangle) \rightarrow (\text{does } p, vX) \mid v \in T_{Vpl} - T_{VA}, p \in T_{PPn} \right\}$$

is applied at the beginning of the derivation. It inserts *does* at the beginning of the sentence, converts the predicator *gets* to its plain form *get*, and puts X behind it. Next, productions from

$$\left\{ (\langle x \rangle, X) \rightarrow (x, X) \mid x \in T - T_V \right\}$$

rewrite every word $\langle w \rangle \in I$ appearing in front of the predicator to $w \in T$. Note that they do not rewrite verbs — in this way, the grammar verifies that the first verb in the sentence was previously selected as the predicator. For instance, in the sentence

He has been working for hours.,

has must be selected as the predicator; otherwise, the derivation is unsuccessful. Finally, the grammar rewrites all words behind X and erases X in the last step, as in the previous case.

Based on this intuitive explanation, we can see that the set of all input sentences K described in the introduction of this example is transformed by G into $T(G, K)$, which is the set of all interrogative sentences constructed from K. Therefore,

$$(\langle \text{he} \rangle \langle \text{is} \rangle \langle \text{mowing} \rangle \langle \text{the} \rangle \langle \text{lawn} \rangle, \text{is he mowing the lawn}) \in T(G, K),$$
$$(\langle \text{she} \rangle \langle \text{usually} \rangle \langle \text{gets} \rangle \langle \text{up} \rangle \langle \text{early} \rangle, \text{does she usually get up early}) \in T(G, K).$$

Question tags

Question tags are special constructs that are primarily used in spoken language. They are used at the end of declarative clauses, and we customarily use them to ask for agreement or confirmation. For instance, in

Your sister is married, isn't she?

isn't she is a question tag, and we expect an answer stating that she is married. The polarity of question tags is always opposite the polarity of the main clause — if the main clause is positive, the question tag is negative, and vice versa. If the predicator is an auxiliary verb, the question tag is formed by the same auxiliary verb. For lexical verbs, the question tag is made by using *do* as

He plays the violin, doesn't he?

There are some special cases that have to be taken into account. First, the verb *be* has to be treated separately because it has more paradigms than other verbs, and the question tag for first-person singular is irregular:

I am always right, aren't I?

Second, for the verb *have*, the question tag depends on whether it is used as an auxiliary verb or a lexical verb. In the first case, *have* is used in the question tag as

He has been working hard, hasn't he?

In the latter case, the auxiliary *do* is used as

They have a dog, don't they?

To explain the basic concepts as simply as possible, we omit the special cases of the verb *have* in the following transformational scattered context grammar G, which supplements a canonical clause with a question tag. For the same reason, we only sketch its construction and do not mention all the created productions explicitly. In addition, we suppose that the subject is represented by a personal pronoun.

Set $G = (V, T, P, I)$, where $V = T \cup I \cup \{X, Y\}$ (X, Y are new symbols such that X, $Y \notin T \cup I$), and P is defined as follows:

$$
\begin{aligned}
P = \Big\{ &\big(\langle p\rangle, \langle \text{will}\rangle, \langle x\rangle\big) \rightarrow (p, \text{will } X, Yx \text{ won't } p),\\
&\big(\langle p\rangle, \langle \text{won't}\rangle, \langle x\rangle\big) \rightarrow (p, \text{won't } X, Yx \text{ will } p),\\
&\ldots \mid p \in T_{PPn}, x \in T\Big\}\\
\cup \Big\{ &\big(\langle \text{I}\rangle, \langle \text{am}\rangle, \langle x\rangle\big) \rightarrow (\text{I}, \text{am } X, Yx \text{ aren't I}),\\
&\big(\langle \text{you}\rangle, \langle \text{are}\rangle, \langle x\rangle\big) \rightarrow (\text{you}, \text{are } X, Yx \text{ aren't you}),\\
&\ldots \mid x \in T\Big\}\\
\cup \Big\{ &\big(\langle p\rangle, \langle v\rangle, \langle x\rangle\big) \rightarrow (p, vX, Yx \text{ doesn't } p),\\
&\big(\langle q\rangle, \langle v\rangle, \langle x\rangle\big) \rightarrow (q, vX, Yx \text{ don't } q) \mid\\
&p \in \{\text{he, she, it}\}, q \in T_{PPn} - \{\text{he, she, it}\}, v \in T_V - T_{VA}, x \in T\Big\}\\
&\vdots\\
\cup \Big\{ &\big(\langle x\rangle, X\big) \rightarrow (x, X),\\
&\big(X, \langle y\rangle, Y\big) \rightarrow (X, y, Y) \mid x \in T - T_V, y \in T\Big\}\\
\cup \big\{ &(X, Y) \rightarrow (\varepsilon, \varepsilon)\big\}.
\end{aligned}
$$

First, we describe the generation of question tags for clauses whose predicator is an auxiliary verb:

$$
\begin{aligned}
&\langle \text{I}\rangle\langle \text{am}\rangle\langle \text{always}\rangle\langle \text{right}\rangle\\
\Rightarrow_G\ &\text{I am } X\langle \text{always}\rangle Y \text{ right aren't I}\\
\Rightarrow_G\ &\text{I am } X \text{ always } Y \text{ right aren't I}\\
\Rightarrow_G\ &\text{I am always right aren't I.}
\end{aligned}
$$

Here, the production

$$\big(\langle \text{I}\rangle, \langle \text{am}\rangle, \langle \text{right}\rangle\big) \rightarrow (\text{I}, \text{am } X, Y \text{ right aren't I})$$

initiates the derivation. When it finds *I am* at the beginning of the sentence, it generates the question tag *aren't I* at its end. In addition, it adds X behind *I am* and Y in front of *right aren't I*. Next, it rewrites all words from $\langle w\rangle \in I$ to $w \in T$. It makes sure that the predicator was chosen properly by productions from

$$\Big\{\big(\langle x\rangle, X\big) \rightarrow (x, X) \mid x \in T - T_{\rm V}\Big\},$$

similarly to the previous example. In addition, productions from

$$\Big\{\big(X, \langle y\rangle, Y\big) \rightarrow (X, y, Y) \mid x \in T - T_{\rm V}, y \in T\Big\}$$

check whether the question tag was placed at the very end of the sentence. If not, there remains some symbol from the input vocabulary behind Y that cannot be rewritten. Finally, the last production removes X and Y from the sentential form.

When the predicator is a lexical verb in present, the question tag is formed by *does* or *do*, depending on the person in which the predicator occurs:

$$\begin{aligned}
&\langle \text{he}\rangle\langle \text{plays}\rangle\langle \text{the}\rangle\langle \text{violin}\rangle \\
\Rightarrow_G\ & \text{he plays } X\langle \text{the}\rangle Y \text{ violin doesn't he} \\
\Rightarrow_G\ & \text{he plays } X \text{ the violin } Y \text{ doesn't he} \\
\Rightarrow_G\ & \text{he plays the violin doesn't he.}
\end{aligned}$$

The rest of the derivation is analogous to the first case.

Based on these derivations, we can see that the set of all input sentences K described in the introduction of this example is transformed by G into $T(G, K)$, which is the set of all sentences constructed from K that are supplemented with question tags. Therefore,

$$\big(\langle \text{I}\rangle\langle \text{am}\rangle\langle \text{always}\rangle\langle \text{right}\rangle, \text{I am always right aren't I}\big) \in T(G, K),$$
$$\big(\langle \text{he}\rangle\langle \text{plays}\rangle\langle \text{the}\rangle\langle \text{violin}\rangle, \text{he plays the violin doesn't he}\big) \in T(G, K).$$

Generation of grammatical sentences

The purpose of the following discussion, which closes this section, is six-fold, (1)–(6), stated as follows:

(1) We want to demonstrate that ordinary scattered context grammars, discussed in the previous chapters of this book, can be seen as a special case of transformational scattered context grammars, whose applications are discussed in the current section.

(2) As pointed out in the notes following the general definition of a transformational scattered context grammar (see Definition 12.1), there exists a close relation between ordinary scattered context grammars and transformational scattered context grammars. That is, for every scattered context grammar $G = (V, T, P, S)$, there is a transformational scattered context grammar $H = \big(V, T, P, \{S\}\big)$ satisfying

$$L(G) = \Big\{y \mid (S, y) \in T\big(H, \{S\}\big)\Big\},$$

and in this way, $L(G)$ is defined by H. Next, we illustrate this relation with a specific example.

(3) From a syntactical point of view, we want to show that scattered context grammars can generate an infinite non-context-free grammatical subset of the English language in a very succinct way.

(4) In terms of *morphology* — the area of linguistics that studies the structure of words and their generation — we demonstrate how to use transformational scattered context grammars to create complicated English words within English sentences so that the resulting words and sentences are grammatically correct.

(5) As stated at the beginning of this section, so far we have assumed that the set of common English words is finite. Next, we want to demonstrate that, from a strictly theoretical point of view, the set of all possible well-formed English words, including extremely rare words in everyday English, is infinite. Indeed, L, given as follows, includes infinitely many words of the form $(great\text{-})^i grandparents$, $(great\text{-})^i grandfathers$, and $(great\text{-})^i grandmothers$, for all $i \geq 0$, and purely theoretically, they all represent well-formed English words. Of course, most of them, such as

great-great-great-great-great-great-great-great-great-grandfathers

cannot be considered common English words because most people never use them during their lifetime.

(6) We illustrate that the language generation based upon scattered context grammars may have significant advantages over the generation based upon classical grammars, such as context-sensitive grammars.

Without further ado, consider the language L consisting of these grammatical English sentences:

Your grandparents are all your grandfathers and all your grandmothers.

Your great-grandparents are all your great-grandfathers and all your great-grandmothers.

Your great-great-grandparents are all your great-great-grandfathers and all your great-great-grandmothers.

$\vdots$

In brief,

$$L = \big\{\text{your } \{\text{great-}\}^i\text{grandparents are all your } \{\text{great-}\}^i\text{grandfathers and all your } \{\text{great-}\}^i\text{grandmothers } \mid i \geq 0\big\}.$$

Introduce the scattered context grammar $G = (V, T, P, S)$, where

$$T = \{\text{all}, \text{and}, \text{are}, \text{grandfathers}, \text{grandmothers}, \text{grandparents}, \text{great-}, \text{your}\},$$

$V = T \cup \{S, \#\}$, and P consists of the following three productions:

$$(S) \rightarrow (\text{your } \#\text{grandparents are all your } \#\text{grandfathers and all your } \#\text{grandmothers}),$$
$$(\#,\#,\#) \rightarrow (\#\text{great-}, \#\text{great-}, \#\text{great-}),$$
$$(\#,\#,\#) \rightarrow (\varepsilon, \varepsilon, \varepsilon).$$

Obviously, this scattered context grammar generates L; formally, $L = L(G)$. Consider the transformational scattered context grammar $H = \big(V, T, P, \{S\}\big)$. Note that

$$L(G) = \Big\{y \mid (S, y) \in T\big(H, \{S\}\big)\Big\}.$$

Clearly, L is not context-free, so its generation is beyond the power of context-free grammars. It would be possible to construct a context-sensitive grammar that generates L. However, a context-sensitive grammar like this would have to keep traversing across its sentential forms to guarantee the same number of occurrences of *great-* in the generated sentences. Compared to this awkward way of generating L, the scattered context grammar G generates L in a more elegant, economical, and effective way.

In this chapter, we have illustrated how to transform and generate grammatical sentences in English by using transformational scattered context grammars, which represent a very natural linguistic apparatus straightforwardly based on scattered context grammars. However, from a more general perspective, we can apply these grammars basically to any area of science that formalizes its results by strings containing some scattered context dependencies. This general perspective brings us to the concluding chapter of this book, in which we make remarks about some selected scientific areas that involve a formalization of scattered context, and we also suggest how to make use of scattered context grammars in them.

12.2 Musicology

The fundamental purpose of this brief section is to bring a challenging interdisciplinary scientific area to the attention of both computer scientists and musicologists. That is, it explains how to characterize selected musical structures in terms of the apparatus of formal language theory, with a focus on scattered context grammars. The chapter only gives general ideas that explain how to grammaticize selected musical structures. It pays special attention to the formalization of structures that can be properly and elegantly described by scattered context rules, but they are indescribable by ordinary context-free rules. Therefore, just like most natural languages, these structures represent languages that belong to $\mathbf{L}_{\mathbf{CSG}} - \mathbf{L}_{\mathbf{CFG}}$.

The style of this section is even more informal than that used in the previous section of this chapter. It only sketches all the musical structure formalizations in an utterly descriptive way without any exact mathematical definitions. Indeed, the section always presents only the crucially important grammatical rules without presenting grammars in their entirety. To make this presentation even more simpler, in all the grammatical rules given in this section, we automatically assume that X represents a nonterminal, while all the other symbols

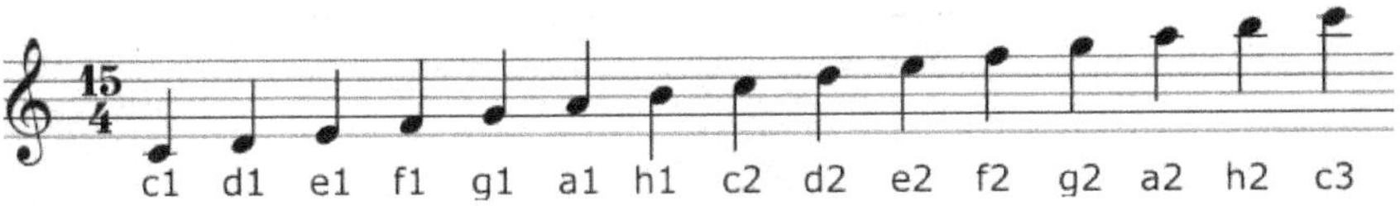

Figure 12.1. Notes and their names.

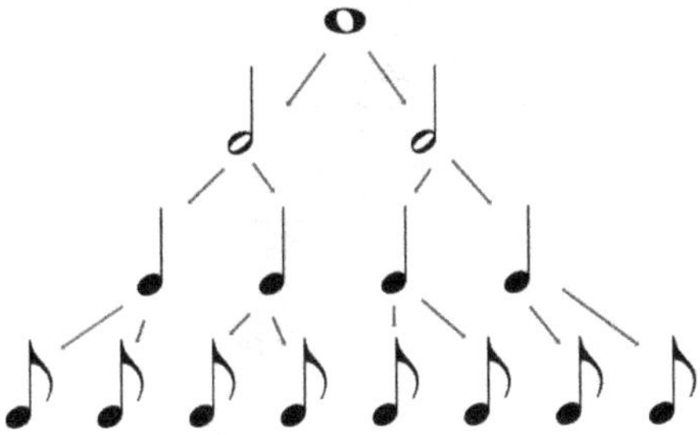

Figure 12.2. Hierarchy of note lengths.

are terminals; as a matter of fact, X is the only nonterminal that occurs in this section.

Basics of musical notation

Concerning the musical notation, this chapter starts from scratch in order to make its discussion, especially the examples given in the following, readable and understandable to all computer scientists, including those who are not familiar with this notation at all.

The music staff consists of five lines with four gaps between them, each of which represents a different pitch. Tones are specified by notes (see Figure 12.1).

The notes are organized into measures, whose lengths are specified by time signatures. In Figure 12.2, the root of the tree represents the whole note, which has a length of four beats. From the whole note, we obtain two half two-beat notes. On the third level, there are quarter notes, each of which has the length of one beat. The last level presents eighth notes, whose duration is the eighth of a whole note. Musical compositions are expressed by notes. As a rule, the heart of a composition consists of a major musical theme, whose variations, which are based on modifying the rhythm or melody of

Figure 12.3. Arch dependency.

Figure 12.4. Serial dependency.

the main theme, are subsequently repeated throughout the composition. Of course, between these repetitions, there exist contextual dependencies, whose formalization represents the subject of the current chapter. The chapter narrows its attention only to two kinds of these dependencies: *arch dependency* (see Figure 12.3) and *serial dependency* (see Figure 12.4). The current section sketches how to formalize the former and the latter with context-free and scattered context grammars, respectively.

Context-free music: Arch dependencies

In essence, an arch musical form is a sectional structure for a piece of music written in such a way that each section is based on a musical passage followed by its repetition in reverse order. Most of these musical sections are symmetrical around a central movement. All the repetitions of these sections may not be verbatim, but they usually share major thematic material. At first glance, this form of music may appear static, thus denying any progress. In fact, however, the symmetric pairs of sections create a unidirectional process with the center. As a result, this process engenders a strongly expressive psychological power, which might otherwise be unavailable for the musical work as a whole.

Béla Bartók wrote several pieces of music purely based on the arch form, including his fourth and fifth string quartets, Concerto for Orchestra, Music for Strings, Percussion and Celesta, the second

Figure 12.5. A retrogradation.

piano concerto, and the second violin concerto. Shostakovich's String Quartet No. 8 in C minor is based on the arch form too.

Retrogradation

As an example of an arch musical form, consider retrogradation, usually consisting of a melodic line that is the reverse of a previous line. An exact retrograde includes both the pitches and rhythms in reverse. An even more exact retrograde reverses the physical contour of the notes themselves, but this utterly strict reversion is possible only in electronic music. In live music, most compositions usually choose to subject either the pitches or the rhythms of a musical line to retrograde. In twelve-tone music, a reversal of the pitch classes alone is regarded as a retrograde.

Arch musical dependencies are usually describable by using context-free grammars based on context-free rules of the form $X \rightarrow xXx^R$. For instance, Figure 12.5 gives a retrogradation based on a mirrored-structure variation that consists of a sequence of notes, which represents an increasing melody. Specifically, tones $g1$, $a1$, $h1$, and $c2$ are followed by their reservations $c2$, $h1$, $a1$, and $g1$, respectively. This retrogradation can be captured by the following context-free rules:

$$(X) \rightarrow (g_{quarter}Xg_{quarter}), (X) \rightarrow (a_{quarter}Xa_{quarter}),$$
$$(X) \rightarrow (h_{quarter}Xh_{quarter}), (X) \rightarrow (c2_{quarter}Xc2_{quarter}).$$

Scattered context music: Serial dependencies

In musical compositions, serial dependencies occur whenever several musical passages, such as a major theme and its variations, reflect each other in series throughout the composition. Based on whether these passages overlap each other, musicology distinguishes two important kinds of such dependencies. *Cross-serial dependencies,* also known as crossing dependencies, occur when these passages cross over each other. Scattered serial dependencies take place when these dependent passages are scattered throughout the whole composition without any overlapping of each other. As a rule, in a musical composition, the latter kind of serial dependencies usually consists of a major musical theme gradually followed by many of its variants. The current chapter only deals with scattered serial dependencies in what follows.

Ravel's Boléro is perhaps the best example of scattered serial dependencies in which the main musical passage and its straightforward variants follow one after another. Indeed, beginning *pianissimo* and rising in a *continuous crescendo* to *fortissimo possibile* built over an unchanging ostinato rhythm successively played 169 times. However, the overwhelming majority of classical music works contain many passages based on scattered serial dependencies that take place far away from each other. To illustrate, take Mahler's Third. Most of its musical themes are introduced in the first long movement, while their variants occur later on in the remaining shorter movements. The current chapter concentrates its discussions on this kind of scattered serial dependencies. Specifically, it covers its most frequently used variants — transposition, sequence, contrary motion, augmentation, and diminution, whose proper formalization usually goes far beyond the power of context-free grammars. On the other hand, scattered context grammars can capture them easily and elegantly, as demonstrated in the following.

Transposition

In music, transposition refers to the process or operation of moving a collection of notes up or down in pitch by a constant interval. It shifts a melody or a harmonic progression to another key while maintaining the same tone structure, i.e., the same succession of whole tones and semitones as well as remaining melodic intervals. A transposition may shift a tone row or an unordered collection of pitches, such as a chord, so that it begins on another pitch. In fact, an entire piece of music can be transposed into another key.

Consider a simple transposition that moves a theme to another pitch level upward, as illustrated in Figure 12.6. In this figure, there are notes h_1, g_1, and a_1, which are then shifted to another pitch level

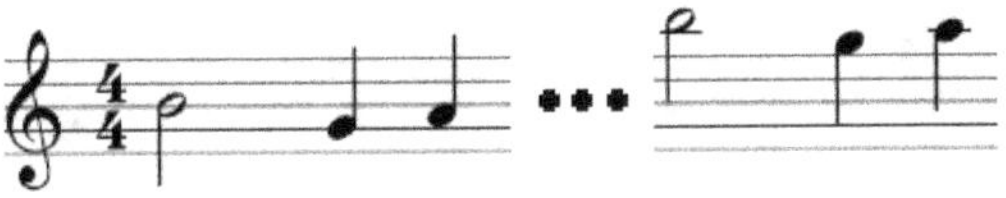

Figure 12.6. A transposition.

in the range from c_2 to h_2. Specifically, in Figure 12.6, these notes are h_2, g_2, and a_2, respectively, whose length is changed too. Specifically, $h1$ is represented as a half note, which lasts two beats, while g_1 and a_1 are quarter notes with a length of one beat. This transposition can be expressed by the following scattered context rules:

$$(X, X) \to (h_{half}X, h2_{half}X), (X, X) \to (g_{quarter}X, g2_{quarter}X),$$
$$(X, X) \to (a_{quarter}X, a2_{quarter}X).$$

Sequence

In music, a sequence is the restatement of a motif or longer musical passage at a higher or lower pitch, usually in one direction, while all the tones of the moved passage continue with the same interval distance. A constant sequence growth represents a special kind of this modification, with a constant growth every tone either higher or lower.

In Figure 12.7, the first measure consists of four quarter notes: g_1, e_1, a_1, and f_1. In the second measure, a sequence takes place so that the previous four-note melody is decreased by one. The scattered context rules that describe this change are as follows:

$$(X, X) \to (g_{quarter}X, f_{quarter}X), (X, X) \to (e_{quarter}X, d_{quarter}X),$$
$$(X, X) \to (a_{quarter}X, g_{quarter}X), (X, X) \to (f_{quarter}X, e_{quarter}X).$$

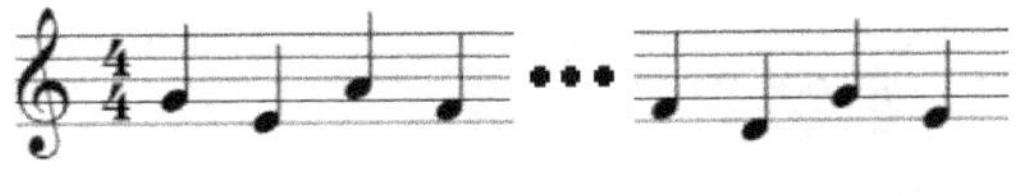

Figure 12.7. A sequence.

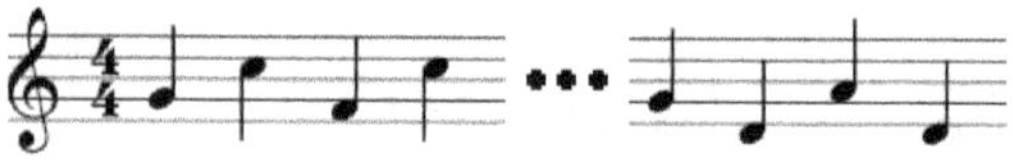

Figure 12.8. A contrary motion.

Contrary motion

As its name suggests, a contrary motion moves a musical passage in the opposite direction. That is, to put it in terms of music staff, when one of the lines moves up, the other line inversely moves down. This modification is said to be a strict contrary motion if the moved passages always move at the same intervals in opposite directions.

Figure 12.8 presents a contrary motion with tones g_1, c_2, f_1, and c_2. It shows the distance specified as 3, −1, and 3, which define how many times the tones are moved to obtain the new tones. Thus, the resulting variation will have distances −3, 1, and −3, respectively. This contrary motion can be captured by the following scattered context rules:

$$(X,X) \to (g_{quarter}X, g_{quarter}X), (X,X) \to (c2_{quarter}X, d_{quarter}X),$$
$$(X,X) \to (f_{quarter}X, a_{quarter}X), (X,X) \to (c2_{quarter}X, d_{quarter}X).$$

Augmentation

In music, augmentation means the lengthening of a note or the widening of an interval. Consequently, it represents a compositional device where a melody, theme, or motif is presented in longer note values compared to its previous use.

In Figure 12.9, the augmentation changes the rhythm so that the length of the tones is doubled. More precisely, the augmentation of the tones e_1, a_1, h_1, and c_2 results in the doubled length of e_1 from the half note to the whole note, so these tones have changed their length from the eighth note to the half note. The other tones change their length from the eighth note to the half note too. The rules that can describe this augmentation can be defined as follows:

$$(X,X) \to (a_{half}X, a_{full}X), (X,X) \to (e_{quarter}X, e_{half}X),$$
$$(X,X) \to (a_{quarter}X, a_{half}X).$$

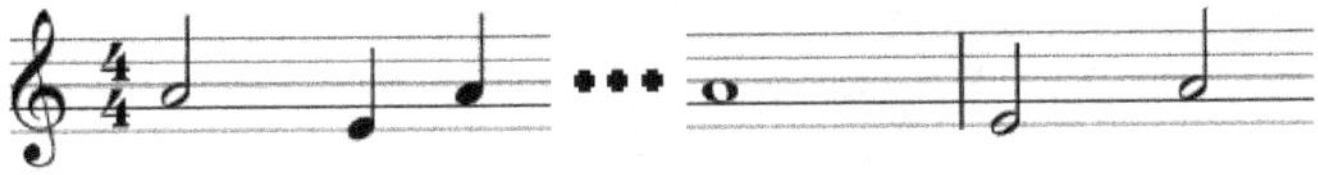

Figure 12.9. An augmentation.

Figure 12.10. A diminution.

Diminution

Thus, diminution is the opposite of augmentation, where note values are prolonged. Diminution is a form of embellishment in which a long note or a series of long notes is divided into shorter values. Figure 12.10 presents the same notes, but they have halved length from the half notes to the quarter notes. This diminution can be described by the rules given as follows:

$$(X, X) \rightarrow (a_{half}X, a_{quarter}X), (X, X) \rightarrow (c_{half}X, c_{quarter}X).$$

Part 5

Conclusion

This concluding part consists of two brief chapters. Chapter 13 summarizes this book. Chapter 14 places all its material in the historical and bibliographical context and recommends further reading to the serious reader.

Chapter 13

Summary

This book covers a variety of automata as mathematical models that formalize procedures in the intuitive sense. It explains the reason why they are central to computer science when answering such important questions as what is computable and what is not. It approaches them from two essential standpoints: automata as language models and automata as models of computation. At the same time, it demonstrates that both approaches are nothing but two different but literally inseparable viewpoints on the same thing. Indeed, based upon a computationally founded motivation, it always conceptualizes and studies a new type of automata as language models; then, in turn, it establishes the consequences following from this study in terms of computation and its limits.

Primarily, this book gives an account of the basic theory of automata and computation. Secondarily, it describes its modern trends and applications. The text maintains a balance between theoretical and practical approaches to this subject. From a theoretical viewpoint, it covers all rudimentary topics concerning automata. The book demonstrates the most fundamental properties of languages defined by automata. Concerning applications, it explains how these models underlie computer-science engineering techniques for language processing, such as lexical and syntactic analyses. From a more theoretical viewpoint, this book applies automata, especially Turing machines, to computation in general and, thereby, demonstrates the fundamentals underlying the theory of computation, including computability, decidability, and computational complexity.

The entire text of this book is divided into five parts.

Part 1, consisting of Chapters 1 and 2, gives an introduction to this book. Chapter 1 reviews the basic mathematical notions used throughout the book. Chapter 2 explains the essential reason why automata are central to computer science when answering such important questions as what is computable and what is not. It demonstrates two fundamental approaches to them: automata as language models and automata as models of computation. At the same time, it explains that both approaches are nothing but two different but literally inseparable viewpoints on the same thing. Indeed, based upon a computationally founded motivation, it conceptualizes and studies a new type of automata as language models; then, in turn, it establishes the consequences following from this study in terms of computation and its limits.

Part 2 consists of Chapters 3–7. Chapter 3 is devoted to *finite automata* as strictly finitary models of computation. Each of its components is of a fixed and finite size, and none of them can be extended during the course of computation. Chapter 4 extends finite automata by adding a potentially infinite workspace always organized as stacks, referred to as pushdowns, hence the name of these automata — *pushdown automata.* Chapter 5 introduces *Turing machines*, named after their originator, Alan Turing, as quite general models for languages and computation. This chapter treats these machines primarily as language models while leaving their discussion as general models of computation to Chapter 7. Chapter 6 covers selected grammars, which generate their languages by finitely many grammatical rules. It concentrates this coverage primarily on grammars with the same language-expressive power as automata discussed earlier in Part 2, so these grammars actually represent counterparts to the corresponding equally powerful automata in terms of their language-defining capability. Chapter 7 returns to the Turing machines by reconsidering them as ultimately general computational models, thus underlying the theory of computation in its entirety. It explains that this theory actually represents a metaphysics of computation that states what is computable and what is not. Analogically, it demonstrates that some problems are algorithmically decidable while others are not. Concerning decidable problems, the theory of computation takes a finer look at some of them by studying their time and space computational complexity. It distinguishes problems whose computation takes a reasonable amount of time from intractable problems whose solutions require an unmanageable amount of time. Thus, apart from a theoretical significance, results of this kind are obviously crucially important to most application-oriented areas of computer science. Considering all these crucially important theoretical results together with their practical consequences, Chapter 7 definitely crowns the discussion of Part 2 as a whole.

Part 3 gives an account of selected modern trends in automata theory. It consists of Chapters 8–10. Chapter 8 presents *regulated automata*, whose behavior is controlled by additional simple languages. Chapter 9 modifies classical finite automata into *jumping automata* for discontinuous computation. Chapter 10 studies *deep pushdown automata* that can modify their stacks under their tops in a limited way.

Part 4, consisting of Chapters 11 and 12, sketches the application of language models discussed earlier in this book. Chapter 11 explains how to use them in programming language processing. Chapter 12 sketches grammatical applications in natural language processing and musicology.

Part 5 is the current two-chapter part. Chapter 13 sums up the book, after which Chapter 14 makes remarks about it from their historical perspective and recommends five excellent texts for further reading.

Chapter 14

Historical and Bibliographical Remarks

First, this chapter places the subject of this book into a historical and bibliographical context of the studies published during the previous century. It simultaneously sketches this history from two slightly different but literary inseparable perspectives: automata as computational models and automata as language models. It restricts its attention only to the crucially important papers and books published at that time. Then, this chapter recommends five excellent books published in the current century for further reading.

The Twentieth Century

As computational models, Turing (1936) opened the subject of this book by proving that the *TM-Halting* problem is undecidable (see Problem 7.13 in Chapter 7). As language models, most of the early important works on automata appeared in the 1960s, when the first high-level programming languages, such as FORTRAN and ALGOL 60, were developed, because these models were used in the compilers of these languages as models of their components, such as finite and pushdown automata as models of scanners and parsers, respectively. Their introduction, however, considerably preceded this compiler-related period, as pointed out in the following.

Finite automata first emerged as models of neural nets in the work of Mcculloch and Pitts (1943). As a formalization of switching circuits, they were introduced by Harrison (1965). Moore (1956) proved the decidability of all the problems concerning finite automata covered in the current book. Rabin and Scott (1959) continued with this subject by investigating many more decision problems for finite automata, and this work also established the equivalence of deterministic and non-deterministic finite automata. Johnson *et al.* (1968) used the theory of finite automata to design lexical analyzers of compilers for the first time.

The notion of a pushdown was introduced by Burks *et al.* (1954). Pushdown automata were first conceptualized by Oettinger (1961). The equivalence between these automata and context-free grammars was demonstrated by Chomsky (1962) and Evey (1963). Schützenberger (1963) defined and investigated deterministic versions of pushdown automata. Cantor (1962), Floyd (1962), and Chomsky and Schützenberger (1963) proved that the *CFG-Ambiguity* problem is undecidable (see Problem 7.26 in Chapter 7). Based on pushdown automata, Aho and Ullman (1972a) gave an excellent coverage of parsing methods achieved before 1972.

As already pointed out, concerning Turing machines, Turing (1936) started the subject of this book, and this crucially important paper was followed by many more studies on this topic. Almost all of these studies were summarized by Davis (1965), who paid principal attention to the results on decidability and computability. Hoare and Allison (1972) provided a very readable introduction to computability published in the previous century.

During the past three decades of the previous century, the basic knowledge concerning automata was summarized in a great number of books, including those by Aho and Ullman (1972a, 1972b, 1977), Aho *et al.* (2007), Alblas and Nymeyer (1996), Appel (1998), Bergmann (1994), Elder (1994), Fischer (1991, 1999), Fraser (1995), Gries (1971), Haghighat (1995), Hendrix (1990), Holmes (1995), Holub (1990), Hunter (1999), Kiong (1997), Lemone (1992a, 1992b), Lewis *et al.* (1976), Louden (1997), Mak (1996), Morgan (1998), Muchnick (1997), Parsons (1992), Pittman and Peters (1992), Sampaio (1997), Sorenson and Tremblay (1985), Waite (1993), Wilhelm (1995), and Wirth (1996). All of them are still useful and readable.

The Twenty-First Century

Numerous books about automata have been published during this century. Rather than attempt to be encyclopedic, we recommend only the following five excellent texts out of them. Compared to the way automata are explained in the current text, each of these books approaches them from a slightly different perspective.

I. Hopcroft, J. E., Motwani, R. and Ullman, J. D. (2006)

Authors:	John E. Hopcroft, Rajeev Motwani, and Jeffrey D. Ullman
Title:	Introduction to Automata Theory, Languages, and Computation
Publisher:	Pearson Education; 3rd edition, Addison Wesley, 2006
ISBN:	0321486811

This famous book is a revised and updated version of its two predecessors. It covers all the major topics in the theory of automata and computation, and it is a splendid reference for research in these areas. This book contains hardly any applications.

II. Sipser, M. (2014)

Authors:	Michael Sipser
Title:	Introduction to the Theory of Computation
Publisher:	Course Technology Inc; 3rd edition; 2014
ISBN:	8131771865

This book contains all the material needed for an advanced course on the theory of computation and complexity. Its author is a talented writer, and thanks to his writing style, it nicely presents all the knowledge concerning the theory of automata and computation in an effective way. Proofs of theorems are almost always given in the following two-phase way: First, the book gives the idea that lies behind the proof, and second, it gives the proof itself in a rigorous way. Regarding automata considered as formal language models, the book restricts its attention to the models needed in the theory of computation. However, just like the book by Hopcroft *et al.* (2006), this one lacks any applications.

III. Kozen, D. C. (2012)

Authors: Dexter C. Kozen
Title: Automata and Computability
Publisher: Springer; 2007
ISBN: 1461273099

As its title indicates, this book provides an introduction to the theory of automata and computability. It contains a reasonable selection of essential material concerning the theory of automata and computation. The presentation is somewhat unusual because it consists of lectures rather than chapters. Apart from the basic lectures, the text adds eleven supplementary lectures that cover special and advanced topics on the subject. Frequently, this book makes seemingly difficult-to-understand topics easy to grasp. Regarding applications, it contains a single lecture about parsing, explained in a rather theoretical way.

IV. Martin, J. C. (2010)

Authors: John Martin
Title: Introduction to Languages and the Theory of Computation
Publisher: McGraw-Hill Higher Education; 4th edition, 2010
ISBN: 0071289429

This book is a mathematically oriented survey of some important fundamental concepts in the theory of automata and computation and covers a wider range of topics than most other introductory books on the subject. It contains worked-out proofs of every major theorem, but the reader needs a fairly high level of mathematical sophistication to fully grasp these proofs. It is strictly theoretically oriented. As a result, it is not designed for undergraduate students. Although it contains some algorithms and examples, their presentation is theoretical, too.

V. Sudkamp, T. A. (2005)

Authors: Thomas A. Sudkamp
Title: Languages and Machines: An Introduction to the Theory of Computer Science
Publisher: Pearson Education; 3rd edition, 2005
ISBN: 0321315340

This book provides the reader with a mathematically sound presentation of the theory of automata and computation. The theoretical concepts are presented in such a way that they are preceded by an intuitive understanding of the concepts through numerous examples and illustrations. It contains a good selection of topics concerning computational complexity. Parsing based upon LL and LR grammars is included to lay the groundwork for the study of compiler design, but its presentation is so theoretical that ordinary readers can hardly see how to implement parsers based on these grammars.

About the Supplementary Website

As already pointed out in the Preface section of this book, the text is supported by this website

http://www.fit.vutbr.cz/~meduna/books/atta

Some of its parts are protected by a password. If prompted, enter **amtk** as the password.

Bibliography

Aho, A. V., Lam, M. S., Sethi, R., and Ullman, J. D. (2007). *Compilers: Principles, Techniques, and Tools*, 2nd edn. (Addison-Wesley, Boston).

Aho, A. V., and Ullman, J. D. (1972a). *The Theory of Parsing, Translation and Compiling, Volume I: Parsing* (Prentice-Hall, New Jersey).

Aho, A. V., and Ullman, J. D. (1972b). *The Theory of Parsing, Translation and Compiling, Volume II: Compiling* (Prentice-Hall, New Jersey).

Aho, A. V., and Ullman, J. D. (1977). *Principles of Compiler Design* (Addison-Wesley).

Alblas, H., and Nymeyer, A. (1996). *Practice and Principles of Compiler Building with C* (Prentice Hall, London).

Appel, A. W. (1998). *Modern Compiler Implementation in ML* (Cambridge University Press, Cambridge).

Backus, J. W., Bauer, F. L., Green, J., Katz, C., McCarthy, J., Perlis, A. J., Rutishauser, H., Samelson, K., Vauquois, B., Wegstein, J. H., van Wijngaarden, A., Woodger, M., and Naur, P. (1960). Report on the algorithmic language algol 60. *Communications of the ACM* **3**, 5, pp. 299–314.

Baeza-Yates, R., and Ribeiro-Neto, B. (2011). *Modern Information Retrieval: The Concepts and Technology behind Search*, 2nd edn. (Addison-Wesley Professional, Boston).

Bar-Hillel, Y., Perles, M., and Shamir, E. (1961). On formal properties of simple phrase structure grammars. *STUF - Language Typology and Universals* **14**, 1–4, pp. 143–172.

Barnett, M. P., and Futrelle, R. P. (1962). Syntactic analysis by digital computer. *Communications of the ACM* **5**, 10, pp. 515–526.

Bergmann, S. (1994). *Compiler Design: Theory, Tools, and Examples* (W.C. Brown, Oxford).

Bouchon-Meunier, B., Coletti, G., and Yager, R. R. (eds.) (2006). *Modern Information Processing: From Theory to Applications* (Elsevier Science, New York).

Buettcher, S., Clarke, C. L. A., and Cormack, G. V. (2010). *Information Retrieval: Implementing and Evaluating Search Engines* (The MIT Press, Cambridge).

Cantor, D. G. (1962). On the ambiguity problem of backus systems. *Journal of the ACM* **9**, 4, pp. 477–479.

Chomsky, N. (1962). Context-free grammars and pushdown storage. *Quarterly Progress Report*, pp. 187–194.

Chomsky, N., and Schützenberger, M. P. (1963). The algebraic theory of context-free languages*. *Studies in Logic and the Foundations of Mathematics* **35**, pp. 118–161.

Conway, M. E. (1963). Design of a separable transition-diagram compiler. *Communications of the ACM* **6**, 7, pp. 396–408.

Courcelle, B. (1977). On jump deterministic pushdown automata. *Mathematical Systems Theory* **11**, pp. 87–109.

Dassow, J., and Păun, G. (1989). *Regulated Rewriting in Formal Language Theory* (Springer, New York).

Davis, M. E. (1965). *The Undecidable: Basic Papers on Undecidable Propositions, Unsolvable Problems and Computable Functions* (Raven Press, Hewlett, New York).

de Bakker, J. W. (1969). *Semantics of Programming Languages* (Springer US, Boston, MA), pp. 173–227. ISBN 978-1-4899-5841-9.

Elder, J. (1994). *Compiler Construction: A Recursive Descent Model* (Prentice Hall, London).

Evey, R. J. (1963). Application of pushdown-store machines, in *Proceedings of the November 12–14, 1963, Fall Joint Computer Conference*, AFIPS'63 (Fall) (Association for Computing Machinery, New York, NY, USA), pp. 215–227. ISBN 9781450378833.

Fischer, C. N. (1991). *Crafting a Compiler with C* (Benjamin/Cummings Publishing Co., Redwood City).

Fischer, C. N. (1999). *Crafting a Compiler Featuring Java* (Addison-Wesley, Harlow).

Fischer, P. C., and Rosenberg, A. L. (1968). Multitape one-way nonwriting automata. *Journal of Computer and System Sciences* **2**, pp. 38–101.

Floyd, R. W. (1962). On ambiguity in phrase structure languages. *Communications of the ACM* **5**, 10, p. 526.

Fraser, C. W. (1995). *A Retargetable C Compiler: Design and Implementation* (Addison-Wesley, Harlow).

Gilbert, É., and Conklin, D. (2007). A probabilistic context-free grammar for melodic reduction, in *Proceedings of the International Workshop on Artificial Intelligence and Music, 20th International Joint Conference on Artificial Intelligence (IJCAI)*, pp. 83–94.

Ginsburg, S., Greibach, S. A., and Harrison, M. (1967). One-way stack automata. *Journal of the ACM* **14**, 2, pp. 389–418.

Ginsburg, S., and Rice, H. G. (1962). Two families of languages related to algol. *Journal of the ACM* **9**, 3, pp. 350–371.

Ginsburg, S., and Spanier, E. H. (1968). Control sets on grammars. *Theory of Computing Systems* **2**, 2, pp. 159–177.

Greibach, S. A. (1969). Checking automata and one-way stack languages. *Journal of Computer and System Sciences* **3**, pp. 196–217.

Gries, D. (1971). *Compiler Construction for Digital Computers* (John Wiley & Sons, New York).

Grossman, D. A., and Frieder, O. (2004). *Information Retrieval: Algorithms and Heuristics*, 2nd edn. (Springer, Berlin).

Haghighat, M. R. (1995). *Symbolic Analysis for Parallelizing Compilers* (Kluwer Academic, Boston).

Harrison, M. (1978). *Introduction to Formal Language Theory* (Addison-Wesley, Boston).

Hartmanis, J., and Stearns, R. E. (1965). On the computational complexity of algorithms. *Transactions of the American Mathematical Society* **117**, pp. 285–306.

Hendrix, J. E. (1990). *A Small C Compiler* (Prentice Hall International, London).

Hoare, C. A. R., and Allison, D. C. S. (1972). Incomputability. *ACM Computing Surveys* **4**, 3, pp. 169–178.

Holmes, J. (1995). *Building Your Own Compiler with C++* (Prentice Hall International, London).

Holub, A. I. (1990). *Compiler Design in C* (Prentice Hall International, London).

Hopcroft, J. E., Motwani, R., and Ullman, J. D. (2006). *Introduction to Automata Theory, Languages, and Computation*, 3rd edn. (Addison-Wesley, Boston).

Huddleston, R., and Pullum, G. (2002). *The Cambridge Grammar of the English Language* (Cambridge University Press).

Huddleston, R. and Pullum, G. (2005). *A Student's Introduction to English Grammar* (Cambridge University Press).

Hunter, R. (1999). *The Essence of Compilers* (Prentice Hall, London).

Ibarra, O. H. (1970). Simple matrix languages. *Information and Control* **17**, pp. 359–394.

Irons, E. T. (1961). A syntax directed compiler for algol 60. *Communications of the ACM* **4**, 1, pp. 51–55.

Johnson, W. L., Porter, J. H., Ackley, S. I., and Ross, D. T. (1968). Automatic generation of efficient lexical processors using finite state techniques. *Communications of the ACM* **11**, 12, pp. 805–813.

Jurish, B. (2004). Music as a formal language, online.

Kasai, T. (1970). An hierarchy between context-free and context-sensitive languages. *Journal of Computer and System Sciences* **4**, pp. 492–508.

Kiong, D. B. K. (1997). *Compiler Technology: Tools, Translators, and Language Implementation* (Kluwer Academic Publishers, London).

Kolář, D., and Meduna, A. (2000). Regulated pushdown automata. *Acta Cybernetica* **2000**, 4, pp. 653–664.

Kurki-Suonio, R. (1969). Notes on top-down languages. *BIT* **9**, 3, pp. 225–238.

Kučera, J., Meduna, A., and Soukup, O. (2015). Absolutely unlimited deep pushdown automata, in *Proceedings of the 10th Doctoral Workshop on Mathematical and Engineering Methods in Computer Science (MEMICS 2015)* (Ing. Vladislav Pokorný - Litera), pp. 36–44. ISBN 978-80-214-5254-1.

Landin, P. J. (1965). Correspondence between algol 60 and church's lambda-notation: Part i. *Communications of the ACM* **8**, 2, pp. 89–101.

Lemone, K. A. (1992a). *Design of Compilers: Techniques of Programming Language Translation* (CRC Press, Boca Raton).

Lemone, K. A. (1992b). *Fundamentals of Compilers: An Introduction to Computer Language Translation* (CRC Press, Boca Raton).

Lewis, H. R., and Papadimitriou, C. H. (1981). *Elements of the Theory of Computation* (Prentice-Hall, New Jersey).

Lewis, I., P. M., Rosenkrantz, D. J., and Stearns, R. E. (1976). *Compiler Design Theory* (Addison-Wesley, Reading, Massachusetts).

Lewis, P. M., and Stearns, R. E. (1968). Syntax-directed transduction. *Journal of the ACM* **15**, 3, pp. 465–488.

Louden, K. C. (1997). *Compiler Construction: Principles and Practice* (PWS Publishing, London).

Mak, R. (1996). *Writing Compilers and Interpreters* (John Wiley & Sons, New York).

Manning, C. D., Raghavan, P., and Schütze, H. (2008). *Introduction to Information Retrieval* (Cambridge University Press, New York).

McCarthy, J. (1960). Recursive functions of symbolic expressions and their computation by machine, part i. *Communications of the ACM* **3**, 4, pp. 184–195.

McCarthy, J., and Painter, A. J. (1966). Correctness of a compiler for arithmetic expressions (American Mathematical Society).

Mcculloch, W. S., and Pitts, W. (1943). A logical calculus of ideas immanent in nervous activity. *Bulletin of Mathematical Biophysics* **5**, pp. 127–147.

Meduna, A. (2000). *Automata and Languages: Theory and Applications* (Springer, London).

Meduna, A. (2003). Simultaneously one-turn two-pushdown automata. *International Journal of Computer Mathematics* **2003**, 80, pp. 679–687.

Meduna, A. (2014). *Formal Languages and Computation* (Taylor & Francis Informa plc, New York).

Meduna, A., and Zemek, P. (2014). *Regulated Grammars and Automata* (Springer US). ISBN 978-1-4939-0368-9.

Moore, E. F. (1956). *Gedanken-Experiments on Sequential Machines* (Princeton University Press, Princeton), pp. 129–154. ISBN 9781400882618.

Morgan, R. C. (1998). *Building an Optimizing Compiler* (Butterworth-Heinemann, Oxford).

Muchnick, S. S. (1997). *Advanced Compiler Design and Implementation* (Morgan Kaufmann Publishers, London).

Nisan, N., and Schocken, S. (2005). *The Elements of Computing Systems: Building a Modern Computer from First Principles* (The MIT Press, Cambridge).

Oettinger, A. G. (1961). Automatic syntactic analysis and the pushdown store (American Mathematical Society).

Parsons, T. W. (1992). *Introduction to Compiler Construction* (Computer Science, Oxford).

Pittman, T., and Peters, J. (1992). *The Art of Compiler Design: Theory and Practice* (Prentice Hall).

Rabin, M. O., and Scott, D. (1959). Finite automata and their decision problems. *IBM Journal of Research and Development* **3**, 2, pp. 114–125.

Rego, S. K. C. (2009). Rhythmically-Controlled Automata Applied to Musical Improvisation, Ph.D. thesis, Instituto Nacional de Matemática Pura e Aplicada.

Roads, C., and Wieneke, P. (1979). Grammars as representations for music, online.

Rosebrugh, R. D., and Wood, D. (1973). A characterization theorem for n-parallel right linear languages. *Journal of Computer and System Sciences* **7**, pp. 579–582.

Rosebrugh, R. D., and Wood, D. (1975). Restricted parallelism and right linear grammars. *Utilitas Mathematica* **7**, pp. 151–186.

Rozenberg, G., and Salomaa, A. (eds.) (1997). *Handbook of Formal Languages, Vol. 1: Word, Language, Grammar* (Springer, New York).

Russell, S., and Norvig, P. (2002). *Artificial Intelligence: A Modern Approach*, 2nd edn. (Prentice-Hall, New Jersey).

Sakarovitch, J. (1981). Pushdown automata with terminating languages, in *Languages and Automata Symposium, RIMS 421, Kyoto University*, pp. 15–29.

Salomaa, A. (1973). *Formal Languages* (Academic Press, London).

Sampaio, A. (1997). *An Algebraic Approach to Compiler Design* (World Scientific, London).

Schulze, W. (2009). A Formal Language Theory Approach to Music Generation. Master's thesis, University of Stellenbosch, Matieland 7602, South Africa.

Schützenberger, M. (1963). On context-free languages and push-down automata. *Information and Control* **6**, 3, pp. 246–264.

Shan, M.-K., Chiang, M.-F., and Kuo, F.-F. (2008). Relevance feedback for category search in music retrieval based on semantic concept learning, online.

Siromoney, R. (1969). Studies in the Mathematical Theory of Grammars and Its Applications. Ph.D. thesis, University of Madras, Madras, India.

Siromoney, R. (1971). Finite-turn checking automata. *Journal of Computer and System Sciences* **5**, pp. 549–559.

Sorenson, P. G., and Tremblay, J. P. (1985). *The Theory and Practice of Compiler Writing* (McGraw Hill, New York).

Stearns, R., Hartmanis, J., and Lewis, P. (1965). Hierarchies of memory limited computations, pp. 179–190.

Turing, A. (1936). On computable numbers, with an application to the entscheidungs problem. *Proceedings of the London Mathematical Society* **42**, 2, pp. 230–265.

Valiant, L. (1989). The equivalence problem for deterministic finite turn pushdown automata. *Information and Control* **81**, pp. 265–279.

Waite, W. (1993). *An Introduction to Compiler Construction* (HarperCollins, New York).

Wijngaarden, A., Mailloux, B., Peck, J., Koster, C., Sintzoff, M., Lindsey, C., Meertens, L., and Fisker, R. (1975). Revised report on the algorithm language algol 68. *Acta Inf.* **5**, pp. 1–236.

Wilhelm, R. (1995). *Compiler Design* (Addison-Wesley, Workingham).

Wirth, N. (1996). *Compiler Construction* (Addison-Wesley, Harlow).
Wood, D. (1973). Properties of n-parallel finite state languages. Technical report, McMaster University.
Wood, D. (1975). m-parallel n-right linear simple matrix languages. *Utilitas Mathematica* **8**, pp. 3–28.
Wood, D. (1987). *Theory of Computation: A Primer* (Addison-Wesley, Boston).
Zuidema, W., Hupkes, D., Wiggins, G., Scharff, C., and Rohrmeier, M. (2019). Formal models of structure building in music, language and animal song.

Index

Q

R

Printed in the USA
CPSIA information can be obtained
at www.ICGtesting.com
JSHW011931241023
50596JS00003B/34

9 789811 278129